Digital Control
Applications
Illustrated with MATLAB®

Digital Control Applications

Illustrated with MATLAB®

Hemchandra Madhusudan Shertukde

UNIVERSITY OF HARTFORD, CT, USA

CRC Press

Taylor & Francis Group

Boca Raton   London   New York

CRC Press is an imprint of the
Taylor & Francis Group, an **informa** business

CRC Press
Taylor & Francis Group
6000 Broken Sound Parkway NW, Suite 300
Boca Raton, FL 33487-2742

First issued in paperback 2021

© 2015 by Taylor & Francis Group, LLC
CRC Press is an imprint of Taylor & Francis Group, an Informa business

No claim to original U.S. Government works

Version Date: 20150106

ISBN 13: 978-0-367-77904-7 (pbk)
ISBN 13: 978-1-4822-3669-9 (hbk)

Visit the Taylor & Francis Web site at
http://www.taylorandfrancis.com

and the CRC Press Web site at
http://www.crcpress.com

To all my professors at the Indian Institute of Technology, Kharagpur, India, who taught me

control theory and electrical power engineering in my formative years in the early 1970s and later

during my graduate study days at the University of Connecticut, Storrs, in the early 1980s.

Finally, to my lovely wife, Rekha; my dear and accomplished kids, Dr. Amola,

Karan, and Rohan; and, as always, my adopted loyal dog, Sheri

Above all this book is dedicated to the memory of my beloved parents; Late Shri Madhusudan

Gajanan Sherukde and Late Smt. SUlabha Madhusudan Shertukde; both were excellent

teachers and as in Mahbharata my father taught me how to tell stories in four different ways!

May their souls rest in peace!

Contents

Preface

My association with the theory of controls in continuous time started during my studies at the Indian Institute of Technology, Kharagpur, India, in 1974 as an undergraduate student in the Controls and Power program. The initial introduction by Professors Kesavamurthy, Y. P. Singh, and Rajagopalan laid the foundation for a good basic understanding of the subject matter. This pursuit and further advanced study in the field of digital controls continued during my days as a graduate student in the Electrical and Systems Engineering Department at the University of Connecticut in Storrs, from 1983 to 1988.

The fundamentals and advances in control theory is a vast field to study. During my graduate studies, the knowledge imparted by Professors Charles Knapp, Peter Luh, Yaakov Bar-Shalom, David Jordan, and David Kleinman has been invaluable and priceless. Further my classmates Krishna Pattipati and Daniel Serafaty have guided and helped considerably in this arduous task.

In addition, the urge to start a graduate program was initiated in the Electrical and Computer Engineering Department at the University of Hartford to include controls as a specialty along with power in the graduate degree program has been a great motivation above all.

Since 1988, after being hired as an assistant professor in the Electrical and Computer Engineering (ECE) at the University of Hartford, I have personally pursued this dream of teaching every aspect of controls such as advanced controls, digital controls, optimal controls, and stochastic controls. The dream finally came true in 1992 when I started the graduate program as the director of Graduate Studies for the College of Engineering in 1993 and the chair of the ECE Department in 1994. For the last 22 years, this teaching process continued, and in 2010 the desire to write books in different areas of controls and power started very successfully, first with Verlag-Dr. Mueller company and later with CRC Press in 2013.

The book *Digital Controls and Applications Illustrated with MATLAB®* will be published by CRC Press with the support and agreement offered to me by Ms. Nora Konopka, publisher for Taylor & Francis Group. I am thankful to her for providing me this opportunity. This is my second book with this publisher and hope that this mutual understanding will continue for posterity's sake for years to come.

This book can be used as a text for two semester courses on digital controls at the senior undergraduate level or introductory level at the graduate level and the second one for an advanced level in digital controls with appropriate prerequisites at each level. Chapters 1 through 5 can be used for the first introductory course and Chapters 5 through 9 for the advanced course in digital controls. Chapter 1 explains the process of digital control, followed by a review of Z-transforms, feedback control concepts, and s to z plane conversions, mappings, signal sampling, and data reconstruction. This is followed by mathematical representations of discrete systems affected by use of advances in computing methodologies and the advent of computers.

Chapter 4 illustrates state-space representations and construction of transfer functions and their corresponding discrete equivalents. Chapter 5 then deals with the design approach and related design processes, followed by performance criteria evaluations through simulations and review of classical designs for comparison in Chapter 6.

Advances in the design of compensators using its discrete equivalent are studied in Chapter 7 and stability tests using transformations are illustrated in Chapter 8. Further, Chapter 9 deals with the computational aspects handled by the present PCs. With the advent of modern computers, hardware and software packages such as FEEDBACK≪® and MATLAB, both of which have been extensively used, make the understanding of the theory and its application more hands-on than the days when we toyed with dated equipment. I hope the use of such packages makes the simulation and hands-on experience fruitful.

I would like to acknowledge the cooperation of my publisher Ms. Nora Konopka of Taylor & Francis, Florida and the patience she has exhibited in the completion of this book. Further, the help and guidance of others in the project team, now Ms. Laurie Schlags and earlier Ms. Kate Gallo, is greatly appreciated.

My first comical introduction to control theory was inspired by a famous cartoon by Dr. S. M. Joshi of the Dynamic Systems and Control Branch, NASA Langley Research Center, Hampton, Virginia, as shown below.

"Nice artwork, kiddo! I've got a gut feeling that a great many people will make a living off that third line some day!"

Observe carefully the third line as the cartoonist points out. Over the past five to six decades many have definitely made a mark and, in the process, made some money using it intelligently. In this effort, I have tried to make the complicated theory in digital controls a little more palatable and understandable. My comical impressions and slightly amorous bend to control theory can be seen from my cartoon below, which I carefully crafted to deliver a message on a popular "Maximum Likelyhood Estimate (MLE)" algorithm, by a name sake cartoon as illustrated. This was my bold opening slide at my Doctoral Defense at the University of Connecticut, Storrs on April 28, 1989. The MLE algorithm took a comical form as illustrated in my depiction of maximal love entrapment of my own life.

I would like to thank the Book Program at MathWorks Inc. for allowing me to adopt some of the MATLAB material. Similarly thanks to Professor Patricia Mellodge of the Electrical and Computer Engineering Department at the University of Hartford for permitting me to use the pictures of assemblies of workstations in the laboratory (D-319) using FEEDBACK≪ equipment. Also thanks to Mr. Iman Salehi, a graduate student in our department for helping me to understand the software and hardware complexities with this equipment. Finally, I would like to thank my wife, Rekha, for her patience in tolerating my mood changes during the completion of my book and that of my children, Amola, Karan, Rohan, and my loyal dog, Sheri.

MATLAB® is a registered trademark of The MathWorks, Inc. For product information, please contact:

The MathWorks, Inc.
3 Apple Hill Drive
Natick, MA 01760-2098 USA
Tel: 508 647 7000
Fax: 508-647-7001
E-mail: info@mathworks.com
Web: www.mathworks.com

FEEDBACK≪® is a registered trademark of Feedback Inc in UK. For the product information please contact:

Feedback Instruments Limited
5 & 6 Warren Court
Crowborough
East Sussex, TN6 2QX UK
Telephone: +44 (0)1892 653322
Fax: +44 (0)1892 663719

Author

Hemchandra Madhusudan Shertukde, SM'92, IEEE, was born in Mumbai, India, on April 29, 1953. He graduated from Indian Institute of Technology, Kharagpur, India with a BTech (high honors) with distinction in 1975. He earned his MS and PhD in electrical engineering with a specialty in controls and systems engineering from the University of Connecticut, Storrs, in 1985 and 1989, respectively. He joined the faculty in the Electrical and Computer Engineering Department at the University of Hartford in 1988. Since 1995, he has been a full professor in the Department of Electrical and Computer Engineering in the College of Engineering, Technology, and Architecture (CETA) at the University of Hartford (West Hartford, Connecticut). Since fall 2011, he has been a senior lecturer at the School of Engineering and Applied Sciences (SEAS), Yale (New Haven, Connecticut). He is the principal inventor of two commercialized patents (US Patent 6,178,386 and US Patent 7,291,111). He has published several articles in *IEEE Transactions* and has written three solo books namely, *Transformers*, *DPV Grid Transformers* and *Target Tracking*, respectively.

The book *DPV Grid Transformers* by CRC Press/Taylor & Francis Group has received international recognition and will be published in simple Chinese by China Machine Press in August 2016.

1

Digital Control Introduction and Overview

1.1 Overview of Process Control: Historical Perspective

In recent times, over the past several decades, interest in digital control has been stimulated by two major factors. Thus for:

- *Process control*: where industry needs for efficient operation of a large plant warrant a constant requirement in the advances of computer technology: with costs decreasing, size decreasing, reliability increasing and plant power increasing it is imperative to seek involvement of computers with human and machine interface (HMI).
- Consider the example of a conventional power plant, shown in Figure 1.1. In the forward loop, fuel is injected by the fuel oil pump into the boiler, where water is heated in the water drum and converted into steam. The steam from the boiler is directed into a high-pressure turbine through the main steam pipe controlled by the main throttle valve. The shaft of the high-pressure turbine is connected to the synchronous generator with voltage control to generate the nominal voltage and mega watt (MW) of power as compared to the command from the dispatch center. Simultaneously, the process of generating voltage and power can be achieved by using a low-pressure turbine as well.

EXAMPLE 1.1: CONVENTIONAL POWER PLANT

A conventional power plant is illustrated in Figure 1.1. In the feedback loop, steam is extracted and fed onto the main condenser from where water is further guided back to the boiler by the main feed pump.

- As seen in Figure 1.1, the overall system is a combination of numerous subsystems:
  - Boiler, feedwater, high-pressure or/and low-pressure turbines and generator
- Each subsystem is characterized by desired setpoints or reference levels as
  - Steam drum pressure and water level
  - Main feed pump discharge pressure
  - Generator synchronous speed in rpm (3600 rpm for a salient two-pole synchronous generator with a voltage signal frequency of 60 Hz)
  - Synchronous generator voltage, MW, mega volt amp reactive (MVAR) outputs, and related mega volt amps (MVA)

Desired setpoints are computed, for example, on the basis of economics for normal operation, safety for shutdown, time to bring the generator back to its nominal voltage and load, etc.

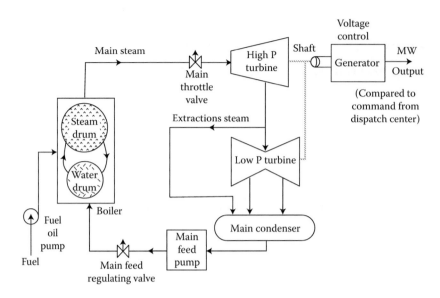

FIGURE 1.1
Conventional power plant with its subsystems at desired setpoints.

Setpoints are usually recomputed every few minutes (or more often, depending on the speed of the plant dynamics) by a central computer. With the recent advances in electronics as well as supervisory control and data acquisition (SCADA) methods, these setpoints are computed more frequently every few seconds by the central computer as shown in Figure 1.2.

Computer-generated setpoints can be, and often are, overridden by the system operator. This creates the scenario of human involvement in the operations called as the "cyberlab" system.

Virtually, all large, complex systems are handled in this manner; that is, a central computer considers global plant operation and determines the setpoints {n} using static models and global optimization criteria, generally economic.

These are often computed by solving a large nonlinear programming problem subject to (safety) constraints, and physical relations f_i and r_i as shown below:

$$\min_{r_1, r_2, \dots} \quad J(r_1, r_2, \dots r_i \dots, r_q) \tag{1.1}$$

$$\text{Subject to } f_i(r_1, r_2, \dots, r_q) < \gamma_i \tag{1.2}$$

Assuring that the output γ_i of a given subsystem accurately follows its desired setpoint than it is a dynamic control problem. It is this class of problems that we will

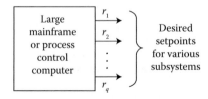

FIGURE 1.2
System showing large computer with desired setpoints.

consider in this book. These are generally classified as problems of local computer control in a given system.

1.2 Feedback Control Structures for Continuous Systems: Mathematical Representation of (Sub) System Dynamics

In this book we consider nth order, single input (u), and single output (y) (SISO) systems:
Systems in this book are classified as Causal Linear and Time Invariant (CLTI).
Here, the system dynamics are continuous, and clearly not discrete.
Some salient aspects of the mathematical nature of such systems are delineated in the following equations:

1. The differential equation model for an "n" dimensional system is shown in Equation 1.3a

$$y^{(n)}(t) + a_1 y^{(n-1)}(t) + \cdots + a_n y(t) = b_0 u^{(n)}(t) + b_1 u^{(n-1)}(t) + \cdots + b_n u(t) \tag{1.3a}$$

where $y^i(t)$ and $u^i(t)$; $i = 1, 2, , \ldots, n$ are the output of the system and inputs to the n-dimensional system.

a. Generally, $b_0 = 0$ for most systems

b. With initial conditions

$$y^{(n)}(0), \ldots, y(0)$$

$$u^{(n)}(0), \ldots, u(0)$$

and; $u(0)$ is generally $= 0$; since control is applied at $t = 0$.

State-space model (with vectors as columns), representation Equation 1.3a is given as follows:

$$X = \text{state vector}$$

$$X(t) = AX(t) + Bu(t); \quad X(0) = \text{initial state vector} \tag{1.3b}$$

$$y(t) = CX(t) + du(t) \tag{1.3c}$$

Note: Here Equation 1.3b and 1.3c are alternate representation of Equation 1.3a.

One can recall in the s-domain, the transfer function is given by $G(s) = C(sI - A)^{-1}B + d$; the state vector is given by a column vector with n-components as

$$\mathbf{X}(t) = [x_1(t) \ldots x_n(t)]'$$

where
$A = n \times n$ system matrix
$B = n \times 1$ control matrix
$C = 1 \times n$ output matrix
$d =$ scalar

for the generation of $G(s)$ using A, B, C and D.

2. Transfer function representation in the s-domain is shown in Equation 1.4

$$\frac{Y(s)}{U(s)} = G(s) = \frac{(b_0 s^n + b_1 s^{n-1} + \cdots + b_n)}{(s^n + a_1 s^{n-1} + \cdots + a_n)} \tag{1.4}$$

where
$s =$ Laplace transform variable or operator
   Then in general for an n-dimensional SISO
$G(s) =$ is defined as the Laplace transform of the impulse response of the system = £[impulse response]

This representation in the s-domain as in Equation 1.4 is equivalent to Equation 1.3a if initial conditions are equal to zero.

1.3 Basic Feedback Control Loop: Single Input Single Output (SISO) System

Consider a particular subsystem under local computer control as shown in Figure 1.3. $r(t)$ is the desired setpoint, $u(t)$ is the control input to the subsystem with known dynamics in the continuous time domain, and $y(t)$ is the output of the subsystem.

1.3.1 Goal

The feedback control problem is to design the controller to have $y(t) \approx r(t) =$ reference setpoint, even in the face of external disturbances, imprecise models of dynamics, and so on.
   Error, $e(t)$, is defined as

$$e(t) \triangleq r(t) - y(t) \tag{1.5}$$

- We define $u(t) =$ control signal produced by the controller. This is the input signal to the subsystem (e.g., valve opening, fuel flow; field excitation to generator, etc.). If only measurements of the system are variable, then the resulting system is defined as

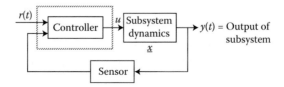

FIGURE 1.3
A particular subsystem under local computer control.

FIGURE 1.4
Samples of strep and/or ramp input.

$$u(t) = H[r(t), y(t)]; \quad \text{Output Feedback} \tag{1.6}$$

- Often we have measurements of the system state X available for feedback, in which case:

$$u(t) = H[r(t), x(t)]; \quad \text{State Variable Feedback (SVFB)} \tag{1.7}$$

- When $r(t) = 0$ we desire $y \approx 0$. This is a regular problem, where we wish to bring the system to the rest state (e.g., reduce spin of a satellite).
- When $r(t) \neq 0$ we have an output command problem, where we wish to reduce $e(t) \to 0$. Typically, $r(t)$ is a step or ramp command as shown in Figure 1.4.

1.4 Continuous Control Structures: Output Feedback

As shown in Figure 1.5, continuous control structures with output feedback can be discussed with the set of equations shown below:

$$T(s) = \frac{G(s)H(s)}{1 + G(s)H(s)} = \frac{y(s)}{r(s)} \tag{1.8}$$

The input to the system $u(s)$ is given by

$$u(s) = H(s) \cdot e(s) = H(s)[r(s) - y(s)] \tag{1.9}$$

Here, the design objective is to determine the transfer function $H(s)$ so that $y(t) \to r(t)$ "nicely" and the closed loop has desirable stability/transient response. Usually, $H(s)$ has a simple form, for example,

$$H(s) = K(1 + s/\alpha\omega_1)/(1 + s/\omega_1) \ \alpha > 1; \quad \text{Lag Compensator} \tag{1.10}$$

$$H(s) = K(1 + S/\omega_2)/(1 + s/\beta\omega_2) \ \beta > 1; \quad \text{Lead Compensator} \tag{1.11}$$

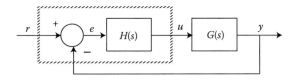

FIGURE 1.5
A simple continuous control structure with output feedback.

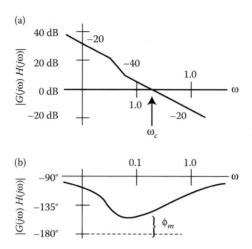

FIGURE 1.6
(a) The Bode plot—magnitude of GH versus frequency plot. (b) The Bode plot—angle of GH versus frequency plot.

Equations 1.10 and 1.11 can sum together to form the Lag-Lead compensator. In addition,

$$H(s) = K_0 + K_1/s + K_2 s/(\tau_2 s + 1); \quad \text{PID Compensator (~Lag-Lead)} \tag{1.12}$$

Selection of design parameters like $(K, \alpha, \beta, \omega_1, \omega_2, ...)$ can be achieved via either root locus or the Bode plot methods—which entail classical design methods using properties of loop gain CG, that is, *GH*—is shown in Figures 1.6a and b, respectively, where:

ω_c = crossover frequency, place at which $|GH|_{j\omega_c} = 1$; and ω_c is a measure of closed-loop (CL) system bandwidth

Φ_m is defined as the phase margin = π+ angle of $GH|_{\omega_c}$

Φ_m is a measure of the CL stability and tolerance to loop delays

$$\text{Recall angle of } e^{-s\tau}\Big|_{s=j\omega c} = -\omega_c \tau$$
$$\Rightarrow \tau_{\max} = \Phi_m / \omega_c$$

A good design will have a $\Phi_m \approx 45°$ to 60°. This aspect is always kept in mind in the "classical design" methodology.

1.5 Continuous Control Structures: State Variable Feedback

In the continuous time domian the system transfer function is defined as

$$G(s) = C(sI - A)^{-1}B\{+d\} \tag{1.13}$$

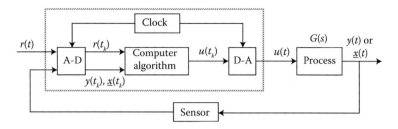

FIGURE 1.7
Continuous control structure with state variable feedback.

$$u(t) = K_r r(t) - K_c \underline{X}(t); \quad K_c = 1 \times n \text{ gain matrix} \tag{1.14}$$

In this continuous control structure as shown in Figure 1.7, the design objective is to have $y(t) \rightarrow r(t)$, especially when

$$r(t) = \text{step input}$$

and to have a desirable closed-loop stability and transient response.
CL system dynamics are

$$X(t) = (A - BK_c) \, \underline{X}(t) + BK_r r(t) \tag{1.15}$$

$$y(t) = C \, \underline{X}(t) + \{du(t)\}$$

- The selection of feedback gains K_c is done so that eigenvalues of $A - BK_c$ are in suitable locations on the left-hand s-plane, which implies a pole placement problem.

While the crossover frequency and phase margin are evaluated by examining the Bode plot of,

$$\text{Loop gain (LG)} = K_c \, (sI - A)^{-1} B \big|_{s=j\omega} \tag{1.16}$$

This implies that continuous controllers require continuous (i.e., analog) feedback of $y(t)$ and/or $x(t)$ and implementation using analog components (e.g., circuits, op-amps, analog chips).

1.6 Digital Control Basic Structure

We are now dealing not with continuous signals in the controller, but with samples of these signals. Usually, $t_k = kh$, where h is the sample time interval. The real-time clock maintains synchronism. The basic structure is shown in Figure 1.8 and the digitized signals that are compared between desired and output are shown in Figure 1.9. The next steps are described below.
  * Primary steps in computing $u(t)$:

      1. Wait for interrupt at time t_k

(A-D)   2. Sample $r(t)$ and $y(t)$ to obtain

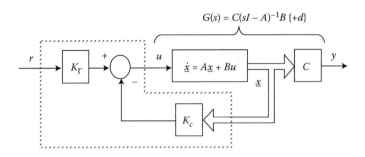

FIGURE 1.8
Basic structure of a digital control system.

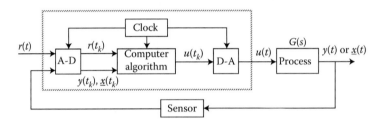

FIGURE 1.9
Digitized signals between output, and reference signals using A/D, D/A and persoal computer (PC).

$y(k)$ = value of $y(t)$ at time $t = t_k$

$r(k)$ = value of $r(t)$ at time $t = t_k$

3. Compute $u(k)$

$U(k) = H[y(k), y(k-1), r(k), \dots, u(k-1), u(k-2), \dots]$

(D-A)   4. Output $u(k)$ through the D-to-A converter to give $u(t)$

If the D-A is a hold circuit, then $u(k)$ = value of the control <u>over</u> the time interval $[t_k + \epsilon, t_k + h\,\epsilon]$, where ϵ is the computational delay at step 3 due to A-D characteristics; which are generally a few nanoseconds with present nanotechnology the delays could be as low as picoseconds.

5. Precompute any variables needed for the next cycle

6. Return to step 1 with $k = k + 1$

Note: All operations in steps 2 through 5 must be done in less than h sec!, to avoid any further complications like aliasing to be discussed later.

1.7 Relationship of Time Signals and Samples

Due to signals obtained at different samples $t_k = kh$, we need to study the relationship of these signals and related samples. There is often a computational delay at step 3 depending

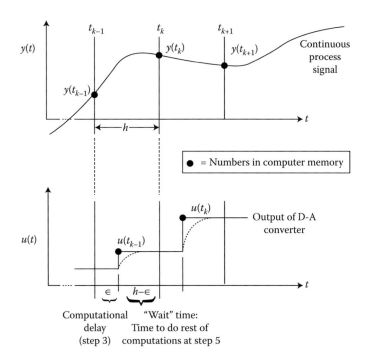

FIGURE 1.10
Continuous process signal and output of a D-A converter.

on the complexity of the computations performed. This process of time sampling is shown in Figure 1.10.

The delay ϵ is like a delay in $y(t)$ or in the process, that is, $G(s) \to G(s)\, e^{-\epsilon s}$.

Subsequently as a disadvantage, delays reduce Φ_m.

It is important to note here to minimize the delay at step 3 by arranging things so as to do the least amount of computations at step 3, while shifting the rest of the calculations to step 5.

1.8 A Typical Algorithm for *H*

$$e(k) = r(k) - y(k) = \text{error at time } t_k \tag{1.17}$$

$$u(k) = -\alpha_1\, u(k-1) - \alpha_2\, u(k-2) - \cdots -\alpha_m\, u(k-m) + \beta_0\, e(k) + \beta_1\, e(k-1) + \cdots + \beta_m e(k-m) \tag{1.18}$$

This is a difference equation, that is, a relationship between a sequence of values. An alternate way of writing the algorithm is via a discrete transfer function.

$$\text{Notationally, } u(k) \to u(Z),\, u(k-i) \to z^{-1}\, u(z)$$

$$u(z) = H(z)\, e(z) \tag{1.19}$$

$$H(z) = (\beta_0 + \beta_1 z^{-1} + \cdots + \beta_m z^{-m})/(1 + \alpha_1 z^{-1} + \cdots + \alpha_m\, z^{-m}) \tag{1.20}$$

Referred to as an "*m*th order compensator."
Implementation of Equation 1.18:

1. Directly as shown at Step 3 in Section 1.6. This would involve ~(2 *m* + 1) Multiplication and ADDS (MADDS)

   or

2. Compute

$$u(k) = \text{WI} + \beta_0\, e(k) \text{ at step 3}$$

where

$$\text{WI} = -\sum_{i=1}^{m} \alpha i\, u(k - i) + \sum_{i=1}^{m} \beta i\, e(k - i) \tag{1.21}$$

was computed at step 5 during the previous time step. Requires only 1 MADD!

Clearly this is a clever organization of the algorithm that can help to reduce ϵ, the time delay as required.

1.9 Differences in Digital versus Analog Control Methods

1. In continuous control, control design is based on samples of $y(t)$, $r(t)$, and $x(t)$.
2. In digital control, control input to the system is piecewise constant over intervals of length h (assuming D-A is a hold circuit).
3. Additional computational delays are encountered due to digitization of signal.
4. Further controlling a continuous system, $G(s)$, using a discrete algorithm, $H(z)$
   ⇨ A mix of continuous and discrete elements constitutes the feedback (FB) loop.
5. Most analysis will need to be performed using z-transforms and working in the z-plane with digitized signals of $y(k)$, $r(k)$, and $x(k)$.
6. Computer-aided design software becomes much more necessary for analysis, design, and eventual evaluation.
7. Effects of roundoff error in computations exist due to finite word length.
8. Quantization error in A-D conversion relevant to micro and nanotechnology used in the related hardware.

1.10 Computing the Time Response of a Linear, Time Invariant, Discrete Model to an Arbitrary Input

Consider the time response of a Linear Time Invariant (LTI) discrete model as shown in Figure 1.11. System response is considered only at times when $t = kh$.

We assume; the system is initially at rest at $t = 0$.

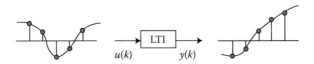

FIGURE 1.11
Time response of a linear time-invariant discrete model.

The objective is to find response y at time "kh" knowing the system input as seen in Figure 1.11

$$u(i) \quad \text{for } i = 0, 1, \ldots, k \tag{1.22}$$

First, consider the system's response to a unit pulse (see Figure 1.12):

$$\delta_0(k) = 1 \forall k = 0; \quad 0 \text{ otherwise}; \quad y(k) = g(k)$$
$$\triangleq \text{unit pulse response}; \quad 0 \text{ otherwise for } k < 0 \tag{1.23}$$

Then

$$y(k) = \sum_{i=0}^{k} u(i)g(k-i)$$

Now, if $u(.)$ is an arbitrary input (see Figure 1.13).

This is called the convolution sum formula. Its derivation requires assumptions of linearity and time invariance. This is cumbersome to evaluate.

Note: Continuous time analog is given by

$$y(t) = \int_0^t u(\sigma)g(t-\sigma)d\sigma$$

FIGURE 1.12
System response to a unit pulse.

$$y(k) = u(0)\,g(k) + u(1)\,g(k-1) + \cdots$$

Output response to $u(0)\,\delta_0(k)$ Output response to $u(1)\,\delta_0(k-1)$

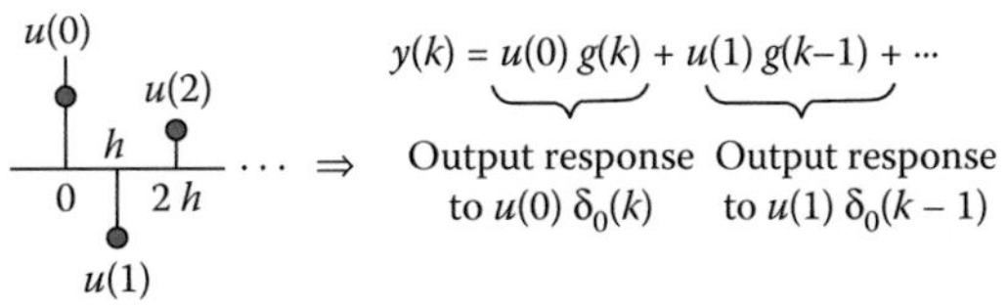

FIGURE 1.13
System response to an arbitrary input.

1.11 Review of *z*-Transforms for Discrete Systems

Z-Transforms is a Tool to Help in the Computation of System Response.
One-sided *z*-transform of a time sequence $f(k)$ is defined as

$$F(Z) \triangleq \sum_{k=0}^{\infty} f(k)z^{-k} \text{ (region of convergence } |z| > 1)$$

$$= f(0) + f(1)z^{-1} + \cdots \tag{1.24}$$

The coefficient of z^{-k} is $f(k)$.

Note that $F(z)$ is like the Laplace transform of a sequence of impulses as shown in Figure 1.14

$$f(0)\delta(t) + f(1)\delta(t - h) + \cdots \text{ with } e^{sh} \text{ to } z \tag{1.25}$$

$$z^{-1} = e^{-sh} \sim h \text{ sec delay}$$

$F(z)$ would not be useful if it were not possible to obtain a closed-form expression for the series definition, for example, if

$$f(k) = \sigma_1(k) = \text{unit step:}$$
$$F(z) = 1 + z^{-1} + z^{-2} + \cdots = \frac{1}{1 - z^{-1}}$$

Advantage for systems analysis: Taking the *z*-transform of the convolution sum yields

$$y(z) = G(z)u(z) \tag{1.26}$$

The first term on the right-hand side (RHS) is the *z*-transform of $g(k)$ and the second term is the *z*-transform of $u(k)$, respectively, thus

$G(z)$ = discrete transfer function of the system as $G(s)$ is the transfer function in the continuous time domain.

1.12 Some Useful Results for One-Sided *z*-Transforms

If $f(k)$ has transform $F(z)$, then:

FIGURE 1.14
Unit step in digital domain.

TABLE 1.1

z-Transform Table

$f(k)$	$F(Z) \triangleq \sum\limits_{k=0}^{\infty} f(k)z^{-k}$
$\delta_0(k)$ = unit pulse	1
$\delta_1(k)$ = unit step	$1/(1 - z^{-1}) = z/(z - 1)$
k (unit ramp)	$z/(z - 1)^2$
a^k (exponential)	$z/(z - a)$
ka^k	$az/(z - a)^2$
$a^k \cos(kb)$	$[z(z - a \cos b)]/[z^2 - 2az \cos b + a^2]$
$a^k \sin(kb)$	$[za \sin b]/[z^2 - 2az \cos b + a^2]$

$$kf(k) \rightarrow -z \; d/dz \; F(z)$$
$$f(k - n) \rightarrow z^{-n}[F(z) + f(-1)z + f(-2)z^2 + \cdots + f(-n)z^n]$$
$$f(k - n) \rightarrow z^n[F(z) - f(0)z - f(1)z^{-1} - \cdots - f(n - 1)z^{-(n-1)}]$$

For example: $f(k + 1) \rightarrow z[F(z) - f(0)]$
Final value theorem: ($z = 1$ is steady state, i.e., $s = 0$)

$$\lim f(k) \big|_{k \to \infty} = (1 - z^{-1})F(z) \big|_{z=1}$$

The one-sided z-transforms are shown in Table 1.1 for some commonly used discrete time functions.

1.13 How to Find $y(k)$ Using z-Transforms

a. Compute the z-transform of $u(k)$ by using Equation 1.24, or use Table 1.1
b. Obtain $y(z) = G(z) \, u(z)$
c. Compute inverse transform of $y(z)$ to get $y(k)$

There are several possibilities that exist for step c above:

1. Divide denominator of $y(z)$ into numerator to get

$$y(z) = c_0 + c_1 z^{-1} + c_2 z^{-2} + \cdots \tag{1.25}$$

and then obtain $y(k)$ from coefficients. This method is good only if you want one or two values of y.
2. Use tables to obtain $y(k)$. This method is good when $y(z)$ is not complex and low order is less than 2.

3. Do a partial fraction expansion of $y(z)$ and use tables on the simpler factors

$$y(z) = \frac{(2z^2 - z)}{(z^2 + 2z + 0.75)};$$

Do partial fraction expansion (PFE) of $y(z)/z$ as

$$y(z)/z = (2z - 1)/(z^2 + 2z + 0.75)$$
$$y(z)/z = 4/(z + 3/2) - 2/(z + 1/2)$$
$$y(z) = 4z/(z + 3/2) - 2z/(z + 1/2)$$

This yields: $y(k) = 4\left(-\frac{3}{2}\right)^k 2\left(-\frac{1}{2}\right)^k \quad k = 0,1,\ldots$

4. Use complex contour integration formulae.
   Note: If you ever have

   $$y(z) = c_{-1}z + c_0 + c_1 z^{-1} + \cdots.$$
   $\Rightarrow$ non-causal system with response@$k = -1 \Rightarrow$ not in the scope of this material in this book.

1.14 Use of z-Transforms to Solve nth Order Difference Equations

Consider an nth order difference equation with input u and output y:

$$y(k) + a_1 y(k - 1) + \cdots + a_n y(k - n) = b_0 u(k) + b_1 u(k - 1) + \cdots + b_n u(k - n) \qquad (1.26)$$

Generally, $b_0 = 0$ when Equation 1.26 represents a model of a dynamic system. To compute $y(k)$, $k \geq 0$, one needs to know the following:

n past values of y (stored in pushdown vector)
n past values of u (stored in pushdown vector) and
$u(k)$ = current value of input

Start with $k = 0$, if you know $y(-1)$, ..., $y(-n)$ (zero if system was initially at rest), and $u(-1)$, ..., $u(-n)$ (zero if input was applied at $k = 0$), and $u(0)$, then you can compute $y(0)$. Store $u(0)$, $y(0)$ in stack.
Next, compute $y(1)$, etc.
This procedure is quite suitable for computer simulation.to obtain $y(k)$ analytically, given $u(z)$ and the initial conditions $y(-1)$, $y(-n)$, $u(-1)$, ..., $u(-n)$, we take the z-transforms of both sides of Equation 1.26 using the relation for $i > 0$,

$$Z\{f(k-1)\} = z^{-i}[F(z) + f(-1)z + f(-2)z^2 + \cdots + f(-i)z^i] \tag{1.27}$$

Apply to each term in Equation 1.26

$$Z\{y(k)\} = y(z); \quad Z\{u(k)\} = u(z)$$

to obtain:

$$y(z) = \frac{\beta_0 + \beta_1 z^{-1} + S + \cdots + \beta_n z^{-n}}{(1 + \alpha_1 z^{-1} + \cdots + \alpha_n z^{-n})} u(z) + A(z) \tag{1.28}$$

The coefficient function of the first term on the RHS is $G(z)$ = transfer function of the system, and the second term involves all of the initial condition terms (which results into a mess):

$$y(z) = \text{Forced response} + \text{Free response} = B(z) + A(z)$$
$$G(z) = (b_0 z^n + b_1 z^{n-1} + \cdots + b_n)/(z^n + \alpha_1 z^{n-1} + \cdots + \alpha_n) \tag{1.29}$$
$$= n(z)/p(z)$$

If a system is initially at rest at $k = 0$, and if $u(k) = 0$ for $k < 0$, then $Y(z) = G(z)\,u(z)$ as seen before. But if not, one must use Equation 1.28:

$$y(k) + 1.5y(k-1) - y(k-2) = u(k-2)$$

Initial conditions are $y(-1) = 1.0$, $y(-2) = -1.5$ step input applied at time $k = 0 \rightarrow u(z) = z/(z-1)$. Take z-transforms:

$$y(z) = 1.5z^{-1}\left[y(z) + y(-1)z\right] - z^{-2}[y(z) + y(-1)z + y(-2)z^2]$$
$$= z^{-2}\left[u(z) + u(-1)z + u(-2)z^2\right]$$

$$\nwarrow \quad \nearrow$$
$$0 \cdots 0$$

The coefficients of the last two terms of the right-hand side (RHS) inside the braces $u(-1)$ and $u(-2)$ are zero, thus

$$y(z) = 1.5z^{-1}[y(z) + z] - z^{-2}[y(z) + z - 1.5z^2] = z^{-2}u(z)$$

which yields

$$y(z) = \frac{1}{z^2 + 1.5z - 1} u(z) + \frac{-3z^2 + z}{z^2 + 1.5z - 1} \text{ solution per Equation 1.28}$$

Forced response:
Let us now consider a forced response, from earlier approach

$$B(z) = 1/(z^2 + 1.5z - 1)z/(z-1);$$

Thus,

$$B(z) = (2/3\,z)/(z-1) - (4/5\,z)/(z-1/2) + 2/15(z/(z+2))$$

Therefore the forced response $= 2/3 - 4/5(1/2)^k + 2/15\,(-2)^k\ \ k \geq 0$

Free response A(z): Similarly $A(z)$ can be evalua   ted from earlier approach as

$$\frac{A(z)}{z} = \frac{-3z + 1}{(z - (1/2))(z + 2)} = \frac{-1/5}{z - 1/2} + \frac{-14/5}{z + 2}$$

Now using partial fraction expansion (PFE)

$$A(z) = -\frac{1}{5}\frac{z}{z - 1/2} - \frac{14}{5}\frac{z}{z + 2}$$

Thus,

$$\text{Free response} = -\frac{1}{5}\left(\frac{1}{2}\right)^k - \frac{14}{5}(-2)^k\ \ k \geq 0$$

Thus, the total response:

$$y(k) = \frac{2}{3} - \left(\frac{1}{2}\right)^k - \frac{8}{3}(-2)^k\ \ k \geq 0$$

which yields for different values of $k \geq 0$

$$y(k) = \{-3.\ 5.5,\ -10.25,\ 21.875,\ -42.06\ \ldots\}$$

Response grows without bounds as $k \to \infty$ due to the term $(-2)^k$ in $y(k)$, as shown in Figure 1.15.

What has been learned so far regarding the stability of response is evident from the previous example. Thus if the denominator of $G(z)$ has a root (eigenvalue) λ, $y(k)$ will have a component λ^k.

We now proceed to study the stability of the time response as in Section 1.15.

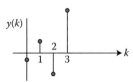

FIGURE 1.15
Output response $y(k)$ in the digital domain k.

1.15 Stability of the Time Response

$p(z)$ = Denominator of transfer function $\triangleq$ Characteristic polynomial (CP)

$$= z^n + a_1 z^{n-1} + \cdots + a_n$$

$$= (z - \lambda_1)(z - \lambda_2)\cdots(z - \lambda_n); \quad \lambda_i = \text{Roots of } p(z)$$

If input $u(k)$ does not $\to \infty$, then $y(k)$ does not $\to \infty$ if roots of λ_i of the characteristic polynomial of $|\lambda_i| < 1$ λ_i are called the poles of the system. We say a system is stable iff (if and only if) λ_i lies within the unit circle (see Figure 1.16).

Usually, the difference in Equation 1.26 is a discrete model of an underlying continuous process or continuous signal. We often have a good mental picture of the response of $g(t)$ given the poles s_i of the continuous transfer function $G(s)$ (see Figure 1.17).

How do we develop the same insight in the z-plane? This is the building block of the digital domain analysis in the sequel, using the mapping in the previous section Figure 1.17.

1.16 Continuous versus Discrete Relationships

$z = e^{sh}$ gives relationship between Laplace (s) and z domains $\hspace{2em}$ (1.30)

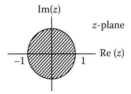

FIGURE 1.16
z-plane plot of roots of characteristic polynomial. The hatched region is the unit circle where stability of the system can be predicted.

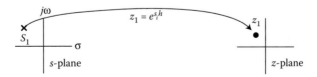

FIGURE 1.17
Impulse response $g(t)$ for various locations of poles in the continuous time domain.

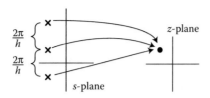

FIGURE 1.18
Many to one plotting from s to z plane.

EXAMPLE 1.2

If $f(t)$ has a Laplace transform with poles at s_1, s_2, ... Then the sampled $f(kh)$ has a z-transform with poles at $z_1 = e^{s_1 h}$, and so on, as shown

$$z_1 = e^{s_1 h}, \quad z_2 = e^{s_2 h}, ...$$

This mapping from s-plane to z-plane is shown in Figure 1.18.
One should now focus on a nonuniqueness property:

$$e^{(\sigma + j\omega)h} = e^{(\sigma + j\omega \pm j(2\pi/h)N)h} \quad N = \text{integer}$$

Figure 1.18 shows that

$$\Rightarrow s = \sigma + j\omega \quad \text{and} \quad \sigma + jw \pm j\frac{2\pi}{h}N \text{ give same } z$$

This is not 1-1 mapping → one has 1 primary point and rest aliases in this example. Accordingly a similar s-plane to z-plane mapping is shown in Figures 1.19 and 1.20.

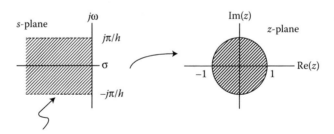

FIGURE 1.19
s-plane to z-plane plotting, Note: if $\text{Re}(s_i) < 0$, $|z_i| < 1$.

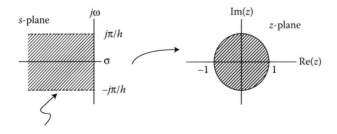

FIGURE 1.20
s-plane to z-plane mapping onto a unit circle.

The primary hatched strip in the *s*-plane will fill the unit circle. Points outside this strip will be duplicates.

1.17 *s*-to-*z* Plane Mappings

As shown in Figure 1.21, various *s* to *z*-plane mapping is displayed in (a) circles of radius $e^{\sigma h}$, (b) increasing ω, and (c) line of constant damping ζ. Figure 1.21 shows some useful mapping regions of *s*-plane into regions of *z*-plane unit circle.

1.18 LOCI of Constant Damping Ratio (ζ) and Natural Frequency (ω_n) in *s*-Plane to *z*-Plane Mapping

These loci of ζ versus ω_c are shown in Figure 1.22 for *s*-plane to *z*-plane mapping.

EXAMPLE 1.3

Figure 1.23 shows the *s*-plane to *z*-plane mapping for a strip in the *s*-plane to a unit circle with hatched regions equivalence.

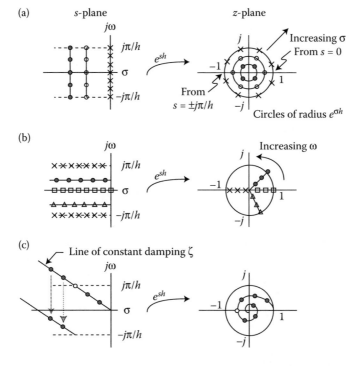

FIGURE 1.21
Various *s*-plane to *z*-plane mapping is shown in (a) circles of radius *eσh*, (b) increasing ω, and (c) line of constant damping ζ.

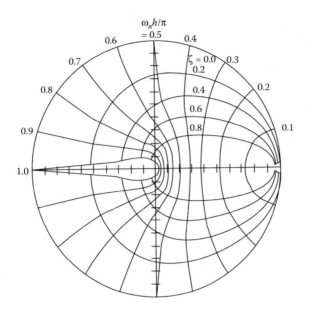

FIGURE 1.22
Loci of ζ versus ω_n curves for s-plane to z-plane mapping.

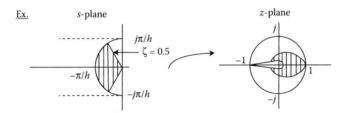

FIGURE 1.23
s-plane to z-plane mapping for a strip in the s-plane to a unit circle with hatched regions equivalence. Note: The semi-circle of radius π/h fills the entire unit circle minus the small "airfoil" area on z-axis.

Note: As seen the semi-circle of radius π/h fills the entire unit circle minus the small "airfoil" area on z-axis.

1.19 Signal Sampling and Data Reconstruction

One can examine the sampling process from a mathematical viewpoint as seen in Figure 1.24.

Where,

h = sampling period or time step

f_s = sampling frequency = number of samples/s = $1/h$

$\omega_s = 2\pi/h$

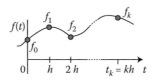

FIGURE 1.24
Sampled value of $f(t)$: $f_k = f(kh)$, for $t = kh$.

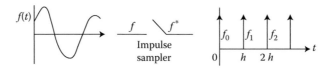

FIGURE 1.25
Mathematical model of an impulse sampler.

The problem here is that sampling a signal loses information, especially for the points between $(k-1)h$ and kh.

Thus, if we sample too slowly we lose information and if we sample too fast we overwork the computer. This leads to some major questions, that is,

1. How should sample occur so as not to lose information?
2. How should we approach reconstruction of the signal $f(t)$, or an approximation, from (f_k)?

This leads to the understanding of "impulse" sampling as a mathematical model (Figure 1.25):

Here one needs to understand that the area of impulse is f_k.

1.20 Impulse Sampling

$$f^*(t) = f_0\delta(t) + f_1\delta(t-h) + f_2\delta(t-2h) + \cdots \tag{1.31}$$

Equation 1.27 can be written as

$$f^*(t) = f(t) \cdot m^*(t) \tag{1.32}$$

where $m^*(t)$ is a periodic train of unit impulses $-\infty < t < \infty$.

The signal $f^*(t)$ is not "real," but when an impulse sampler is followed by a suitable transfer function $H_0(s)$, one can model any practical sampling situation.

This is illustrated in Figure 1.26.

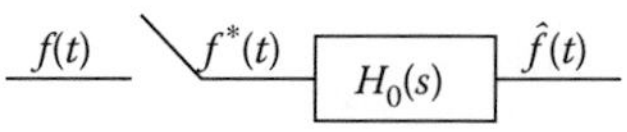

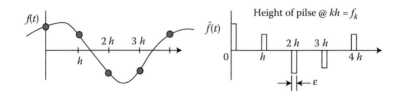

FIGURE 1.26
Model of a practical sampling condition for $f(t)$ with a suitable $H_0(s)$.

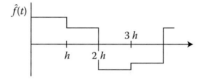

FIGURE 1.27
Impulse response of $H_0(s)$ as a pulse train.

FIGURE 1.28
Sample and hold as the most common form of sampling plus data reconstruction.

EXAMPLE 1.4

$$\text{If } H_0(s) = \frac{1}{s} - \frac{e^{-\epsilon s}}{s} = \frac{1}{s}(1 - e^{-\epsilon s}) \tag{1.33}$$

If Equation 1.33 is the impulse response of H_0, we get an output pulse train as shown in Figure 1.27 for an arbitrary input $f(t)$.

If $\epsilon = h$, the transfer function $H_0(s)$ is shown in Equation 1.33.

Figure 1.28 is a model of "sample and hold," which is the most common form of a sampling process that also results in a good data reconstruction of $f(t)$.

1.21 Laplace Transform of a Sampled Signal

Let us take Laplace transform of $f^*(t)$ as

$$\text{Take Laplace Transform of } f^*(t) \triangleq F^*(s)$$

$$F^*(s) = \int_0^\infty f^*(t)e^{-st}dt \tag{1.34}$$

$$= f_0 + f_1 e^{-sh} + f_2 e^{-2sh} \cdots$$

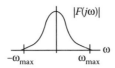

FIGURE 1.29
Magnitude of $F(j\omega)$ versus ω.

One can also see that since $z^{-1} = e^{-sh}$, then using the series expansion of e^{-sh}

$$F^*(s) = \sum_{k=0}^{\infty} f_k z^{-k} \big|_{z=e^{sh}} = F(z)\big|_{z=e^{sh}} \tag{1.35}$$

Where also $F(z) = z$-transform of the sampled sequence $\{f_k\}$, notationally,

$$F^*(s) = Z\{f(kh)\}\big|_{z=e^{sh}} \tag{1.36}$$

We wish to examine the relationship between $F^*(s)$ and $F(s) = $ Laplace transform of $f(t)$, and between

$$S_F(j\omega) = \text{Spectrum or Power Spectral Density (PSD) off } (t) = |F(j\omega)|^2 \tag{1.37}$$

and

$$S_{F^*}(j\omega) = \text{Spectrum or Power Spectral Density (PSD) off}^*(t) = |F^*(j\omega)|^2 \tag{1.38}$$

The spectrum or PSD indicates where a signal has power as shown in Figure 1.29. For example, a sine wave has impulses at $\pm\omega_0$.

To find Laplace transform of $[f(t) \cdot m^*(t)]$, one must first use the Fourier series to get a different way to write $m^*(t)$. To analyze this further recall, if a signal $x(t)$ is periodic with period h, then

$$x(t) = \frac{1}{h}\sum_{n=-\infty}^{\infty} c_n e^{jn\omega_s t}, \quad \text{where } \omega_s = \frac{2\pi}{h} \tag{1.39}$$

where the Fourier coefficients are given as below:

$$c_n = \int_0^h x(t)e^{-jn\omega_s t}dt \tag{1.40}$$

1.22 Nyquist Theorem

To understand the previous Equation 1.40, apply Fourier series to $x(t) = \delta(t)$.

$$\rightarrow c_n = \int_0^h \delta(t)^{-jn\omega_s t} dt = 1 \forall \text{ all } n \tag{1.41}$$

So, an alternate representation of $m^*(t)$ is

$$m^*(t) = \frac{1}{h} \sum_{n=-\infty}^{\infty} e^{jn\omega_s t} \tag{1.42}$$

and

$$f^*(t) = \frac{1}{h} \sum_{n=-\infty}^{\infty} f(t) e^{jn\omega_s t} \tag{1.43}$$

Thus,

$$F^*(s) = \mathcal{L}[f^*(T)] = \frac{1}{h} \sum_{n=-\infty}^{\infty} \mathcal{L}\left[f(t) e^{jn\omega_s t} \right] \tag{1.44}$$

This frequency domain representation is illustrated in Figure 1.30. Using the relation $\mathcal{L}[x(t)e^{at}] = X(s - a)$,

$$F^*(s) = \frac{1}{h} \sum_{n=-\infty}^{\infty} F(s - jn\omega_s) \rightarrow F^*(j\omega) = \frac{1}{h} \sum_{n=-\infty}^{\infty} F(j\omega - jn\omega_s) \tag{1.45}$$

1.22.1 Nyquist Result

If the original signal $f(t)$ does not have any frequency components $> \omega_s/2$, we can (in theory) reconstruct/recover $f(t)$ from $f^*(t)$ using an ideal low-pass filter (LPF).

$$\omega_N = \frac{\omega_s}{2} = \pi/h$$

Equation 1.46 is called the Nyquist frequency. Thus, one must sample $f(t)$ at a rate that is at least twice the highest frequency ω_{max} in the signal:

$$\omega_s > 2\omega_{max} \tag{1.46}$$

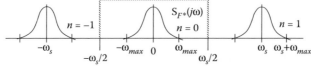

FIGURE 1.30
The Nyquist theorem illustrated using samples representing terms on LHS of Equation 1.44.

or

$$\omega_N > \omega_{max} \tag{1.47}$$

1.23 Recovering $f(t)$ from $f^*(t)$

Assume

$$\hat{F}(s) = H_0(s)F^*(s) \tag{1.48}$$

In an ideal case:

If $H_0(j\omega)|$ is the ideal shape as shown in Figure 1.31, then $\hat{f}(t) = f(t)$ and the signal is recovered from its samples. However, such an $H_0(s)$ is unrealizable.

Suppose, $H_0(s) = (1 - e^{-sh}/s)$, that is, $\hat{f}$ is a sample and hold (zero-order hold)

$$H_0(j\omega) = e^{-j\omega h/2} \left[\frac{e^{(j\omega h/2)} - e^{(j\omega h/2)}}{j\omega} \right] = e^{-j\omega h/2} \cdot h \cdot \left(\frac{\sin(\omega h/2)}{(\omega h/2)} \right) \tag{1.49}$$

$$\Rightarrow |H_0(j\omega)| = h \left| \frac{\sin(\omega h/2)}{(\omega h/2)} \right| \tag{1.50}$$

$$\text{Angle of } H_0(j\omega) = -\frac{\omega h}{2} \left(\text{delay of } \frac{h}{2} \text{ seconds,} \quad \text{for } \omega < \frac{2\pi}{h} = \omega_s \right) \tag{1.51}$$

This is an approximation to an ideal LPF (Figure 1.32).

Still, one can get some high-frequency components in $\hat{f}(t)$. Other signal reconstructors $H_0(s)$ are possible (e.g., polynomial interpolators) but usually are not worth the added

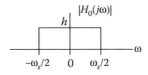

FIGURE 1.31
Ideal shape of $H_0(j\omega)$.

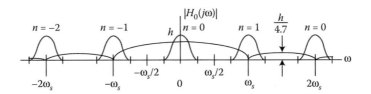

FIGURE 1.32
Magnitude of $H_0(j\omega)$ versus ω, which shows the approximation to an ideal LPF as in Equation 1.48.

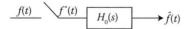

FIGURE 1.33
Digitized signal of $f(t)$.

complexity. The zero-order hold is the most common form of $H_0(s)$. The digitized block-diagram representation of f(t) is shown in Figure 1.33.

1.24 Aliasing

An interesting phenomenon happens when $\omega_s/2 < \omega_{max}$.

In this case, the components of $F(j\omega - jn\omega_s)$ overlap in $S_{F^*}(j\omega)$ and it becomes impossible to recover $f(t)$.

In addition, the sampled signal $f^*(t)$ has power at frequencies not present in the original signal $f(t)$!

EXAMPLE 1.5

$$f(t) = A \sin \omega_0 t \text{ and we sample at } \omega_s < 2\omega_0 \tag{1.52}$$

The original signal is "hidden," and the sampled signal is an "alais." The low-frequency signal does not really exist in $f(t)$, but will exist in see Figures 1.34 and 1.35.

$$\hat{f}(t) = f(t)$$

Since $H_0(s)$ is an LPF.

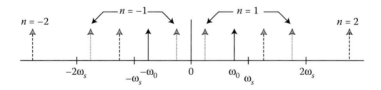

FIGURE 1.34
Low-frequency components in $f^*(t)$ at $(\omega_s-\omega_0)$ in the digital domain.

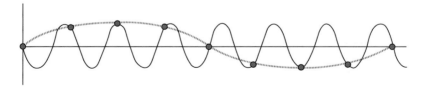

FIGURE 1.35
Low-frequency components as in Equation 1.48.

TABLE 1.2

Sampling of a Signal $f(t)$ with Components of 0.1, 0.8, and 1.4 Hz, Respectively

	$n = 0$	$n = 1$	$n = 1$	$n = 2$	$n = 2$
f_1	0.1	1.9	1.9	3.9	4.1
f_2	0.8	1.2	2.8	3.2	4.8
f_3	1.4	0.6	3.4	2.6	5.4

EXAMPLE 1.6

Sample a signal $f(t)$ that has frequency components at $f_1 = 0.1$ Hz, $f_2 = 0.8$ Hz and $f_3 = 1.4$ Hz using $f_s = 2$ Hz (note Nyquist says $f_s > 2.8$ Hz). What are the first five positive frequency components of the sampled signal?

The results of the sampled signal from Example 1.6 are shown and tabulated in Table 1.2.

1.25 How to Avoid Aliasing

1. There is no way to fix $f^*(t)$ after you have sampled. So, you must assure that the signal to be sampled has no frequencies higher than $\omega_N = \pi/h$.

   But, real signals have power in $[-\infty, \infty]$ (with a neat caveat!). This implies the need to prefilter the signal (f9t) before sampling (antialiasing, Figure 1.36).

$$\text{Typical} \, G_f(s) = \frac{\omega_0^2}{s^2 + 2\zeta\omega_0 s + \omega_0^2} \quad \forall \zeta = 0.707 \text{ (Butterworth filter)} \qquad (1.53)$$

   Usually, pick $\omega_0 \approx \omega_N/2 = \pi/2\,h$ to be safe, but beware of using $G_f(s)$ in a feedback loop due to added negative phase shift that reduces Φ_m as shown in Figure 1.37.

2. Sampling for accuracy: For a single-sine wave, the Nyquist criterion suggests to use more than two (2) samples/period ($\omega_s > 2\,\omega_0$), but the reconstruction error via the zero-order hold is terrible.

If we sample and hold with $N \geq 4$ samples/period, then $h = 2\pi/N\omega_0$, then the maximum relative error $= ((A\sin 2\pi/N)/A) = \sin 2\pi/N$ and maximum relative error with $h/2$ shift is $= \sin \pi/N$.

Usually, one tries for $\omega_s = (10 \rightarrow 30)\, \omega_{max}$.

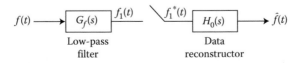

FIGURE 1.36

Prefiltering the signal before sampling, called antialiasing.

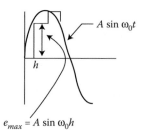

FIGURE 1.37
Sampling of a sinusoid $A \sin \omega_0 t$.

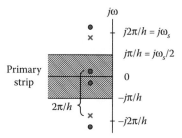

FIGURE 1.38
Aliasing frequencies in the strip resulting in all points $j2\pi N/h$ apart, give same z_i.

1.26 Interpretation of Aliasing in *s*-Plane

After sample and hold (or another type of reconstructor), we pick out predominantly those signals in the primary strip, $-\pi/h < \omega < \pi/h$ (Figure 1.38).

Since the aliased frequencies are not "real," that is, not in the original signal, any controller aimed at reducing the "observed" oscillations will fail.

Finally, aliasing effects will be observed in

- Frequency folding in *s*-plane
- Time response
- Fourier spectrum

1.27 Example of Aliasing in a Control Setting

In the system shown in Figure 1.39, the aliasing problem arises due to the backlash in the valve positioning. This leads to

$$\rightarrow \text{Oscillations in pressure } (P) \text{ and Temperature } (T)$$

Pressure and temperature are coupled and should oscillate at the same frequency! What happened?

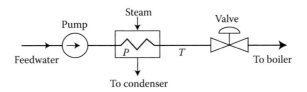

FIGURE 1.39
Feedwater–condenser–boiler system.

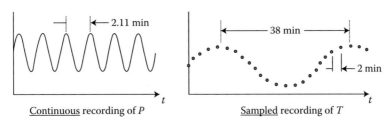

FIGURE 1.40
Continuous signal and its sampled output.

1. Sampling frequency, $\omega_s = 2\pi/2 = 3.14$ rad/min
2. Pressure oscillation frequency, $\omega_p = 2\pi/2.11 = 2.98$ rad/min
3. Lowest aliasing frequency, $\omega_s - \omega_p = 0.16$ rad/min $\rightarrow T = 38$ min

One can conclude that the sampler in this case did not take this course! This aliasing effect and reproduction of the continuous signal in an improper manner is shown in Figure 1.40.

PROBLEMS

P1.1   Consider the open-loop system as shown in Figure P1.1.

The plant is described as

$$\dot{X} = \begin{bmatrix} -1 & 0 \\ 0 & -2 \end{bmatrix} X + \begin{bmatrix} 4 \\ 1 \end{bmatrix} u$$
$$y = \begin{bmatrix} 3 & 5 \end{bmatrix} X$$

Find the transfer function: $T(z) = Y(z)/R(z)$.

P1.2   For the system shown in Figure P1.2, where

$$G(z) = Z[zoh \cdot G(s)] = \frac{k(z + 0.2)}{(z - 1)(z - 0.5)}$$

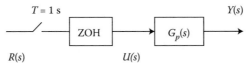

FIGURE P1.1
Simple open-loop system.

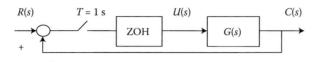

FIGURE P1.2
Simple close loop system with unity feedback.

    a. What is the minimum steady-state error for a unit ramp input that this system can deliver?

    b. What is the type of the system?

P1.3 Given the z-transform:

$$F = 0.792z^2/(z-1)(z^2 - 0.416z + 0.208)$$

    a. Determine the final value of $f(kT)$ using the final-value theorem.

    b. Verify the result by expanding $F(z)$ into a power series in z^{-1} (Figure P1.3).

P1.4 The forward-path transfer function $G(s)$ of the system shown in Figure P1.4 is

$$G(s) = \frac{10}{s(s+5)}$$

The feedback-path transfer function is

$$H(s) = 1$$

The sampling period $T = 0.1$ s.

    a. Find the $G(z)$ of the forward path.

    b. Find the characteristic equation of the closed-loop system.

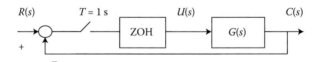

FIGURE P1.3
Sampled-data system with ZOH to find final value of $f(KT)$.

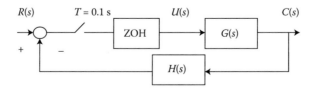

FIGURE P1.4
Sampled-data system with ZOH for a unity feedback system.

P1.5 A digital system is described by the differential equation:

$$c(k + 2) + 2c(k + 1) + 3c(k) = r(k)$$

The state diagram of the system is shown in Figure P1.5.

a. Write the output equation in the discrete domain.

b. Identify the A, B, C, and D components of the system state and measurement models and find the transfer function $T(z) = C(z)/R(z)$.

P1.6 Find the z-transform of

$$f(t) = te^{-at}$$

using the partial differential equation.

P1.7 The block diagram of a sampled-data system is shown in Figure P1.7.

a. Find the forward-path and closed-loop transfer functions of the system in the z-domain.

b. Use sampling time = 0.1 s.

c. Plot the unit-step response $y(kT)$ for $k = 0$–200.

d. Redo the problem for a sampling time of 0.05 s.

e. Use MATLAB®/Simulink® to illustrate your answers.

P1.8 The block diagram of a sampled-data system is shown in Figure P1.8.

a. Find the error constants K_p^*, K_v^*, and K_a^*.

b. Derive the transfer function $Y(z)/E(z)$ and $T(z) = Y(z)/R(z)$.

c. For $T = 0.1$ s, find the critical value of K for system stability.

d. Compute the unit step response $y(kT)$ for $k = 0$–100 for $T = 0.05$ s and $K_t = 5$.

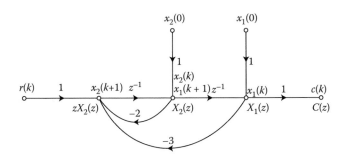

FIGURE P1.5
State diagram for differential equation of Problem P1.5.

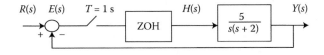

FIGURE P1.7
Sampled-data system with a single integrator.

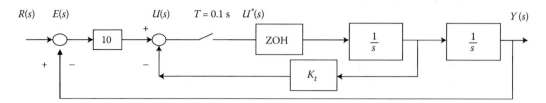

FIGURE P1.8
Sampled-data system with two integrators.

    e.   Repeat part (d) for $T = 0.1$ s and $K_t = 1$.

    f.   Use MATLAB/Simulink to illustrate your answers.

P1.9   A dc-motor is incorporated in a digital control system as shown in Figure P1.9a.

    A microprocessor collects the information from the encoder and computes the angular velocity ω_m information. This generates the sequential information $\omega(kT)$, $k = 0, 1, 2, 3,\ldots$. The microprocessor then generates the error signal $e(kT) - r(kT) - \omega(kT)$. The digital control is modified by the block diagram as in Figure P1.9b.

    For the following parameters:

$K_s = 1$ V/rad/s

$K - 10$

$K_b = 0.0706$ V/rad/s

$J = J_h + J_m = 0.1$ oz-in-s^2

$K_i = 10$ oz-in./A

$J_R = 0.05$ oz-in.-s^2

$K_D = 1$ oz-in.-s

$R_a = 1\ \Omega$

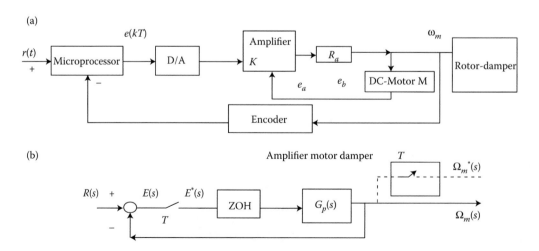

FIGURE P1.9
DC-motor incorporated in a digital control system.

a. Find the transfer function $\Omega(z)/E(z)$ with $T = 0.1$ s.

b. Find the CL transfer function $\Omega(z)/R(z)$. Find the characteristic equation (CE) and its roots. Locate these poles in the z-plane.

c. Is the system stable?

d. Repeat parts for $T = 0.01$ and 0.001 s.

e. Use MATLAB to show the performance of these systems.

f. Find the error constants K_p^*, K_v^*, and K_a^*.

g. Find the steady-state error $e(kT)$ for a unit step, unit-ramp, and a sinusoid input function.

h. Plot $e(kT)$. Use MATLAB to illustrate these results.

P1.10 A sampled data system is shown in Figure P1.10.

The Laplace transform of the output of the system is written in terms of input to the zero-order hold as

$$Y(s) = \frac{1 - e^{-Ts}}{s} \frac{1}{s + 1} U^*(s)$$

1. Find the z-transform of $Y(s)$.

2. Find the discrete dynamic equation of the system.

3. Draw the state diagram.

P1.11 The block diagram of a sampled-data system is shown in Figure P1.11.

1. Find the discrete-state equations of the open-loop system, assuming output $y(t)$ is not connected to the summing junction with $r(t)$.

2. Find the discrete-state equations of the closed-loop system, when the output $y(t)$ is connected to the summing junction with $r(t)$.

P1.12 The position control system is described in Figure P1.12.

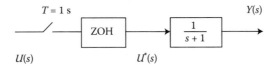

FIGURE P1.10
A block diagram representation for a sampled data system.

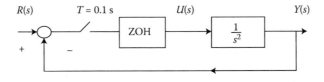

FIGURE P1.11
Sampled-data with zero order hold for a double integrator

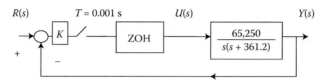

FIGURE P1.12
Position control sampled-data with unity feedback.

For $K = 14.5$, the transfer function of the controlled process is

$$G_p(s) = \frac{65,250}{s(s + 361.2)}$$

1. Find the forward-path transfer function for a sampling time $T = 0.001$ s.
2. Find the closed-loop transfer function of the system.
3. For a unit step input $R(z) = z/(z - 1)$, find the output transform $Y(z)$.
4. Plot the unit-step responses of the continuous and discrete systems using MATLAB.

P1.13 The state model is given by

$$X(k + 1) = \begin{bmatrix} 0.967144 & 0.14841 \\ -0.29682 & 0.521909 \end{bmatrix} X(k) + \begin{bmatrix} 0.016429 \\ 0.148411 \end{bmatrix} u(k)$$

The measurement model is given by

$$y(k) = [1 \quad 0]X(k)$$

Find the transfer function $G(z)$ for $h = 0.2$ s.

P1.14 The observable canonic form of the Problem 1.13 is given by

$$\begin{bmatrix} x_1(k + 1) \\ x_2(k + 1) \end{bmatrix} = \begin{bmatrix} 1.48905 & 1 \\ -0.548811 & 0 \end{bmatrix} \begin{bmatrix} x_1(k) \\ x_2(k) \end{bmatrix} + \begin{bmatrix} 0.01643 \\ 0.013452 \end{bmatrix} u(k)$$
$$y(k) = [1 \quad 0]x(k)$$

Find the transfer function $G(z)$ and confirm it is the same as that in Problem 1.13.

P1.15 A discrete time system is given by the state model as

$$\text{State model:} \begin{bmatrix} x_1(k + 1) \\ x_2(k + 1) \end{bmatrix} = \begin{bmatrix} \dfrac{1}{2} & \dfrac{1}{8} \\ \dfrac{1}{8} & \dfrac{1}{2} \end{bmatrix} \begin{bmatrix} x_1(k) \\ x_2(k) \end{bmatrix} + \begin{bmatrix} 1 & 0 \\ 0 & 1 \end{bmatrix} \begin{bmatrix} u_1(k) \\ u_2(k) \end{bmatrix}$$

$$\text{Measurement model: } y(k) = [1 \quad 2] \begin{bmatrix} x_1(k) \\ x_2(k) \end{bmatrix}$$

Find $y(k)$ if $x_1(k) = -1$, $x_2(0) = 3$.

P1.16 The state equations of a system are

$$\begin{bmatrix} x_1(k+1) \\ x_2(k+1) \end{bmatrix} = \begin{bmatrix} 0 & 1 \\ -3 & -2 \end{bmatrix} \begin{bmatrix} x_1(k) \\ x_2(k) \end{bmatrix} + \begin{bmatrix} 0 \\ 1 \end{bmatrix} r(k)$$

Output equation is

$$c(k) = \begin{bmatrix} 1 & 0 \end{bmatrix} \begin{bmatrix} x_1(k) \\ x_2(k) \end{bmatrix}$$

a. Draw the state diagram.
b. Find the CE.
c. Find the $(C(z)/R(z))$ for this system.

P1.17 Consider the difference equation:

$$y(k+1) + 2y(k) = 4^k; \quad k = 0,1,\ldots$$

with $y(0) = 0$

Show that the $y(k)$ resulting after applying the z-transform yields an expression composed of two delayed geometric sequences: one with ratio of 4 and the other with ratio -2, respectively.

P1.18 Find the pulse transfer function of the sampled-data analog system in Figure P1.18 consisting of an A/D converter with zoh in closed loop and an analog plant of transfer function:

$$G(s) = \frac{(s+1)}{[(s+2)(s+3)]}$$

P1.19 Consider the open-loop system as shown in Figure P1.19.

The plant is described by

$$\frac{dx_1}{dt} = -x_1 + 4u$$

$$\frac{dx_2}{dt} = -2x_2 + u$$

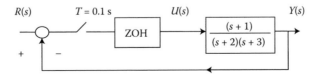

FIGURE P1.18
Unity feedback sampled-data analog system.

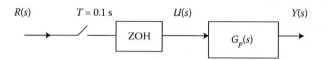

FIGURE P1.19
Open-loop system with an A/D and a ZOH.

The measurement model is described by

$$y = 3x_1 + 5x_2$$

Find the transfer function $T(z) = Y(z)/R(z)$ with $T = 0.5$ s.

2

Mathematical Models of Discrete Systems

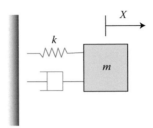

(See color insert.)

2.1 Discrete Time System Representations

2.1.1 Difference Equation Form

Consider an n-th order difference equation model of a system that we wish to control given by Equation 2.1

$$y(k) + a_1 y(k-1) + \cdots + a_n y(k-n) = b_0 u(k) + b_1 u(k-1) + \cdots + b_n u(k-n) \tag{2.1}$$

where we assume the input $u(k) = 0$ for $k < 0$.

The transfer function, $G(z)$, of the system defined by Equation 2.1 is given by Equations 2.2a and 2.2b as

$$G(z) = (b_0 + b_1 z^{-1} + \cdots + b_n z^{-n})/(1 + \alpha_1 z^{-1} + \cdots + \alpha_n z^{-n})$$
$$= y(z)/u(z) \tag{2.2a}$$

which yields

$$G(z) = \frac{b_0 z^n + b_1 z^{n-1} + \cdots + b_n}{z^n + \alpha_1 z^{n-1} + \cdots + \alpha_n} \tag{2.2b}$$

Generally, for most dynamical systems, $b_0 = 0$, that is, there exists some delay before the control applied at time kh appears at the output y.

Further, to use Equation 2.2 to find the time response $y(k)$, it is necessary to assume zero initial conditions for $y(-1)$, $y(-2)$, ..., $y(-n)$, or else one can follow the approach using z-transforms in Equation 2.1.

Unfortunately, neither difference representation in Equation 2.1 nor transfer function representations in Equations 2.2a and 2.2b are well-suited to the analysis of system/model performance via digital computer.

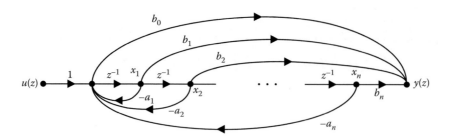

FIGURE 2.1
Signal flow diagram illustrating Mason's rule.

Toward this end, we will develop state-variable models; to describe discrete time systems for digital control applications with the advent of the micro-computers.

2.1.2 Signal Flow Diagram and Analysis

See Figure 2.1 that illustrates Mason's rule for:

$$\frac{y(z)}{u(z)} = \frac{b_0 + b_1 z^{-1} + \cdots}{1 + \sum_{i=1}^{n} a_i z^{-i}}$$

But what is important here is the interpretation of the rule. z^{-1} = unit (1 time step h) delay is seen to be a key element for the synthesis of a discrete model.

An n-th-order discrete system can be synthesized using n delay elements. The state of such a system is the stored energy in each delay element (i.e., the minimum amount of knowledge needed at time step k to predict the future $y(\bullet)$, knowing the future u).

Define $x_i(k)$ = output value of delay element i at time k. Our objective is to obtain an equation for the x_i values at time $k + 1$ in terms of $x_i(k)$ and $u(k)$.

Since a signal at a node = Σ of all incoming signals,

$$x_1(z) = z^{-1}\left[-a_1 x_1(z) - a_2 x_2(z) - \cdots - a_n x_n(z) + u(z)\right]$$

$$x_2(z) = z^{-1}[x_1(z)]$$
$$\vdots$$
$$x_n(z) = z^{-1}[x_{n-1}(z)]$$

2.2 State Equations from Node Equations

Since $f(z) = z^{-1}g(z) \Rightarrow f(k + 1) = g(k)$ in time domain,

$$x_1(k + 1) = -a_1 x_1(k) - a_2 x_2(k) - \cdots - a_n x_n(k) + u(k)$$
$$x_2(k + 1) = x_1(k)$$
$$\vdots$$
$$x_n(k + 1) = x_{n-1}(k)$$

These are the time-recursive equations for the states x_i. The output,

$$y(k) = b_1 x_1(k) + b_2 x_2(k) + b_n x_n(k) + b_0 \left[u(k) - a_1 x_1(k) - a_2 x_2(k) - a_n x_n(k) \right]$$

or

$$y(k) = \tilde{b}_1 x_1(k) + \tilde{b}_2 x_2(k) + \tilde{b}_n x_n(k) + b_0 u(k), \quad \text{where } \tilde{b}_i = b_i - b_0 a_i$$

Note that we have replaced one n-th-order difference equation representation of a system with n first-order difference equations.

What has been gained? The ability to use a matrix-vector description, as defined in Equation 2.3

Define:

$$X(k) = \begin{bmatrix} x_1(k) \\ x_2(k) \\ \vdots \\ x_n(k) \end{bmatrix} = \text{State vector} \tag{2.3}$$

Note: Henceforth, all state vectors will be bold faced.

Then,

$$X(k + 1) = AX(k) + Bu(k) \tag{2.4}$$

where

$A = n \times n$ matrix; and B is $n \times 1$ matrix

$$y(k) = CX(k) + du(k) \tag{2.5}$$

and

$$C = 1 \times n \text{ matrix}$$

and we let $x(0) =$ initial state at time $k = 0$.

2.3 State Variable Forms: I

$$A = \begin{bmatrix} -a_1 & -a_2 & & -a_n \\ 1 & 0 & \cdots & 0 \\ 0 & 1 & & 0 \\ \vdots & & \ddots & \vdots \\ 0 & & \cdots & 10 \end{bmatrix} \quad B = \begin{bmatrix} 1 \\ 0 \\ \vdots \\ 0 \end{bmatrix} \tag{2.6}$$

$$C = \begin{bmatrix} \tilde{b}_1 & \tilde{b}_2 & \cdots & \tilde{b}_n \end{bmatrix} \quad d = b_0, \quad \text{where } (\tilde{b}_i = b_i - b_0 a_i)$$

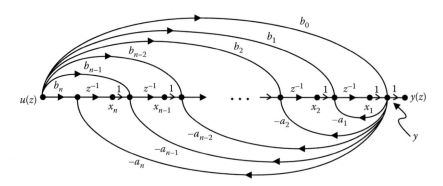

FIGURE 2.2
Standard observable form (SOF)—alternate form of $G(z)$.

This is called the Standard Controllable Form (SCF) or Control Canonical Form (CCF) as shown in Figure 2.1. Its advantages are

1. Easily written directly from $G(z)$ or difference equation
2. System characteristic polynomial by inspection

$$p(z) = z^n + a_1 z^{n-1} + \cdots + a_n$$

3. Minimal number of non-zero parameters in $\{A, B, C, d\}$ good model for system identification

But, the states are not related 1:1 with the physical variables, and mathematical operations using A are often numerically ill-conditioned.

An alternate form of $G(z)$ synthesis (see Figure 2.2):

$$\frac{y(z)}{u(z)} = \frac{b_0 + b_1 z^{-1} + \cdots}{1 + \displaystyle\sum_{i=1}^{n} a_i z^{-i}} = G(z) \tag{2.7}$$

2.4 State Variable Forms

Once again, define $x_i(k) =$ output of the delay element i:

$$x_1(z) = z^{-1}\left[b_1 u(z) - a_1 y(z) + x_2(z)\right]$$

$$x_2(z) = z^{-1}\left[b_2 u(z) - a_2 y(z) + x_3(z)\right]$$

$$\vdots$$

$$x_n(z) = z^{-1}\left[b_n u(z) - a_n y(z)\right]$$

In the time domain noting $y(k) = x_1(k) + b_0 u(k)$

$$x_1(k + 1) = -a_1 x_1(k) + x_2(k) + (b_1 - b_0 a_1)u(k)$$
$$x_2(k + 1) = -a_2 x_1(k) + x_3(k) + (b_2 - b_0 a_2)u(k)$$
$$x_n(k + 1) = -a_n x_1(k) + (b_n - b_0 a_n)u(k)$$

$$A = \begin{bmatrix} -a_1 & 1 & 0 & & -a_n \\ -a_2 & 0 & 1 & \cdots & 0 \\ & \vdots & & \ddots & \vdots \\ & & & & 1 \\ -a_n & 0 & & \cdots & 0 \end{bmatrix} \quad B = \begin{bmatrix} \tilde{b}_1 \\ \tilde{b}_2 \\ \vdots \\ \tilde{b}_n \end{bmatrix} \quad (2.8)$$

Using vector/matrix form $(\tilde{b}_i = (b_i - b_0 a_i))$

$$C = \begin{bmatrix} 1 & 0 & \cdots & 0 \end{bmatrix} \quad d = b_0$$

This is called the standard observable form (SOF) or observable canonical form (OCF). It has much the same properties as the SCF, but with $y = x_1 + b_0 u$.

Note:

$$A_{\text{SOF}} = A'_{\text{SCF}}$$
$$B_{\text{SOF}} = C'_{\text{SCF}}$$
$$C_{\text{SOF}} = B'_{\text{SCF}}$$

Note: There are other state variable forms that can be easily written, knowing the coefficients in $G(z)$.

2.5 Transfer Function of a State-Space Model

Quite often, the physical equations that describe a system mathematically are already available in state variable form, where the states are some set of physical variables (e.g., displacements, velocities, pressures, etc.); that is, we start with

State model: $X(k + 1) = A\ X(k) + B\ u(k)$ \hfill (2.9)

Observation model: $y(k) = C\ X(k) + \{d\ u(k)\}$

$X(0)$ = initial state; second term of the observation model is generally equal to 0 for most physical systems.

Suppose we want transfer function $G(z) = y(z)/u(z)$ for this model; we can go from $G(z)$ state representation easily enough. Incidentally, one wonders if the reverse direction can happen. If one takes z-transform of Equation 2.9 with

$$Z\{X(k)\} \triangleq \begin{bmatrix} x_1(z) \\ x_2(z) \\ \vdots \\ x_n(z) \end{bmatrix} = X(z)$$

and note

$$\begin{aligned} Z\{X(k+1)\} &= zX(k) - zX(0) \\ zX(z) - zX(0) &= AX(z) - zX(0) \\ zX(z) - zX(0) &= AX(z) + Bu(z) \end{aligned}$$

$$\Rightarrow \text{State vector: } X(z) = (zI - A)^{-1}zX(0) + (zI - A)^{-1}Bu(z) \tag{2.10}$$

and,

$$\text{Output: } y(z) = C(zI - A)^{-1}zX(0) + \left[C(zI - A)^{-1}B + d \right]u(z)$$

$$\text{Free response: } C(zI - A)^{-1}zX(0) \tag{2.11}$$

$$\text{Transfer function: } G(z) = \left[C(zI - A)^{-1}B + d \right]$$

The denominator of $G(z) = |(zI - A)| \triangleq p(z)$ is the characteristic polynomial (CP) of the

system $\Rightarrow$ Poles of system $\equiv$ Eigenvalues of λ_i of the system matrix

2.6 State Variable Transformation

We have seen that state variable descriptions of a system are not unique (e.g., SCF and SOF representations of the system). In fact, given any state variable (SV) model of a system, for example, Equations 2.9 through 2.11, it is possible to define a new state representation with state vector

$$V = T^{-1}X$$

The $n \times n$ matrix T (which must have an inverse) is called the "transformation matrix" (TM).

State equations for $V(k)$:

$$V(k+1) = T^{-1}ATV(k) + T^{-1}Bu(k) \tag{2.12}$$

where

$$A_v = T^{-1}AT \text{ and } B_v = T^{-1}B$$
$$y(k) = CTV(k) + du(k)$$

(2.13)

where

$$C_v = CT$$

Since the states are internal characterizations of a system, the external characterization between $u \Rightarrow y$, that is, the transfer function, should not change.

$$G_v(z) = C_v(zI - A_v)^{-1}B_v + d$$

$$= CT[zT^{-1}T - T^{-1}AT]^{-1}T^{-1}B + d$$

$$= CT[T^{-1}(zI - A)T]^{-1}T^{-1}B + d$$

$$= C(zI - A)^{-1}B + d = G(z)$$

It is possible to find the transformation matrix T such that the "new" model is in SCF, or SOF or "modal" (diagonal) form, and so on.

Note: Henceforth, and as in the previous two sections, all state vectors will be denoted in upper case bold, all matrices will be upper case, and all scalars will be lowercase.

2.7 Example

System model:

$$\begin{bmatrix} x_1(k+1) \\ x_2(k+1) \end{bmatrix} = \begin{bmatrix} 1.0 & 0.5 \\ 0.2 & 1.0 \end{bmatrix} \begin{bmatrix} x_1(k) \\ x_2(k) \end{bmatrix} + \begin{bmatrix} 1.2 \\ 0.5 \end{bmatrix} u(k)$$

where

$$X(k+1) = \begin{bmatrix} x_1(k+1) \\ x_2(k+1) \end{bmatrix}; \quad A = \begin{bmatrix} 1.0 & 0.5 \\ 0.2 & 1.0 \end{bmatrix}; \quad X(k) = \begin{bmatrix} x_1(k) \\ x_2(k) \end{bmatrix}; \quad B = \begin{bmatrix} 1.2 \\ 0.5 \end{bmatrix}$$

Measurement model:

$$y(k) = \begin{bmatrix} 1.0 & 0.0 \end{bmatrix} X(k) + 0.2u(k)$$

where

$$C = \begin{bmatrix} 1.0 & 0.0 \end{bmatrix} \quad d = 0.2$$

Transfer function:

Transfer function: $G(z) = \left[C(zI - A)^{-1}B + d \right]$

$$G(z) = \begin{bmatrix} 1 & 0 \end{bmatrix} \begin{bmatrix} z - 1.0 & -0.5 \\ -0.2 & z - 1.0 \end{bmatrix}^{-1} \begin{bmatrix} 1.2 \\ 0.5 \end{bmatrix} + 0.2$$

$$= \frac{\begin{bmatrix} 1 & 0 \end{bmatrix} \begin{bmatrix} z - 1.0 & 0.2 \\ 0.2 & z - 1.0 \end{bmatrix} \begin{bmatrix} 1.2 \\ 0.5 \end{bmatrix}}{(z - 1)^2 - 0.1} + 0.2$$

$$G(z) = \frac{0.2z^2 + 0.8z - 077}{z^2 - 2z + 0.9}$$

If we define a "new" set of SVs,

$$\begin{aligned} v_1 &= x_1 - x_2 \\ v_2 &= x_1 + x_2 \end{aligned} \Rightarrow V = \begin{bmatrix} 1.0 & -1.0 \\ 1.0 & 1.0 \end{bmatrix} X = T^{-1}X$$

$$A_v = T^{-1}AT = \begin{bmatrix} 1.0 & -1.0 \\ 1.0 & 1.0 \end{bmatrix} \begin{bmatrix} 1.0 & 0.5 \\ 0.2 & 1.0 \end{bmatrix} \begin{bmatrix} 0.5 & 0.5 \\ -0.5 & 0.5 \end{bmatrix} = \begin{bmatrix} 0.65 & 0.15 \\ 0.15 & 1.35 \end{bmatrix}$$

$$\text{and} \quad B_v = T^{-1}B = \begin{bmatrix} 1.0 & -1.0 \\ 1.0 & 1.0 \end{bmatrix} \begin{bmatrix} 1.2 \\ 0.5 \end{bmatrix} = \begin{bmatrix} 0.7 \\ 1.7 \end{bmatrix}$$

$$C_v = CT = \begin{bmatrix} 1 & -1.0 \end{bmatrix} \begin{bmatrix} 0.5 & 0.5 \\ -0.5 & 0.5 \end{bmatrix} = \begin{bmatrix} 0.5 & 0.5 \end{bmatrix}$$

$$d = 2.0$$

Computing $G(z)$ using A_v, B_v, C_v, and d will yield the same results as above.

2.8 Obtaining the Time Response $X(k)$

Thus, stepwise:

1. Recursively in the time domain, a numerical procedure:

initialize $X = X(0)$

then $X \Leftarrow AX + Bu(i)$ for $i = 0, 1, \dots, k$

via analytic expression,

$$X(1) = AX(0) + Bu(0)$$

$$X(2) = AX(1) + Bu(1) = A^2X(0) + ABu(0) + Bu(1)$$

$$X(3) = AX(2) + Bu(2) = A^3X(0) + A^2Bu(0) + ABu(1) + Bu(2)$$

$$\vdots$$

$$X(k) = A^kX(0) + \sum_{i=0}^{k-1} A^{k-1-i}Bu(i) \tag{2.14}$$

The first term on the RHS of Equation 2.14 is the free response, the second term is the forced response, and

$$y(k) = CX(k) + du(k) \tag{2.15}$$

Equation 2.14 is a closed expression for $X(k)$ in terms of $X(0)$ and the control applied over the time interval $[0, k-1]$. While "nice" Equation 2.14 is rarely used in practice, it is used mostly for concept development.

2. Via inverse z-transforms of $X(z)$

   This is quite laborious, but inverse transforms must be taken term by term.

   It is okay for low-order, or sparse systems

   $$X(z) = (zI - A)^{-1} zX(0) + (zI - A)^{-1}B\, u(z)$$

3. A heuristic z^{-1}

   1. Scalar case $z/(z - a) \to a^k \Rightarrow (zI - A)^{-1}z \to A^k$
   2. Scalar case $f(z)g(z)$

   $$f(z)g(z) \to \sum_{i=0}^{k-1} f(k - i)g(i)$$

By analogy,

$$f(z) = (zI - A)^{-1} = z^{-1}[(zI - A)^{-1}z] \Rightarrow f(k) = A^{k-1};$$

where z^{-1} = one step delay

$$g(z) = Bu(z) \Rightarrow g(k) = Bu(k)$$

so,

$$(ZI - A)^{-1}Bu(z) \Rightarrow \sum_{i=0}^{k-1} A^{k-1-i}Bu(i)$$

2.9   Computing $G(z)$ from A, B, C, d

Obtaining the coefficients, a_i and b_i, of $G(z)$ is tedious if done by manual calculation.

Leverier Algorithm for $G(z)$, or $a(z - A)^{-1}$

Let $P_i = n \times n$ matrices; $X = n \times n$ scratch matrix;

$$\text{And } tr(X) = \sum_{i=1}^{n} X_{ii}$$

$$P_1 = I \qquad \longrightarrow X = AP_1 \longrightarrow a_1 = -tr(X)/1; \quad \tilde{b}_1 = CP_1B$$

$$P_2 = X + a_1I \quad \longrightarrow X = AP_2 \longrightarrow a_2 = -tr(X)/2; \quad \tilde{b}_2 = CP_2B$$

$$\vdots$$

$$P_n = X + a_{n-1}I \longrightarrow X = AP_n \longrightarrow a_n = -tr(X)/n; \quad \tilde{b}_n = CP_nB$$

$$P_{n+1} = X + a_nI$$

1. $G(z) = \dfrac{\tilde{b}_1 z^{n-1} + \tilde{b}_2 z^{n-2} + \cdots + \tilde{b}_n}{z^n + \alpha_1 z^{n-1} + \cdots + \alpha_n} + d$

$$G(z) = \frac{b_0 z^n + b_1 z^{n-1} + \cdots + b_n}{p(z)}; \quad b_0 = d \text{ and } b_i = \tilde{b}_i + b_0 a_i$$

2. $(zI - A)^{-1} = \dfrac{\left[P_1 z^{n-1} + P_2 z^{n-2} + \cdots + P_n \right]}{p(z)}$ (not often required)

3. $P_{n+1} = 0$ in theory $\rightarrow$ provides an accuracy check. We want to be sure when P_{n+1} is very small with respect to A.

Define,

$$\| A \| = \sqrt{\frac{1}{n} \sum_{i=1}^{n} \sum_{j=1}^{n} a_{ij}^2} \sim \text{average size } \star \text{ of elements in } A$$

Then test $\| P_{n+1} \|$ vs. $\| A \|$.

The Leverier algorithm is one of several available for obtaining $G(z)$.

It is perhaps the simplest, but is often ill-conditioned (especially for matrices A with a widespread in eigenvalue magnitudes). It is usually okay if $n \leq 6$.

* This is not a "proper" norm, but rather a useful one. Other definitions include

$$\max_{i} \sum_{j=1}^{n} |a_{ij}|, \quad \max_{j} \sum_{i=1}^{n} |a_{ij}|, \text{ etc.}$$

2.10 Leverier Algorithm Implementation

(for $G(z)$ coefficients a_i, b_i)

    Initialize $P = I$, $b_0 = d$
    Do for $i = 1, n$
        Printout $P = P_i$ here if want $(z I - A)^{-1}$
        $X = AP$ ⎫
        $a_i = -tr(X)/I$ ⎬ printout and/or store in vector arrays **a** and **b**
        $b_i = CPB$ ⎭
        $P = X + a_i I$; add to diagonal only
    End do
    Compute $||P||$, $||A||$
    Check accuracy $||P|| \leq \epsilon \, ||A||$; $\epsilon = 10^{-8}$
    If $d \neq 0$, then Do for $i = 1, n$
$$b_i = b_i + a_i d$$
    End do

Transfer function evaluation: A necessary tool

Once a transfer function is obtained, what is done with it? Usually, we wish a Bode plot, that is, $|F|$ and angle of F vs. frequency ω (rad/s)

Develop a Subroutine "Bode":

Case 1: continuous model $F(s)|_{s=j\omega}$

Case 2: discrete model $F(z)|_{z=e^{j\omega h} = \cos \omega h + j \sin \omega h}$

$$F(x) = \frac{b_0 x^n + b_1 x^{n-1} + \cdots + b_n}{a_0 x^n + a_1 x^{n-1} + \cdots + a_n}$$

To evaluate F at a particular ω: (X, AA, BB, F is complex)

    If case = 1, $X = 0 + j\omega$
    If case = 2, $X = \cos \omega h + j \sin \omega h$
    Initialize $AA = a_0$; $BB = b_0$
        Do for $i = 1, n$
            $AA = a_i + X * AA$ Polynomial
            $BB = b_i + X * BB$ nesting
        End do
    $F = BB/AA$

Obtain $|F|$ in dB = $20 \log 10 \, |F|$, and angle of $F = \tan^{-1} [\text{Im}(F)/\text{Re}(F)]$ in degrees.

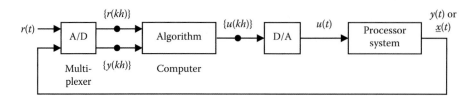

FIGURE 2.3
Standard digital control loop.

2.11 Analysis of the Basic Digital Control Loop

1. The computer algorithm generates a sequence of values $u(kh)$ from the discrete samples $y(kh)$ and $r(kh)$, or from $e(kh) = r(kh) - y(kh)$, then $u(z) = H(z)e(z)$.

2. Process model—continuous inputs and outputs

         transfer function   or   state-space model

$$G(s) \qquad \longleftrightarrow \qquad \dot{X} = AX + Bu, \; y = CX + du$$

3. The computer outputs values $u(kh)$ and at some later time sees the response $y(mh)$. The computer "puts out" samples and "sees" samples, that is, it sees a *discrete system* from $u(kh)$ to

$$y(kh) \Rightarrow \tilde{G}(z)$$

4. Redraw the loop from computer's view (e.g., $u(z) = H(z) \, e(z)$).

See Figure 2.4.

WHY?

1. To enable analysis as a discrete feedback (FB) loop.

2. To enable design of a discrete $H(z)$ vis-a-vis discrete $\tilde{G}(z)$.

3. We are "controlling" $\tilde{G}(z)$ not $G(s)$.

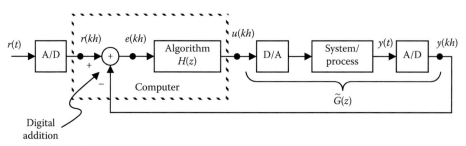

FIGURE 2.4
Redrawing of loop from computer's view.

2.12 Discrete System Time Signals

Typically, there will be the following delays in the loop:

- Computational delays
- Measurement delays
- Process delays

1. All these delays are lumped as some equivalent delay τ.
2. Assume: D/A is a zero-order hold and all A/Ds are synchronized (see Figure 2.5).
3. Via D–A (see Figure 2.6).
4. Via process dynamics (see Figure 2.7).

Definitions:

$$y(k) = y(kh) = \text{sampled value of } y(t) \text{ at time } t = kh$$

$u(k) = u(kh) = $ value of $u(\cdot)$ computed by algorithm using the sample $y(kh)$ and $r(kh)$;

output from computer at time kh^+. If there is no computational delay

$$\Rightarrow u(kh) = \text{value of system input over } [kh^+, (k+1)]$$

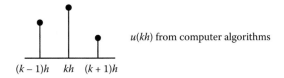

$u(kh)$ from computer algorithms

$(k-1)h \quad kh \quad (k+1)h$

FIGURE 2.5
Signals around the loop $u(kh)$.

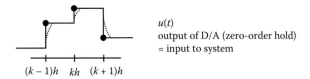

$u(t)$
output of D/A (zero-order hold)
= input to system

$(k-1)h \quad kh \quad (k+1)h$

FIGURE 2.6
$u(t)$ as an output of D/A (zero-order hold) as input to the system.

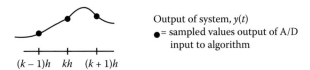

Output of system, $y(t)$
● = sampled values output of A/D
input to algorithm

$(k-1)h \quad kh \quad (k+1)h$

FIGURE 2.7
Output of the system $y(t)$ with sampled values "●" output of A/D input to algorithm.

2.13 Models for Equivalent Discrete System, $\tilde{G}(Z)$

1. System defined by state equations, no delay
2. System defined by transfer function, no delay
3. Modifications to 1 and 2 when $\tau \neq 0$

State-space approach:

$$\dot{X}(t) = AX(t) + Bu(t), \quad y(t) = CX(t) + du(t)$$

$$\Rightarrow G(s) = C(sI - A)^{-1}B + d$$

Compute:

$$X\big[(k + 1)h\big] \triangleq X(k + 1) = \text{value of } X(t) \text{ at } t = (k + 1)h$$

From the knowledge of $X(kh) =$ value of $X(t)$ at $t = kh$ and $u(kh) =$ system input over $(kh, (k + 1)h]$.

Use the state transition equation:

$$X(t_2) = e^{A(t_2 - t_1)}X(t_1) + \int_{t_1}^{t_2} e^{A(t_2 - \xi)}Bu(\xi)d\xi$$

$t_1 = kh, \quad t_2 = (k + 1)h$ and $u(\xi) = u(kh)$ over (t_1, t_2)

$$X\big[(k + 1)h\big] = e^{Ah}X(kh) + \int_{kh}^{(k+1)h} e^{A((k+1)h - \xi)}Bd\xi \cdot u(kh)$$

Let $\sigma = (k + 1)h - \xi$, which yields

$$X\big[(k + 1)h\big] = e^{Ah}X(kh) + \int_{0}^{h} e^{A\sigma}\,d\sigma\, Bu(kh) \tag{2.16}$$

$$\Rightarrow X(k + 1) = \Phi X(k) + \Gamma u(k) \tag{2.17}$$

where

$$\Phi = e^{Ah}; \quad \Psi(h) = \int_{0}^{h} e^{A\sigma}\,d\sigma; \quad \Gamma = \Psi(h)B$$

Output value of system input right at time $t = kh$ (subtle point)

$$y(kh) = C\mathbf{X}(kh) + du[(k-1)h] \qquad (2.18)$$

$$y(k) = C\mathbf{X}(k) + du(k-1) \qquad (2.19)$$

Transfer function: $\tilde{G}(z) = C(zI - \Phi)^{-1}\Gamma + dz^{-1}$

2.14 Computing Φ and Γ (or Ψ)

Note that Φ and Γ are independent of k. Compute once for a given time step h.

Analytic: $e^{Ah} = \mathcal{L}^{-1}[(sI - A)^{-1}]|_{t=h}$ exact value obtained, but very time consuming and not practical for $n > 3$. Then, need to, obtain Ψ by integrating $e^{A\sigma}$ over $[0, h)$.

Note for an interval specification a "[" signifies closed interval beginning and ")" open interval closing.

Numerical: If h is small $\Rightarrow$ Taylor series approximations are good

$$e^{Ah} = I + Ah + \frac{A^2 h^2}{2!} + \cdots \qquad (2.20)$$

To compute $\Psi(h)$ substitute approximation

$$e^{A\sigma} = I + A\sigma + \frac{A^2\sigma^2}{2!} + \cdots$$

$$\Psi(h) = \int_0^h e^{A\sigma}d\sigma = \int_0^h [I + A\sigma + A^2\sigma^2/2! + \cdots]d\sigma$$

$$\Psi(h) \doteq h\left[I + \frac{Ah}{2!} + \frac{A^2 h^2}{3!} + \cdots + \frac{A^M h^M}{(M+1)!}\right] \qquad (2.21)$$

where the number of terms M must be chosen large enough so that the Taylor approximations are valid; that is, we want,

$$\frac{(Ah)^M}{(M+1)!} \ll I \Rightarrow \frac{\|A\|^M h^M}{(M+1)!} < 10^{-6}$$

then

$$\Phi = e^{Ah} = I + A\Psi(h) \qquad (2.22)$$

Algorithm to find M = number of terms in series, given h

$$C_1 = ||A||h/2$$

Do for $M = 2, 20$

$$C_1 = C_1{}^*||A|| \, h/(M + 1)$$

if $C_1 < 10^{-6}$ stop $\Rightarrow r$ return M, if $M < 4$ set $M = 4$

End do

(*Note:* $||A||^{19}/20! \sim 10^{-9}$ if $||Ah|| = \pi$.)
For suitable suggested MATLAB® code for algorithms can be obtained in Appendix III.

2.15 Algorithm for Obtaining $\Psi(h)$ and Φ, Γ

Once M is determined, compute $\Psi(h)$ via series Equation 2.21. Since the magnitude of the higher-order terms in a series decreases as M-grows, sum the series using *reverse nesting*.

$$\Psi(h) = h\left[I...\frac{Ah}{M-2}\left(I + \frac{Ah}{M-1}\left(I + \frac{Ah}{M}\left(I + \frac{Ah}{M+1}\right)\right)\right)\right] \tag{2.23}$$

This assures that very small numbers are never added to much bigger numbers. (See Figure 2.8)

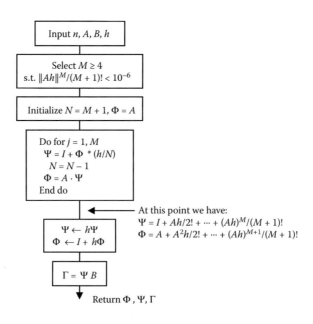

FIGURE 2.8
Flow diagram of a subroutine "Dscrt" for general use.

Then,

$$\tilde{G}(z) = C(zI - \Phi)^{-1}\Gamma + \left\{dz^{-1}\right\} \tag{2.24}$$

Use Leverier algorithm to obtain coefficients.

2.16 Some Discussion on the Selection of *h*

Equivalent discrete time model:

$$X(k + 1) = \Phi X(k) + \Gamma u(k)$$

$$y(k) = CX(k) + \{du(k - 1)\}$$

If $\lambda_1, \lambda_2, ..., \lambda_n$ are the eigenvalues of A then $e^{\lambda_1 h}, e^{\lambda_2 h}, ..., e^{\lambda_n h}$ are the eigenvalues of $\Phi = e^{Ah}$ (Figure 2.9).

To avoid aliasing we must have λ_i within the primary strip in the s-plane, that is, $|\mathrm{Im}(\lambda_i)| < \pi/h$. More manageably,

$$|\lambda_i| < \pi/h$$

$$i = 1, 2, ..., n$$

that is, poles within the circle of radius π/h.

$$\Rightarrow h_{max} = \pi/\left|\lambda_{max}(A)\right|$$

where λ_{max} is the largest eigenvalue of A (generally labeled the spectral radius).

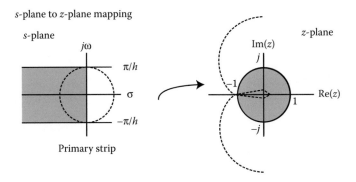

FIGURE 2.9
s-plane to z-plane mapping for the primary strip in the s-plane.

This is too high a limit from a control viewpoint; Instead we seek

$$h \leq \frac{c}{|\lambda_{max}(A)|} \text{ with } c = 0.5 \text{ to } 1.0 \left(\frac{1}{3} \rightarrow \frac{1}{6} \text{ Nyquist rate} \right)$$

An approximation: $|\lambda_{max}(A)| \sim ||A||$
Include warning in subroutine if $||Ah|| > 0.5$

2.17 Examples

EXAMPLE 2.1

If controlling

$$\dot{X}(t) = AX(t) + Bu(t), \quad y(t) = CX(t) + du(t)$$

digitally, what is the reasonable time step h if

$$A = \begin{bmatrix} 1 & 0 & 2 \\ 2 & 1 & -1 \\ 0 & 3 & 1 \end{bmatrix} \quad B = \begin{bmatrix} 1 \\ 0 \\ 1 \end{bmatrix}$$

$$||A|| = \sqrt{\frac{1}{3} \sum_{i=1}^{3} \sum_{j=1}^{3} a_{ij}^2} = \sqrt{\frac{21}{3}} \left(|\lambda_{max}| \right) \sim 2.9$$

Using recommendations from the previous section, we use $h \sim 0.5/||A|| = 0.19$ s (with an upper limit for delay h of 0.38 s).

What is the number of terms needed in series, if $h = 0.3$ s and a desired accuracy in $\Psi = 10^{-4}$?

$$\frac{||A||^M h^M}{(M + 1)!} \leq 10^{-4} \Rightarrow \frac{(0.795)^M}{(M + 1)!} \leq 10^{-4} \Rightarrow M = 6, \text{ which gives upper limit of } 0.5 \times 10^{-4}$$

EXAMPLE 2.2

The equivalent discrete model for the scalar system

$$\dot{x} = -ax + bu, \quad y = x$$

$$G(s) = b/(s + a)$$

$$\Phi = e^{-ah}, \quad \Psi = \int_0^h e^{-a\sigma} d\sigma = -\frac{1}{a}[(e^{-a\sigma})]^h = (1 - e^{-ah})/a$$

$$\Gamma = b\Psi = (1 - e^{-ah})b/a$$

$$x(k + 1) = e^{-ah}x(k) + \left[\frac{1 - e^{-ah}}{a}\right]bu(k)$$

$$y(k) = x(k)$$

$$\tilde{G}(z) = \frac{b(1 - e^{-ah})/a}{z - e^{-ah}} = \frac{z^{-1}(1 - e^{-ah})(b/a)}{1 - z^{-1}e^{-ah}}$$

Note omnipresent one unit (h) delay in $\tilde{G}(z)$ for ($b_0 = 0$).

EXAMPLE 2.3

Model for a motor

$$\begin{bmatrix} \dot{x}_1(t) \\ \dot{x}_2(t) \end{bmatrix} = \begin{bmatrix} 0 & 1 \\ 0 & -a \end{bmatrix}\begin{bmatrix} x_1(t) \\ x_2(t) \end{bmatrix} + \begin{bmatrix} 0 \\ 1 \end{bmatrix}u(t); \quad \lambda_1 = 0 \text{ and } \lambda_2 = -a$$

$$y(t) = \begin{bmatrix} 10 \end{bmatrix}X(k) = x_1(k) \text{ and}; \quad d = 0$$

This a typical model for a motor.
The analytic approach for arbitrary "*a*":

$$\Phi = e^{Ah}; \quad \mathcal{L}^{-1}\left[(sI - A)^{-1}\right]_{t=h} = \mathcal{L}^{-1}\begin{bmatrix} \dfrac{1}{s} & \dfrac{1}{s(s + a)} \\ 0 & \dfrac{1}{s + 2} \end{bmatrix}@_{t=h}$$

This yields Φ, Ψ, Γ as shown in Equations 2.25a, 2.25b, and 2.25c, respectively, as

$$\Phi = \begin{bmatrix} 1 & \dfrac{1}{a}(1 - e^{-ah}) \\ 0 & e^{-ah} \end{bmatrix} \text{ with eigen values } \lambda_1 = 1 \text{ and } \lambda_2 = e^{-ah} \qquad (2.25a)$$

$$\Psi = \int_0^h e^{A\sigma}\,d\sigma = \begin{bmatrix} h & \dfrac{1}{a}\left(h + \dfrac{1}{a}(e^{-ah} - 1)\right) \\ 0 & \dfrac{1 - e^{-ah}}{a} \end{bmatrix} \qquad (2.25b)$$

$$\Gamma = \Psi B = \begin{bmatrix} a^{-1}\left(h + a^{-1}(e^{-ah} - 1)\right) \\ -a^{-1}(e^{-ah} - 1) \end{bmatrix} \qquad (2.25c)$$

$$\tilde{G}(z) = C(zI - \Phi)^{-1}\Gamma + dz^{-1};$$

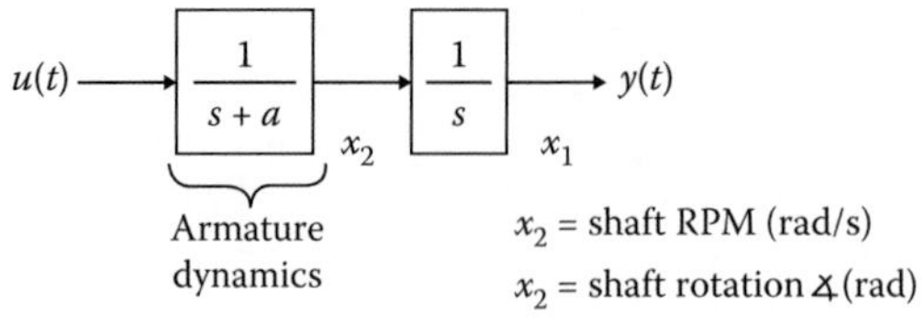
u(t)
1
s + a
x2
1
s
x1
y(t)
Armature
dynamics
x2 = shaft RPM (rad/s)
x2 = shaft rotation ∡ (rad)

FIGURE 2.10
Block diagram representation of a motor.

Note: transfer function of equivalent discrete system is tedius via hand calculation.

$$\tilde{G}(z) = \frac{\left(ah + e^{-ah} - 1\right)\left(\left(z + 1 - e^{-ah} - ahe^{-ah}\right)/\left(ah + e^{-ah} - 1\right)\right)}{a^2(z-1)(z-e^{-ah})} \tag{2.26}$$

Equation 2.26 is the transfer function of an equivalent discrete system for the above example of a motor model in the *s*-domain, and a block diagram form of Figure 2.10.

EXAMPLE 2.4

A special case of Example 2.3 when $a = 0 \quad G(s) = 1/s^2$

We can consider $\overset{\lim}{a \to 0}$ of Equations 2.25 through 2.26 using L'Hospitals's rule, which is generally messy.

Or we can redo the problem for

$$A = \begin{bmatrix} 0 & 1 \\ 0 & 0 \end{bmatrix}; \quad B = \begin{bmatrix} 0 \\ 1 \end{bmatrix}; \quad C = \begin{bmatrix} 1 & 0 \end{bmatrix}$$

$$\Phi = e^{Ah}; \quad \mathcal{L}^{-1}\left[(sI - A)^{-1}\right]\Big|_{t=h} = \mathcal{L}^{-1}\begin{bmatrix} \frac{1}{s} & \frac{1}{s^2} \\ 0 & \frac{1}{s} \end{bmatrix} @_{t=h}$$

$$\Phi = \begin{bmatrix} 1 & h \\ 0 & 1 \end{bmatrix}$$

$$\Psi(h) = \int_0^h e^{A\sigma} d\sigma = \begin{bmatrix} h & \frac{h^2}{2} \\ 0 & h \end{bmatrix}; \quad \Gamma = \Psi B = \begin{bmatrix} h^2/2 \\ h \end{bmatrix}$$

$$\Rightarrow \mathbf{X}(k+1) = \Phi \mathbf{X}(k) + \Gamma u(k)$$

$$\Rightarrow \mathbf{X}(k+1) = \begin{bmatrix} 1 & h \\ 0 & 1 \end{bmatrix} \mathbf{X}(k) + \begin{bmatrix} h^2/2 \\ h \end{bmatrix} u(k)$$

$$\tilde{G}(z) = \begin{bmatrix} 1 & 0 \end{bmatrix} \begin{bmatrix} \frac{1}{z-1} & \frac{1}{(z-1)^2} \\ 0 & \frac{1}{z-1} \end{bmatrix} \begin{bmatrix} \frac{h^2}{2} \\ h \end{bmatrix}$$

$$\tilde{G}(z) = \frac{(h^2/2)}{z-1} + \frac{h^2}{(z-1)^2} = \frac{h^2}{2} \frac{z+1}{(z-1)^2}$$

2.18 Discrete System Equivalents: Transfer Function Approach

See Figure 2.11.

If the process to be controlled is described by a transfer function $G(s)$, we can find $\tilde{G}(z)$ directly.

Indirect approach:

1. Write a state-space model for the process, for example, SCF or SOF
2. Find Φ, Γ using state variable approach
3. Compute

$$\tilde{G}(z) = C(zI - \Phi)^{-1}\Gamma + dz^{-1}$$

Direct approach: Find z-transform of unit pulse response $= \tilde{G}(z)$, between points A and B in Figure 2.11.

Stepwise approach:

First obtain the step response.

1. Let $u(k)$ be a unit step input (see Figure 2.12).

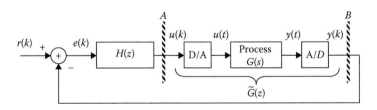

FIGURE 2.11

Transfer function of a discrete system equivalent.

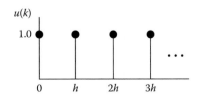

FIGURE 2.12

$u(k)$ as unit step input.

2. If the D/A converter is a zero-order hold, then $u(t)$ will be a pure step, $u(t) = 1$ for $t > 0$:

$$\Rightarrow u(s) = 1/s$$

3. Since the process is continuous, $y(s) = G(s)/s$ and $y(t) = \mathcal{L}^{-1}[(G(s)/s]$.
4. Sampling $y(t)$ and taking the z-transform yields $y(z)$:

$$y(z) = Z\left\{\mathcal{L}^{-1}\left[\frac{G(s)}{s}\right]\right\} = z\text{-transform of step response}$$

usual notation: $Z\{\mathcal{L}^{-1}[F(s)]\} \triangleq Z\{F(s)\}$.

5. If $u(z) = 1$, the response is $(1 - z^{-1})y(z)$:

$$\tilde{G}(z) = (1 - z^{-1})Z\left\{\frac{G(s)}{s}\right\} \tag{2.27}$$

The resulting $\tilde{G}(z)$ must be the same as that obtained via state space.

EXAMPLE 2.5

$$G(s) = \frac{a}{s + a}$$

$$\frac{G(s)}{s} = \frac{a}{s(s + a)}\left[\frac{1}{s} - \frac{1}{s + a}\right]$$

$$\mathcal{L}^{-1}\left[\frac{G(s)}{s}\right] = 1 - e^{-at} = y(t)$$

sampled $y(kh) = 1 - e^{-ahk}$

$$Z\{y(kh)\} = \frac{1}{1 - z^{-1}} = \frac{1}{1 - e^{-ah}z^{-1}}$$

The direct approach gets quite messy for $n > 2$. The preferred method is via state space Φ, Γ then $\tilde{G}(z)$.

- Modifications if $d \neq 0$

$$\tilde{G}(z) = C(zI - \Phi)^{-1}\Gamma + dz^{-1}$$

$$G(z) = \frac{b_1 z^{n-1} + b_2 z^{n-2} + \cdots + b_n}{z^n + a_1 z^{n-1} + \cdots + a_n} + \frac{d}{z} \quad \text{coefficients via Leverier}$$

$$G(z) = \frac{z^{-1}[\tilde{b}_0 z^n + \tilde{b}_1 z^{n-1} + \cdots + \tilde{b}_n]}{z^n + \alpha_1 z^{n-1} + \cdots + \alpha_n}; \; \tilde{b}_i = b_{i+1} + a_i d; \; a_0 = 1, \; b_{n+1} = 0$$

Remember!

1. The computer is controlling a discrete process with transfer function $\tilde{G}(z)$ not a continuous process $G(s)$.

2. Zero-order D/A holds have been assumed (it is possible to re-do state-space approach with first-order holds).

$$\Rightarrow \text{ Of concern is the comparison of } G(s)\big|_{s=j\omega} \text{ vs. } \tilde{G}(z)\big|_{z=e^{j\omega h}}$$

2.19 Relationship between $G(s)$ and $\tilde{G}(z)$

How close is $\tilde{G}(z)\big|_{z=e^{sh}}$ to the original $G(s)$ when $s = j\omega$?
We can expect differences in both magnitude and phase:

$$\tilde{G}(z) = (1 - z^{-1})Z\left\{\frac{G(s)}{s}\right\}$$

$$\tilde{G}(z)\big|_{z=e^{sh}} = (1 - e^{-sh})\left[\frac{G(s)}{s}\right]^{*} \qquad (2.28)$$

{Recall $F^{*}(s)$ is defined as $= F(z)\big|_{z=e^{sh}}$, and relationship between $F^{*}(s)$ and $F(s)$:

$$F^{*}(s) = \frac{1}{h}\sum_{n=-\infty}^{\infty}F(s - jn\omega_{s})\}$$

$$\Rightarrow [G(s)/s]^{*} - \frac{1}{h}\left[\frac{G(s)}{s} + \frac{G(s - j\omega_{s})}{s - j\omega_{s}} + \frac{G(s + j\omega_{s})}{s + j\omega_{s}}\right]$$

If $\omega \ll \omega_{s}/2 = \pi/h$, and $|G(j\omega \pm j\omega_{s})| \ll 1$, then to a first approximation:

$$\Rightarrow [G(s)/s]^{*} \sim \frac{1}{h}\left[\frac{G(s)}{s}\right]$$

and

$$\tilde{G}(z)\big|_{z=e^{sh}} \sim (1 - e^{-sh})\left[\frac{1 - e^{-sh}}{sh}\right]G(s) \qquad (2.29)$$

where

$$\left[\frac{1 - e^{-sh}}{sh}\right]$$

is the sample and hold divided by h:

$$\tilde{G}(z = e^{sh}) \, |_{z=j\omega} = e^{-\frac{j\omega h}{2}} \left(\frac{\sin(\omega h/2)}{(\omega h/2)} \right) G(i\omega) \qquad (2.30)$$

The first term on the RHS is $h/2$ second delay and the second term is magnitude distortion.

By a crude first approximation, the equivalent discrete transfer function is approximately equal to original continuous one with some magnitude distortion and an $h/2$ second delay in the region $\omega \ll \pi/h$.

But "Exact" comparison requires a Bode plot of $G(j\omega)$ versus $\tilde{G}(z = e^{j\omega h})$; and a combination of three subroutines: Discrete, Leverier, and Bode.

2.20 Comparison of a Continuous and Discrete Equivalent Bode Plot

For the motor model as in Example 2.2, the Bode plots for the continuous and discrete equivalent representations are shown for $h = 1.0$ in Figure 2.13.

2.21 Effects of Time Step h on $\tilde{G}(z = e^{j\omega h})$

The Bode plots for a second-order system model with effects of varying sizes of time step h are illustrated in Figure 2.14.

2.22 Anatomy of a Discrete Transfer Function

Examine the Bode plot structure of $G(e^{j\omega h})$ as a function of ω for $\omega > \pi/h$ for any discrete transfer function, $G(z)$ letting $z = e^{j\omega h}$:

$$G^*(j\omega) \triangleq G(e^{j\omega h}) = G\left[e^{-j(2\pi/h-\omega)h} \right] = conj\left\{ G\left[e^{j(2\pi/h-\omega)h} \right] \right\}$$

$$\Rightarrow \left| G^*(j\omega) \right| = \left| G^*\left(\frac{2\pi}{h} - j\omega \right) \right|$$

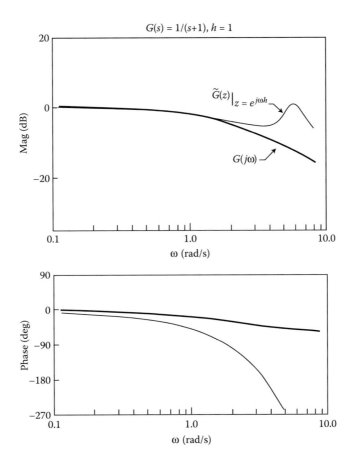

FIGURE 2.13
Continuous and discrete equivalent Bode plots for the motor model with $h = 1.0$.

$$\angle G^* (j\omega) = -\angle G^* \left(\frac{2\pi}{h} - j\omega \right)$$

so, over the interval $[0, 2\pi]$:
$|G^*(j\omega)|$ has even symmetry about $\omega = \pi/h$

$$\angle G^*(j\omega) \text{ has odd symmetry about } \omega = \pi/h$$

$$\left\{ \angle G^* \left(j\pi/h \right) = \text{either } 0° \text{ or } \pm 180° \right\}$$

Over $[2k\pi, 4k\pi]$, $k = 1, 2,...$, $G^*(j\omega)$ is same as that over $[0, 2\pi]$—periodic (Figures 2.15 and 2.16).

Figure 2.15 shows the periodic nature of the magnitude and angle plots of $G^*(j\omega)$.

Note: $(\cdot)^*$ represents the conjugate of a complex function.

$\Rightarrow$ If $G(s)$ has a pole at $s = 0$ then $G^*(j\omega) \to \infty$ for $\omega = 2\pi k/h$, $k = 1, 2,...$.

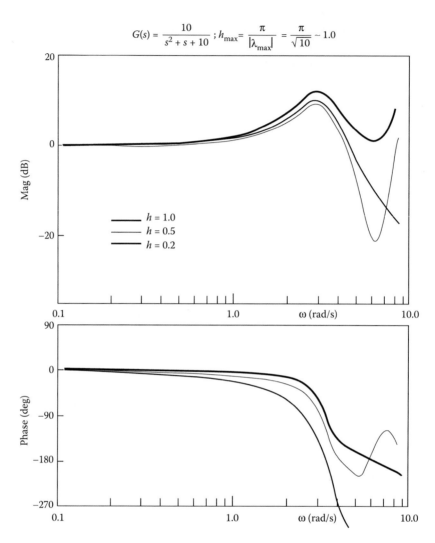

FIGURE 2.14
The Bode plots for a second-order system model with variations in values of h = 1.0, 0.5, and 0.2, respectively.

2.23 Modeling a Process with Delay in Control, $\tau = Mh + \epsilon$

If

$$\dot{X}(t) = A\dot{X}(t) + Bu(t - \tau), \quad y(t) = CX(t) + \{du(t - \tau)\} \qquad (2.31)$$

or

$$G(s) \rightarrow G(s)e^{-s\tau} \qquad (2.32)$$

What is the appropriate discrete equivalent model?

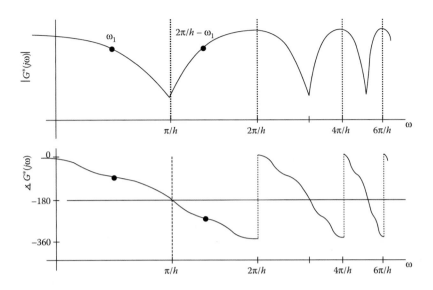

FIGURE 2.15
Magnitude and angle of $G^*(j\omega)$ for different values of $k\pi/h$ for $k = 1, 2, ...$

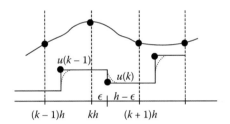

FIGURE 2.16
Mechanism to obtain different states from the previous value and input to system.

Case 1: $M = 0$, $\tau = \epsilon$, and $0 \leq \epsilon < h$ (typical model of computational delay)

Case 2: $M = $ integer ≥ 1; $\tau = Mh + \epsilon$ and $0 < \epsilon < h$ (for cases when there is a large delay)

Consider Case 1 first with a state-space model.

Obtain $X[(k + 1)h]$ from $X(kh)$ and input to the system over $[(kh, (k + 1)h]$

Note the significance of the interval $((\cdot)h, (\cdot)h)$ as open on the left side and closed on the right side of the interval as in: $(\cdot) <$ and $\leq (\cdot)$, respectively. Thus, we have the next state as

$$X[(k + 1)h] = e^{Ah}X(kh) + \int_{kh}^{(k+1)h} e^{A((k+1)h-\xi)}Bu(\xi)d\xi$$

$$= e^{Ah}X(kh) + \int_{kh}^{kh+\varepsilon} e^{A((k+1)h-\xi)}Bd\xi u(k-1) + \int_{kh+\varepsilon}^{(k+1)h} e^{A((k+1)h-\xi)}Bd\xi u(k)$$

where, by using,

$\sigma = (k + 1)h - \xi$; in the second and third term of the previous equation

$$X\big[(k + 1)h\big] = e^{Ah}X(kh) + \int_{h-\epsilon}^{h} e^{A\sigma}d\sigma\, Bu(k - 1) + \int_{0}^{h-\epsilon} e^{A\sigma}d\sigma\, Bu(k)$$

2.24 State Model for a Process with Fractional Delay $\epsilon < h$

Consider a state model as

$$X(k + 1) = \Phi X(k) + \Gamma_1 u(k - 1) + \Gamma_0(k) \tag{2.33}$$

where

$$\Phi = e^{Ah}; \quad \Gamma_0 = \int_{0}^{h-\epsilon} e^{A\sigma}d\sigma B; \quad \Gamma_1 = \int_{h-\epsilon}^{h} e^{A\sigma}d\sigma B = e^{A(h-\epsilon)}\int_{0}^{\epsilon} e^{A\sigma}d\sigma B$$

To compute Φ, Γ_1, and Γ_0:

1. Use a subroutine with (A, B, ϵ): obtain $e^{A\epsilon}$, $\Psi(\epsilon)$
2. Use a subroutine with $(A, B, h - \epsilon)$: obtain $e^{A(h-\epsilon)}$ and $\Psi(h - \epsilon)$
3. $\Gamma_0 = \Psi(h - \epsilon)B$, $\Gamma_1 = eA^{(h-\epsilon)}\Psi(\epsilon)B$, and $\Phi = e^{A(h-\epsilon)} \cdot e^{A\epsilon}$

This yields output equation (as long as $\epsilon < h$)

$$y(k) = CX(k) + \big\{du(k - 1)\big\} \tag{2.34}$$

Define an augmented state model as

$$\chi(k) \triangleq \begin{bmatrix} X(k) \\ u(k - 1) \end{bmatrix}$$

which is an $(n + 1)$ state vector.
Then,

$$\chi(k + 1) \triangleq \begin{bmatrix} X(k + 1) \\ u(k) \end{bmatrix} = \begin{bmatrix} \Phi & \Gamma_1 \\ 0 & 0 \end{bmatrix}\chi(k) + \begin{bmatrix} \Gamma_0 \\ 1 \end{bmatrix}u(k) \tag{2.35}$$

Then the transfer function in general:

$$\tilde{G}(z) = C(zI - \Phi)^{-1}\Gamma + dz^{-1}$$

which yields from Equation 2.33,

$$X(z) = (zI - \Phi)^{-1}[\Gamma_1 z^{-1} + \Gamma_0]u(z)$$

$$y(z) = \tilde{G}(z)u(z) = \frac{1}{z}[C(zI - \Phi)^{-1}(z\Gamma_0 + \Gamma_1) + d]u(z) \tag{2.36}$$

The coefficient of $u(z)$ in Equation 2.36 is the discrete equivalent of the transfer function $\tilde{G}(z)$, which will have a form:

$$\tilde{G}(z) = \frac{c_0 z^n + c_1 z^{n-1} + \cdots + c_n}{z(z^n + a_1 z^{n-1} + \cdots + a_n)} \tag{2.37}$$

2.25 State Model for a Process with Large Delay

$$\tau = Mh + \epsilon; \quad M = \text{integer} \geq 1; \quad 0 \leq \epsilon < h$$

Modeling approach is the same as Case 1 in Section 2.25, but with added M time-step delay:

$$\Rightarrow X(k+1) = \Phi X(k) + \Gamma_1 u(k-1-M) + \Gamma_0 u(k-M) \tag{2.38a}$$

$$y(k) = CX(k) + \{du(k-1-M)\} \tag{2.38b}$$

Augmented state model:

$$\chi(k) \triangleq \begin{bmatrix} X(k) \\ u(k-1-M) \\ u(k-M) \\ \vdots \\ u(k-1) \end{bmatrix} = n+1+M \text{ vector}$$

$$\chi(k+1) = \begin{bmatrix} \Phi & \Gamma_1 & \Gamma_0 & 0 \\ 0 & 0 & 1 & \\ 0 & & & \\ \cdot & \cdot & \cdot & 1 \\ 0 & 0 & 0. & 0 \end{bmatrix} \chi(k) + \begin{bmatrix} 0 \\ 0 \\ \cdot \\ \cdot \\ 0 \\ 1 \end{bmatrix} u(k) \tag{2.39}$$

$$y(k) = \begin{bmatrix} C & d & 0 & \cdots & 0 \end{bmatrix} \chi(k) \tag{2.40}$$

Transfer function:

$$\tilde{G}(z) = \frac{1}{z^M} \cdot \frac{1}{z} \left[C(zI - \Phi)^{-1}(z\Gamma_0 + \Gamma_1) + d \right] \tag{2.41}$$

The second term multiplied by the first term is the previous result for the discrete equivalent transfer function with $M = 0$:

$$\Gamma_0 = \int_0^{h-\epsilon} e^{A\sigma} d\sigma B, \quad \Gamma_1 = e^{A(h-\epsilon)} \int_0^{\epsilon} e^{A\sigma} B$$

2.26 Transfer Function Approach to Modeling a Process with Delay

Since

$$G(s) \rightarrow G(s)\, e^{-(Mh+\epsilon)s},$$

$$\tilde{G}(z) = (1 - z^{-1}) Z \left\{ \frac{G(s) e^{-Mhs} e^{-\epsilon s}}{s} \right\}$$

but $e^{-Mhs} = Z^{-M}$, so pull it out

$$\Rightarrow \tilde{G}(z) = z^{-M}(1 - z^{-1}) Z \left\{ \frac{G(s) e^{-\epsilon s}}{s} \right\} \tag{2.42}$$

Approach:

1. Form $G(s)\, e^{-\epsilon s}/s; \quad 0 \leq \epsilon < h$
2. Take the Laplace inverse:

$$\mathcal{L}^{-1}$$

3. Sample the resulting time signal
4. Take z-transforms

The above steps are messy!

EXAMPLE 2.6

$$G(s) = \frac{1}{s + a} e^{-Mhs} \Rightarrow \dot{x} = -ax + u(t - \tau)$$

$$\Phi = e^{-ah}; \quad \Gamma_0 = \int_0^{h-\epsilon} e^{-a\sigma} d\sigma = \frac{\left[1 - e^{-a(h-\epsilon)}\right]}{a};$$

$$\Gamma_1 = e^{-a(h-\epsilon)} \int_0^{\epsilon} e^{a\sigma} d\sigma = e^{-a(h-\epsilon)}(1 - e^{-a\epsilon})/a$$

$$\tilde{G}(z) = \frac{1}{aZ^{M+1}} \left\{ \frac{\left(1 - e^{-a(h-\epsilon)}\right)z + e^{-ah}(e^{a\epsilon} - 1)}{z - e^{-ah}} \right\}$$

EXAMPLE 2.7

$a = 1.0, M = 2, \epsilon = 0.5, h = 1$

$$\tilde{G}(z) = \frac{1}{z^3} \left\{ \frac{(1 - e^{-0.5})z + e^{-1}(e^{0.5} - 1)}{z - e^{-1}} \right\} = \frac{0.393(z + 0.607)}{z^3(z - 0.368)}$$

Note: In many applications, the time step is dictated by the on-line computational requirements.

⇒ τ is often comparable to h.

PROBLEMS

P2.1 System model is given by

$$\begin{bmatrix} x_1(k+1) \\ x_2(k+1) \end{bmatrix} = \begin{bmatrix} 2.0 & 1.0 \\ 0.4 & 2.0 \end{bmatrix} \begin{bmatrix} x_1(k) \\ x_2(k) \end{bmatrix} + \begin{bmatrix} 2.4 \\ 1.0 \end{bmatrix} u(k)$$

The measurement model is given by

$$y(k) = \begin{bmatrix} 2 & 0 \end{bmatrix} \begin{bmatrix} x_1(k) \\ x_2(k) \end{bmatrix} + 0.4u(k)$$

Find the transfer function: $G(z) = C(zI - A)^{-1}B + d$.

P2.2 We want to control a system digitally, given by

$$\dot{X}(t) = \begin{bmatrix} 2 & 0 & 4 \\ 4 & 2 & -1 \\ 0 & 6 & 2 \end{bmatrix} X(t) + \begin{bmatrix} 2 \\ 0 \\ 2 \end{bmatrix} u(t)$$

What is a reasonable time step "h?"

P2.3 The equivalent discrete model for a scalar system is given by

$$\dot{x} = -2x + 3u; \quad \text{and } y = x$$

1. Find $G(s)$.
2. Find the discrete equivalent transfer function after finding Φ, Ψ, and Γ.

P2.4

$$\begin{bmatrix} \dot{x}_1(t) \\ \dot{x}_2(t) \end{bmatrix} = \begin{bmatrix} 0 & 1 \\ 0 & -3 \end{bmatrix} \begin{bmatrix} x_1(t) \\ x_2(t) \end{bmatrix} + \begin{bmatrix} 0 \\ 1 \end{bmatrix} u(t)$$

$$y(t) = \begin{bmatrix} 1 & 0 \end{bmatrix} \begin{bmatrix} x_1(t) \\ x_2(t) \end{bmatrix}$$

For $t = h$,

1. Find the transfer function of the equivalent discrete system, $C(zI - \Phi)^{-1}\Gamma$.

P2.5 Compare the Bode plots of the following system in continuous and equivalent discrete systems with $h = 1.0$ and using MATLAB.

$$G(s) = \frac{1}{s(s + 1)}$$

Repeat the exercise for $h = 0.5$, 0.3, and 0.1, respectively.

P2.6 A continuous plant is given by

$$\frac{1}{s + a} e^{-Mhs} e^{-\epsilon s}$$

1. Find the state equation.
2. For $a = 2.0$, $M = 4$, $\varepsilon = 0.3$, and $h = 0.5$, find the discrete equivalent transfer function.

P2.7 Mass–Spring–Damper

A mass–spring–damper system is illustrated in Figure P2.1. Write the dynamic equations of motion or the ODE. Use MATLAB to analyze this problem. Also, solve the problem by hand (where required) and compare your hand calculations with the computer results from MATLAB. All MATLAB data printout scans can be easily obtained using the diary command. Be sure your MATLAB printouts include the input data as well as the output data.

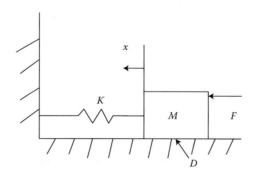

FIGURE P2.1
Mass–spring–damper system.

FIGURE P2.2
Open-loop discrete system with sampling time T and ZOH.

where

F = force applied at $t = 0$

M = mass of body to which F is applied

K = spring constant

D = frictional constant between mass and horizontal surface

P2.8 Type in the coefficients of the numerator and denominator polynomials of the transfer function, $G(s)$, for a linear continuous plant as shown in Figure P2.2.

Also specify a sampling interval, T, for this plant. Then execute the MATLAB function file that will generate the discrete transfer function, which calculates and displays the function of the open-loop systems in the following:

a. $G(s) = \dfrac{5(s + 6)}{(s + 3)(s + 4)}$ with $T = 0.5\,\text{s}$

b. $G(s) = \dfrac{5(s + 6)}{(s + 3)(s + 4)}$ with $T = 1\,\text{s}$

c. $G(s) = \dfrac{5(s + 7)}{s(s + 2)(s + 4)}$ with $T = 0.5\,\text{s}$

d. Submit the MATLAB data printout, hand calculations, and comparisons for these systems in parts (a) through (c).

P2.9 Type in A and B matrices for the state-space model of a linear continuous system. Also, specify a desired sampling interval T, for the systems shown in Figure P2.3. Then execute the continuous-time model to a discrete-time model assuming the system configuration below in the Figure P2.3 and with the A and B matrices shown in the State and Measurement models below.

$$\text{State model: } \dot{X} = \begin{bmatrix} -2 & 3 & 0 \\ 0 & -1 & 0 \\ 5 & -2 & -4 \end{bmatrix} X + \begin{bmatrix} 6 \\ -2 \\ 0 \end{bmatrix} u$$

$$\text{Measurement model: } y = \begin{bmatrix} 0 & 1 & 0 \end{bmatrix} X$$

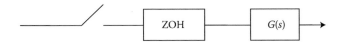

FIGURE P2.3
Open-loop discrete system with sampling time T and ZOH for continuous to discrete configuration execution with MATLAB.

3

Performance Criteria and the Design Process

3.1 Design Approaches and the Design Process

3.1.1 Elements of Feedback System Design

3.1.1.1 Introduction: Series Compensation Design Structure ("Classical")

A standard structure of a series compensator is shown in Figure 3.1. This is one of the "Classical" feedback control structures generally employed.

Given $G(z)$, design a suitable $H(z)$. We start with the equation for input as in Equation 3.1:

$$u(z) = H(z) \, [r(z) - y(z)] \tag{3.1}$$

Thus, the closed-loop (CL) transfer function of the CL system shown in Figure 3.1 is given by

$$\frac{y(z)}{r(z)} = T(z) = \frac{\tilde{G}(z)H(z)}{1 + \tilde{G}(z)H(z)}$$

The denominator: $[1 + \tilde{G}(z)H(z)]$; is the Closed-loop Characteristic Polynomial

The z-transform of the related error function is given by

$$e(z) = \frac{1}{1 + \tilde{G}(z)H(z)} r(z)$$

To study this further, some alternate loop structures employed in feedback control are shown below:

a. Output feedback compensator design is shown in Figure 3.2. This is one of the other alternate loop structures possible in feedback control.

Here,

$$u(z) = r(z) - H(z) \, y(z)$$

and

$$T(z) = \frac{\tilde{G}(z)}{1 + \tilde{G}(z)H(z)}$$

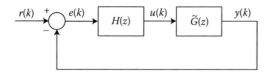

FIGURE 3.1
Series compensator design structure.

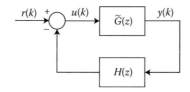

FIGURE 3.2
Output feedback compensator design diagram.

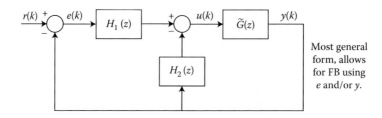

FIGURE 3.3
Mixed series/feedback compensator system model.

 b. Mixed series/feedback compensator design is shown in Figure 3.3 and is yet another one of the alternate loop structures employed in feedback control design. This is one of the most general forms that allows for feedback (FB) using the error "e" and/or output "y."

3.1.2 Elements of FB System Design II

 a. State variable design structure (modern control method)
  Given the state model as in Equation 3.2,

$$X(k + 1) = \Phi X(k) + \Gamma u(k); \quad y(K) = CX(k) \tag{3.2}$$

 We need to design suitable K and K_r to design a suitable negative feedback as in Equation 3.3.

$$u(k) = K_r r(k) - KX(k) \tag{3.3}$$

For a CL system, a suitable format will be explored as follows:

CL:

$$X(k + 1) = (\Phi - \Gamma)X(k) + K_r\Gamma u(k); \quad \bar{\Phi} = (\Phi - \Gamma K)$$

Thus, the transfer function is given in Equation 3.4a:

$$\frac{y(z)}{r(z)} = T(z) = K_rC(zI - \Phi + \Gamma K)^{-1}\Gamma \tag{3.4a}$$

$$= \frac{K_rN(z)}{|zI - \Phi + \Gamma K|} \tag{3.4b}$$

Alternate Formula:

The same can be evaluated with an alternate formula as

$$\frac{K_rC(zI - \Phi)^{-1}\Gamma}{1 + K(zI - \Phi)^{-1}\Gamma}; \quad \tilde{G}(z) = C(zI - \Phi)^{-1}\Gamma \tag{3.4c}$$

Derivation of Equation 3.4c is shown below:

1. $X(z) = (zI - \Phi)^{-1}\Gamma u(z)$

2. $u(z) = K_rr(z) - K(zI - \Phi)^{-1}\Gamma u(z)$

3. $u(z) = [1 + K(zI - \Phi)^{-1}\Gamma]^{-1}K_rr(z)$

4. Substitute into $y(z) = C(zI - \Phi)^{-1}\Gamma u(z)$
   $\rightarrow$ CL characteristic polynomial is thus

$$p(z) = |zI - \Phi + \Gamma|$$

or

$$p(z) = 1 + L(zI - \Phi)^{-1}\Gamma$$

b. Optimal control design ("Classy" method!).
   One method for obtaining K, K_r—by optimizing some criterion.

3.1.3 Closed-Loop System Zeros

a. Here CL transfer function is given by

$$\frac{y(z)}{r(z)} = \frac{K_rC(zI - \Phi)^{-1}\Gamma}{1 + K(zI - \Phi)^{-1}\Gamma} = \frac{K_r \cdot \textit{Open-loop Numerator}}{p(z)}$$

$\rightarrow$ State variable feedback (SVFB) has no effect on system zeros.

b. Also for a *general* state-space model:

$$X(k + 1) = \Phi^* X(k) + \Gamma^* u(k)$$
$$y(k) = C^* X(k) + D^* u(k)$$

where the denominator $= |zI - \Phi^*|$ (values of z, where the system can have a response without any input), and Numerator zeros $=$ values of z where output is always zero, thus

$$\left. \begin{array}{c} \left(zI - \Phi^*\right) X(z) = \Gamma^* {}^* u(z) = 0 \\ C^* X(z) + D^* (z) = 0 \end{array} \right\} \rightarrow \begin{vmatrix} zI - \Phi^* & -\Gamma^* \\ C^* & D^* \end{vmatrix} = 0$$

c. When applied to SVFB system, this yields:

$$X(k + 1) = (\Phi - \Gamma K)X(k) + K_r \Gamma r(k); \quad y(k) = CX(k)$$

We can obtain CL system zeros via

$$\begin{vmatrix} zI - \Phi + \Gamma K & -K_r \Gamma \\ C & 0 \end{vmatrix} = \text{numerator of } \frac{y(z)}{r(z)}$$

To this add K/K_r^* last column to first n columns, yielding

$$\begin{vmatrix} zI - \Phi & -K_r \Gamma \\ C & 0 \end{vmatrix} = K_r \cdot \begin{vmatrix} zI - \Phi & -\Gamma \\ C & 0 \end{vmatrix}$$

$$= K_r \text{ Numerator polynomial of open-loop } (K = 0) \text{ system}$$

$\rightarrow$ CL zeros $=$ Open-loop (OL) zeros.

Thus, for a series compensator design:

$$\text{CL zeros} = \text{OL zeros of } \tilde{G}(z)$$

and those zeros added by $H(z)$.

3.1.4 Design Approaches to Be Considered

a. For series compensation design of $H(z)$,

$$H(z) = \frac{\beta_0 + \beta_1 z^{-1} + \cdots + \beta_m z^{-m}}{1 + \alpha_1 z^{-1} + \cdots + \alpha_m z^{-m}} \tag{3.5}$$

m-th order compensator.

1. Discretization of a continuous design:

$$H(s) \rightarrow \tilde{H}(z)$$

   where $H(s)$ is a series compensator designed for $G(s)$ (will usually be okay when h is very small).
2. Direct design methods for $H(z)$ given $\tilde{G}(z)$.

b. For SVFB design of K, K_r
   1. Discretization of continuous design gains,

$$K \rightarrow \tilde{K}$$
$$K_r \rightarrow \tilde{K}_r$$

   where K, K_r for

$$\dot{X} = AX + Bu$$

   2. Pole placement, direct design methods. Select K so that $|zI - \Phi + \Gamma K|$ has desired roots.
   3. Optimization methods. Find $u(k) = K_r \ \ r(k) - KX(k)$ to optimize some performance:

$$\text{Criterion} \Rightarrow K^*, K_r^*$$

c. Methods for state estimation when $X(k)$ is not directly measureable.

3.2 Performance Measure for a Design Process

a. Mathematical model of system to be controlled is shown in Figure 3.4. This system can be defined by discrete equivalent, or traditional given by

$$\tilde{G}(z) \quad or \quad \{\Phi, \Gamma, C\}$$

b. Performance measures and concerns: These are guided by mathematical criteria that are driven by customer or end-user specifications for behavior of the CL system.
   1. Stability of the CL system:
      - A property of loop dynamics not of $r(k)$ the reference
      - Without stability cannot discuss much else

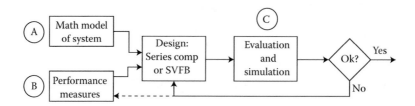

FIGURE 3.4
Block diagram representation of a system to be controlled using digital control techniques.

2. Steady-state accuracy:
   - Does $y(k) \rightarrow r(k)$ as $k \rightarrow \infty$?
   - If $r(k) = 0$ desire $y(k)$ and $X(k) \rightarrow 0$ for all $X(0)$
3. Speed of response/transient:
   - Transient response is linked to CL pole locations
4. Sensitivity/robustness:
   - Ability of CL system to perform with $\Delta \tilde{G}(z)$
   - Feedback desensitizes loop to variations in $\tilde{G}(z)$
5. Optimization

3.2.1 Stability of the Closed-Loop System

- Roots of CL characteristic polynomial $p\text{CL}(z)$ in unit circle, that is, $(|\lambda_i| < 1)$.

$$1 + \tilde{G}(z)H(z) = 0 \quad or \quad \begin{cases} |zI - \Phi + \Gamma K| = 0 \\ 1 + K(zI - \Phi)^{-1}\Gamma = 0 \end{cases} \tag{3.6}$$

- To implement and evaluate its performance, we will need a simple test to determine if a polynomial $p(z)$ has any roots $|\lambda_i| \geq 1$.
  - Similarly, recall the Routh test for whether $p(s)$ has roots in RHP for a continuous time system and related continuous control conditions.
- Phase margin Φ_m is used to give a degree of stability to generally asked questions: "How much more negative phase shift (phase lag) can you put in the FB loop before the system becomes unstable?"—tolerance to time delay needs to be evaluated!
  To determine Φ_m, use the Bode (or Nyquist) plot of loop gain as in Equations 3.7a and 3.7b:

$$-LG_{ain} = \tilde{G}(z)H(z) \text{ series compensation} \tag{3.7a}$$

$$-LG_{ain} = K(zI - \Phi)^{-1}\Gamma \quad \text{for SVFB} \tag{3.7b}$$

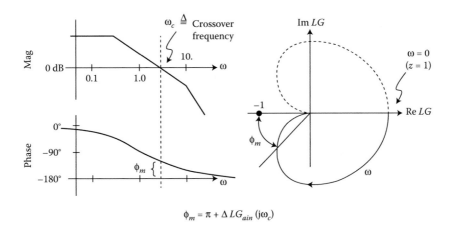

FIGURE 3.5
The Bode and Nyquist plot for loop gain (LG).

A flow diagram to evaluate the stability of a CL system can be seen in Equation 3.7c:

$$\begin{array}{ccccc} Dscrt & & Leverier & & Bode \\ A,B & \rightarrow & \Phi,\Gamma & \rightarrow & \dfrac{N(z)}{D(z)} & \rightarrow & \Phi_m,\omega_c \end{array} \qquad (3.7c)$$

where Dscrt, Leverier, and Bode are algorithms described in the prequel.
Examine loop gain versus ω with $z = e^{j\omega h}$ (option 2 in "Bode").
This is shown in Figure 3.5.

EXAMPLE

Consider a system with a system model and measurement model given by

$$x(t) = 0.5\,u(t), \quad y(t) = x(t)$$

This system is controlled digitally using the algorithm as in Equations 2.3 through 2.5.

$$u(k) = Kr\,(k) - x(k), \quad \text{with time step } h = 0.2 \text{ s}$$

Determine the phase margin. (Note that the open-loop $G(s)$ is unstable.)

a. Discrete equivalent model:

$$\Phi = e^{+0.5h} = 1.1, \quad \Gamma = \left(e^{+0.5h} - 1\right)\left(\frac{0.95}{0.5}\right) = +0.2$$

b. Check stability of closed loop ($K = +1$)

$$\Phi - \Gamma K = 0.9 \Rightarrow \text{stable}$$

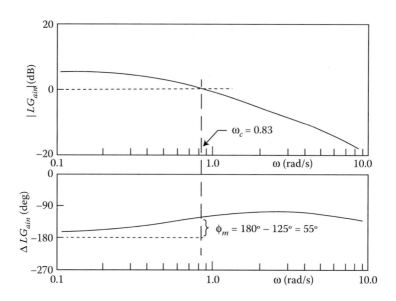

FIGURE 3.6
Magnitude and angle plot of LG_{ain} for this example.

c. Obtain Φ_m via the Bode plot of $K(zI - \Phi)^{-1}\Gamma = 0.2/(z - 1.1)|z = e^{j\omega h}$. The results are shown in Figure 3.6.

Analytic approach: find ω_c where $0.2/(|z - 1.1|) = 1$ at $z = e^{j\omega_c h}$ $0.2 = |(\cos \omega_c h - 1.1) + j \sin \omega_c h \to 0.4 = (\cos \omega_c h\ 1.1)^2 + \sin^2\omega_c h$. Solving gives $\cos \omega_c h = 0.986 = (1/h)$ $\cos^{-1}(0.986) = 0.827$ rad/s.

3.2.1.1 Steady-State Accuracy to a Step Input

a. If $r(k)$ is a step input (e.g., commanded change in setpoint) of value A, we want $y(k) \to A$ in steady state (ss)

Final value theorem for y_{ss}, provided CL system is stable, yields:

$$\lim_{k \to \infty} y(k) = (1 - z^{-1})T(z)r(z)\ |_{z=1} = (1 - z^{-1})T(z)\frac{A}{1 - z^{-1}}\ |_{z=1} = AT(1) \Rightarrow T(1) = 1 \qquad (3.8)$$

b. For series compensation design only

$$T(1) = 1 = \frac{\tilde{G}(1)H(1)}{1 + \tilde{G}(1)H(1)} \Rightarrow \tilde{G}(1)H(1) = \infty \qquad (3.9)$$

Requires loop gain to have a pole at $z = 1$

$$\tilde{G}(z)H(z) = \frac{N(z^{-1})}{(1 - z^{-1})D(z^{-1})}$$

$\to$ Need an integrator in either G (i.e., $\tilde{G}$ or H).

- For SVFB design, achieve $T(1) = 1$ via proper choice of K_r:

$$T(z) = K_r \, C \, (zI - \Phi + \Gamma K)^{-1} \Gamma$$

$$K_r = 1/[C(I - \Phi + \Gamma K)^{-1} \Gamma] \tag{3.10}$$

- If $T(1) \neq 1$, there will be a steady-state error, $A - y_{ss}$:

$$\text{Fractional error} \triangleq \frac{1}{K_p} = \frac{A - y_{ss}}{y_{ss}} = \frac{1 - T(1)}{T(1)}, \quad where \; K_p = \frac{T(1)}{1 - T(1)} \tag{3.11}$$

which is usually large.

This yields a steady-state error as

$$e_{ss} = \frac{A}{1 + K_p} \cong \frac{A}{K_p} \tag{3.12}$$

3.2.1.2 Steady-State Accuracy to a Ramp Input

a. When $r(k)$ is a ramp input, $r(k) = Akh$, we want to command a rate of change in setpoint as shown in Figure 3.7.

Then,

$$y_{ss} \rightarrow \beta kh - \alpha$$

For this result to happen, we need

$$T(1) = 1 \quad \text{for } \beta = A$$

otherwise

$$e_{ss} \rightarrow \infty$$

b. Relative "steady-state" error

$$\frac{e_{ss}}{A} = \frac{\alpha}{A} \triangleq \frac{1}{K_v}$$

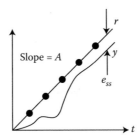

FIGURE 3.7
Ramp input and output response.

In general,

$$K_v = -\frac{1}{h}\left[\frac{dT(z)}{dz}\right]^{-1}\Bigg|_{z=1}$$

One can then show

$$\frac{1}{K_v} = h\left[\sum\frac{1}{1-p_i} - \sum\frac{1}{1-z_i}\right]$$

where

$$p_i = \text{poles}; \quad z_i = \text{zeros of } T(z)$$

c. For series compensation structure, only K_v can be shown to be

$$K_v = \frac{(1-z^{-1})}{h}\tilde{G}(z)H(z)\Bigg|_{z=1} = \frac{N(1)}{hD(1)} = \lim_{s\to 0} s\tilde{G}(e^{sh})H(e^{sh})$$

Thus for a correct evaluation, $\to$ Need at least one integrator in the forward loop gain ($\tilde{G}H$) at low frequency $\tilde{G}H \to (K_v/s)$, that is, K_v is the gain of the low-frequency (LF) asymptote. This provides the criterion for selecting loop gain (LG).

3.2.1.3 Steady-State Error to Sinusoidal Inputs

a. Series compensation design analysis

We have error

$$r(z) - y(z) = e(z) = \frac{1}{1 + \tilde{G}(z)H(z)}r(z) \tag{3.13}$$

and $\to$ want $|\tilde{G}(z)H(z)|$ large over the frequency range of interest
where

$$z = e^{j\omega h}$$

This places lower bounds on ω_c, where $|\tilde{G}(z)H(z)|_{z=e^{j\omega h}=1}$. Simultaneously, we want $|\tilde{G}(z)H(z)|$ small at high frequencies, for noise rejection as

$$y(z) = \frac{\tilde{G}(z)H(z)}{1 + \tilde{G}(z)H(z)}r(z)$$

The above equations provide criteria for selection of $H(z)$, where

$$\omega_c \sim \text{Bandwidth of CL system}$$

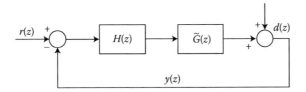

FIGURE 3.8
System model for output disturbance rejection.

b. SVFB design
In the case of SVFB design, bandwidth is determined by CL pole locations. Obtain ω_c via the Bode plot of $K(zI - \Phi)^{-1}\Gamma$
→ Implicit specifications of ω_c are similarly obtainable.

c. Output disturbance rejection is shown in Figure 3.8

$$y(z) = T(z)r(z) + \frac{d(z)}{1 + \tilde{G}(z)H(z)}$$

Figure 3.8 also shows the output evaluated with output disturbance $d(z)$.

3.2.2 Speed of Transient Response

i. Related to the location of CL poles and zeros.
Require some nominal input, that is, speed of response to "what" to be analyzed. Generally, the most common test input is unit step, next suitable is the ramp input. Thus, we now examine step response of a second-order CL system:

- Many systems are interconnections of second-order parts. Many systems have a dominant second-order pair, and have some salient aspects that have to be considered, like roots with the smallest real [s] or largest magnitude of "z," that is, $|z|$.

- Consider $T(s)$, then $s \to z$ plane mapping is used to get $T(z)$ poles.

  The CL transfer function in the z-domain is shown below in Figure 3.8 and conjugate poles plotted in Figure 3.9, in the s-domain.

$$T(s) = \frac{\omega_n{}^2}{s^2 + 2\zeta\omega_n s + \omega_n{}^2}; \quad 0 < \zeta \le 1$$

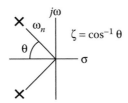

FIGURE 3.9
Pole plot of CL second-order system.

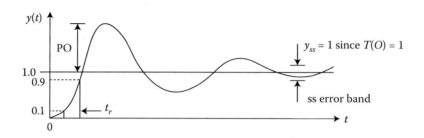

FIGURE 3.10
Output response of a second-order system.

with the roots of the characteristic polynomial at

$$\lambda_{1,2} = -\zeta\omega_n \pm j\omega_n\sqrt{1 - \zeta^2}$$

as shown in Figure 3.10.

The step response of such a second-order system is shown in Figure 3.10 with different figures of merit as listed below.

The several figures of merit of interest are

$$\text{Percentage Overshoot (PO)} = \%\text{overshoot} = 100e^{-\pi\zeta/\sqrt{1-\zeta^2}}, \quad \zeta \le 1$$

As $\zeta \to 0$, system response becomes more oscillatory, with roots on the imaginary axis for continuous time or in the s-domain.

$$t_r = 10\% \text{ to } 90\% \text{ rise time } \sim 2.5/\omega_n$$

- Settling time (TS) = time to get and stay within $\pm x\%$ of steady state (ss).

$$TS_{5\%} \sim 3/\zeta\omega_n; \quad TS_{1\%} \sim 4.7/\zeta\omega_n \ (\zeta\omega_n = [\text{time constant}]^{-1})$$

In terms of steps for z-plane response: it is customary to "Think" in terms of nominal continuous (s-plane) pole locations that are given to start the thought process.

PO and TS specifications. Use a left half plane (LHP) $\to$ unit circle ($s \to z$) map diagram to obtain the desired pole locations in the z-plane.

Results for the second-order continuous system (for a dominant pair of complex roots):

The PO versus damping ratio and phase margin (PM) versus damping ratio (ζ) are illustrated in Figures 3.11a and b for a second-order continuous system.

3.2.3 Sensitivity and Return Difference

Sensitivity to variation in parameter 'x' and its effect on a function $y(x)$ is defined as

$$S_x^y = \frac{\%\text{ change in some } y\,(x)}{\%\text{ change in } x} = \frac{\Delta y/y}{\Delta x/x} - \frac{x}{y}\frac{\partial y}{\partial x} \tag{3.14}$$

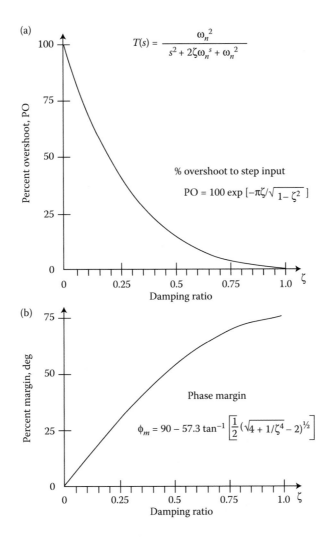

FIGURE 3.11
(a) PO versus ζ for a second-order continuous system. (b) PM versus ζ for a second-order continuous system.

Thus for several FB schemes, specific sensitivity can be evaluated as follows:

a. Series compensation:

$$T(z) = \frac{\tilde{G}(z)H(z)}{1 + \tilde{G}(z)H(z)}$$

$$S_{\tilde{G}(z)}^{T(z)} = \frac{\tilde{G}(z)}{T(z)} \frac{\partial}{\partial \tilde{G}(z)} \left[\frac{\tilde{G}(z)H(z)}{1 + \tilde{G}(z)H(z)} \right] = \frac{1}{1 + \tilde{G}(z)H(z)}$$

The Return Difference (RD) RD $\triangleq 1 + \tilde{G}(z)H(z) = 1 + LG_{ain}(z)$ (3.15a)

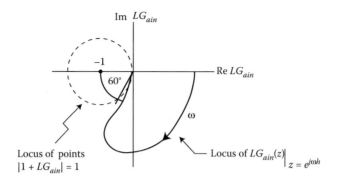

FIGURE 3.12
Locus of points $|1 + LG_{ain}|$ and locus of $LG_{ain}|_{z=e^{j\omega h}}$.

b. SVFB with $u = K_r r - KX$:

$$\text{In this case, Return Difference (RD)} = 1 + K(zI - \Phi)^{-1}\Gamma \qquad (3.15b)$$

For stable operation,

Criteria: Keep $|RD| \gg 1$ over frequency range of interest

c. Relation to Φ_m: The locus of points of the LG_{ain} and $|1 + LG_{ain}|$ in Figure 3.12 show that the $|RD| \geq 1$ for all ω; which also implies $\Phi_m \geq 60°$.

d. It is best to examine the root locus (RL) of CL system poles with respect to individual parameter variations about their nominal values [a_i, b_i, in $G(s)$; a_{ij}, b_{ij}, in A, B; etc.].

3.2.4 Example: Evaluation and Simulation

A sampled output is shown in Figure 3.13 sampled at different times

a. The most time and effort is involved here!
   - To what extent have design specifications been met?
   - Actual Φ_m
   - CL pole locations
   - Effect of different sample times, h
   - Computational lag

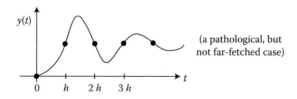

FIGURE 3.13
Output sampled at different sampling times.

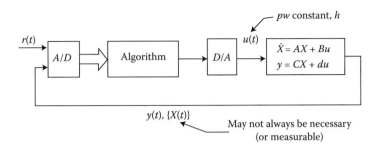

FIGURE 3.14
Simulation structure.

- Root locus with respect to design parameters
b. Time response of CL system to representative command
c. Inputs $r(t)$ and initial conditions
  - Via computer stimulations
  - Must consider response of $y(t)$, $x(t)$ not only at the sample points $t = kh$, but also in between samples too! (As seen in Figure 3.14.)
d. "What if" questions
  - Sensitivity of performance to changes in system parameters, controller parameter
  - Failure modes
  - Control saturation
  - Noise: measurement and/or process noise respectively
  - Unmodelled dynamics, time delays
  - Quantization and other nonlinearities

3.3 Simulation of Closed-Loop Time Response

a. Tool to examine time response:
  - Input (u), output (y), any state (x_i)
  - Obtain response between sample points of the continuous-time variables $y(t)$, $x_i(t)$
  - $u(t)$ is assumed a piecewise constant over intervals of length h
  - Simulate with arbitrary initial conditions (user input)
  - Examine response to representative $r(t)$
b. Need a flexible computer program:
  - Ability to input system dynamics in $G(s)$ or in

$$\dot{X} = AX + Bu, \quad y(t) = CX + (du)$$

Format

→ program will work with a state-space model.

If $G(s)$ format is given, set up A, B, C in standard observable form (SOF)

$$G(s) = \frac{b_0 s^n + b_1 s^{n-1} + \cdots + b_n}{s^n + a_1 s^{n-1} + \cdots + a_n}$$

$$\rightarrow A = \begin{bmatrix} -a_1 & 1 & 0 & & -a_n \\ -a_2 & 0 & 1 & & 0 \\ \vdots & & & \ddots & \vdots \\ & & & & 1 \\ -a_n & 0 & & \cdots & 0 \end{bmatrix}; \quad B = \begin{bmatrix} \tilde{b}_1 \\ \tilde{b}_2 \\ \vdots \\ \tilde{b}_n \end{bmatrix}$$

$$\tilde{b}_i = (b_i - b_0 a_i) \quad C = \begin{bmatrix} 1 & 0 & \cdots & 0 \end{bmatrix} \quad d = b_0$$

- Ability to simulate different control algorithms; option (OPT)
  - OPT = 0; Open-loop response $u(kh) = K_r r(kh)$
  - OPT = 1; State variable feedback control
  - OPT = 2; Series compensation via $H(z)$ (including different implementations)
  - OPT = $i, j...$; Reserve for future control options
- Ability to easily change the control interval, h

3.3.1 Simulation Structure

A basic simulation structure is shown in Figure 3.14 and the corresponding flow chart is shown in Figure 3.15. As shown in Figure 3.15, the steps are as follows:

i. Basic flow:
  1. Obtain $r(t)$ at time t
  2. Sample $r(t)$, $y(t)$ $X(t)$ at $t = kh$ supply $r(kh)$, $y(kh)$, and $X(kh)$ to control algorithm
  3. Obtain $u(kh)$ from control algorithm $u(t) = u(kh)$ for $kh < t \leq (k + 1)h$
  4. Print out info at time t: X, y, u, r
  5. Compute system response $X(t)$, $y(t)$ over $[kh, (k + 1)h]$ at $t = (k + 1)h$; $X[(k + 1)h] = \Phi(h)X(kh) + \Gamma(h)u(kh)$

ii. How to compute $X(t)$ and $y(t)$ at more points in $[kh, (k + 1)h]$; number of steps (NS)
  1. Pick $NS \geq 1$ and let $h_1 = h/NS$
  2. The control algorithm is active every h sec (u is piecewise constant over intervals of length h)
  3. Can compute $x(t)$ at times that are multiples of h_1 while changing u every NS-th multiple of h
  4. Dual-time-scale simulation ($NS = 2 \rightarrow 5$ usually)

Note: Remember!!—even though we simulate the system response using a (small) time step h_1 the control algorithm must have been designed for the actual sample time h.

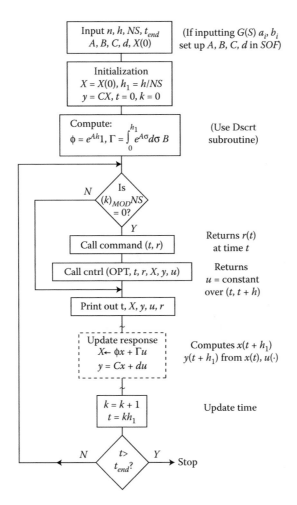

FIGURE 3.15
Flow diagram of a simulation program.

3.3.2 Flow Diagram for Simulation Program for a Control Algorithm

Figure 3.15 shows the flow diagram of a typical simulation program. The result of the simulated quantities are shown in Figure 3.16 for a given input digitized, resulting in different states.

3.3.3 Modifications to Time Delay

History of simulated sequences (Example, $NS = 4$):

a. An understanding facilitates subsequent simulations that will include time delay:
   1. New control computed only at times $k = 0, 4, 8, 12, \ldots$ using the corresponding value of X (or y) at this time. The value of u is not changed at other than these points.
   2. Next, $x(k+1)$ is computed at time $k, k = 0, 1, \ldots$. Using $X(k)$—the previous X and current u. This computation is done at every k.

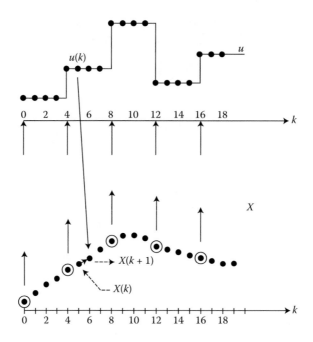

FIGURE 3.16
Simulated sequences for a given input digitized resulting in different states.

$$X[(k + 1)h_1] = \Phi X(kh_1) + Bu(kh_1) \tag{3.16}$$

$$y[(k + 1)h_1] = CX[(k + 1)h_1] + du(kh_1) \tag{3.17}$$

3.3.4 Control (Cntrl) Algorithm Simulation

- Command (t, r) and Cntrl (OPT, t, r, X, y, u) are user-oriented and interactive.
- Command (t, r) returns $r(t)$, for example, $r = 1$, $r = A*t$, and so on.
- Cntrl must distinguish among various options:
  OPT = 0 for open-loop response, $u = K_r r$
  OPT = 1 for SVFB, $u = K_r r - KX$

$$u = K_r r - \sum_{i=1}^{n} K_i * x_i \tag{3.18}$$

where the gain values K_r, $[K_i]$ are read in as input or else set via DATA statement.
OPT = 2 for "standard" series compensation (m-th order):

$$\frac{u(z)}{e(z)} = \frac{\beta_0 z^m + \beta_1 z^{m-1} + \cdots + \beta_m}{z^m + \alpha_1 z^{m-1} + \cdots + \alpha_m} = \frac{\beta_0 + \beta_1 z^{-1} + \cdots + \beta_m z^{-m}}{1 + \alpha_1 z^{-1} + \cdots + \alpha_m z^{-m}} \tag{3.19}$$

- Coefficients $\{\alpha_{ij}\}$, $\{\beta_i\}$, and compensator order "m" would typically be read as input when OPT = 2.

This leads to corresponding discrete algorithm as follows, using the discrete time equation for control as

$$u(k) = \beta_0 e(k) + \beta_1 e(k-1) + \cdots + \beta_m e(k-m)$$
$$- [\alpha_1 u(k-1) + \cdots + \alpha_m u(k-m). \tag{3.20}$$

- To implement $u(k)$ via $H(z)$ we will need (2) storage vectors (each m-dimensional) for the last m values of e and u: paste, pastu (2 storage vectors).

There will be other options to cover different implementations.

EXAMPLE 3.1: ALGORITHM FLOW PROCESS FOR IMPLEMENTING
$H(z)$ COMPENSATOR

0 - enter with t, r, y
1 - if $t = 0$, set pastu $(i) = 0$, paste $(i) = 0$
        For $i = 1, \ldots, m$, $SE = 0$, $SU = 0$, $S = 0$
Now set
2 - $e = r - y$
3 - $u = \beta 0 e + S$ (This is the new value of u.)
4 - (pushdown pastu, paste, if $m > 1$)
        do for $i = 1, m - 1$
                pastu$(m + 1 - i) = $pastu$(m - i)$
                paste$(m + 1 - i) = $paste$(m - i)$
        end do
5 - pastu$(1) = u$
        Paste$(1) = e$ (Store latest u, e)

$$6 - \left. \begin{array}{l} SE = \displaystyle\sum_{i=1}^{m} \beta_i paste(i) \\[4mm] SU = \displaystyle\sum_{i=1}^{m} \alpha_i pastu(i) \end{array} \right\} Setup \ for \ next \ time \ through$$

        $S = SE - SU$
7 - return

We can now use special cases as below:

a. Special case when $m = 1$: $e = r - y$

$$U = \beta_0 e + S$$
$$S = \beta_1 e - \alpha_1 u$$

b. Try to program Cntrl in much the same way as would be done in the real-time implementation. (Note: u can be output at step 3.); permits timing of code, investigation of round-off effects, testing, and so on.
c. The above implementation of $H(z)$ is not the best from a numerical accuracy viewpoint, especially for $m > 2$.

3.3.5 Simulation of Time Delay, τ

We now concentrate on simulation of time delay τ by

a. Lumping all delay in the control (input lag):

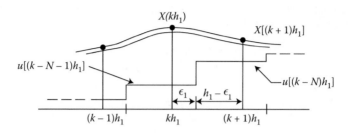

FIGURE 3.17
Simulated sequences for a given input digitized resulting in different states—discrete simulation model with time delay τ.

$$\dot{X}(t) = AX(t) + Bu(t - \tau) \tag{3.21}$$

$$y(t) = CX(t) + du(t - \tau) \tag{3.22}$$

b. Write delay as a multiple of simulation steps, h_1:

$$\tau = Nh_1 + \epsilon_1, \quad 0 \le \epsilon_1 < h_1 \tag{3.23}$$

c. Now the discrete simulation model is as follows, as shown in Figure 3.17.

$$X[(k + 1)h_1] = \Phi X(kh_1) + \Gamma_0 u[(k - N)h_1] + \Gamma_1 u[(k - N - 1)h_1] \tag{3.24}$$

$$y[(k + 1)h_1] = CX(k + 1)h_1] + du[(k - N)h_1] \tag{3.25}$$

Figure 3.18 shows the discrete simulation model with values of state and output for a delay of τ.

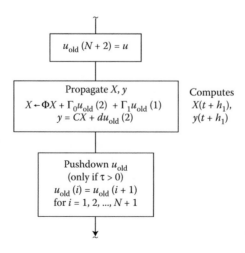

FIGURE 3.18
Update module—discrete simulation model with time delay τ.

Using the state transition matrix and further Γ_0 and Γ_1,

$$\Phi = e^{Ah_1}; \quad \Gamma_0 = \int_0^{h_1 - \epsilon_1} e^{A\sigma} d\sigma B; \quad \Gamma_1 = \int_{h_1 - \epsilon_1}^0 e^{A\sigma} d\sigma B$$

This will need an $(N + 2)$-vector pushdown stack to store past values of u, $u_{old}(i)$, $i = 1, \ldots, N + 1$, and latest value $u_{old}(N + 2)$.

$$u_{old}(i) \equiv u(k - N - 2 + i)$$

Initialize $u_{old}(i)$ 0 at $t = 0$.

Then control algorithm design is based on delay model, Equations 2.34 through 2.39, associated with time step h.

3.3.6 Required Modifications to Simulation Flow Diagram

It is felt that there should be some redefinition to the flow diagram as follows:

i. Initialization:

$$\text{Compute } N = \text{Int}[\tau/h_1]$$

$$\epsilon_1 = \tau - Nh_1$$

$$\text{Set } u_{old}(i) = 0 \quad \text{for} \quad I = 0, \ldots, N + 2$$

$$\text{Compute } \Gamma_0, \Gamma_1$$

$$(\text{Note, } \Gamma_0 = 0 \text{ if } \epsilon_1 = h_1 - \Gamma_1 = 0 \text{ if } \epsilon_1 = 0)$$

ii. A new response update module is shown in Figure 3.18:

- U_{old} stack will be made up of piecewise constant values that change every NS-th point.
- This correctly simulates small delay (when $N = 0$, i.e., $\tau < h_1$).

3.4 Tools for Control Design and Analysis

Thus far, the following are the different tools available for control design and analysis:

- The Bode plots
  - $G(s)$ vs. $\tilde{G}(z), LG_{ain}(z)\big|_{z = e^{j\omega h}}$
- State variable analysis
- Computer programs
  - Leverier, Dscrt, Bode, Simulation, …
- Root locus

…along with a system mode! And with performance specifications in hand, we are now ready to design a digital control algorithm.

But first, let us review classical series compensation design and methods used for continuous time systems.

In the famous control parlance:

"You can't know where you're going if you don't know where you've been!"

3.5 Overview of Classical Design Techniques (Continuous Time)

The stepwise process is as follows in classical design techniques:

1. Graphical methods to pick $H(s)$ via the Bode plot modifications.

2. s-plane methods to pick $H(s)$ via RL, thus improving the shaping.

The above two are trial-and-error methods since frequency domain and s-plane measures are not 1:1 with time-domain measures (e.g., step response), especially for higher-order systems. Let us consider:

a. The Bode plot design, as shown in Figure 3.19.

1. At $\omega \to 0$, $G(s)H(s) \to K_v/s$. Restrictions on steady-state (ss) tracking error to a ramp input will set direct current (DC) gain of GH (recall ss error to ramp command $r(t) = $ At is A/K_v).

2. Since $|e(s)/r(s)| = (1/|1 + G(s)H(s)|)$, restrictions on ss accuracy over mid-frequency range will give lower bound on $|GH|$ (e.g., for < 2%) relative error over $[0, \bar{\omega}]$ $[0, \omega]$, $|GH| > 50$ for $\omega < \bar{\omega}$.

3. At high frequencies, for input noise rejection, we want $|G(s)H(s)|$ to be small (e.g., $|G(s)H(s)| < 0.01$, $\omega > \omega_{max}$).

4. May have restrictions on $\omega_c \sim$ bandwidth. Also may wish $\Phi_m > 45°$ (or as large as possible) via stability criterion (e.g., phase curve of GH).

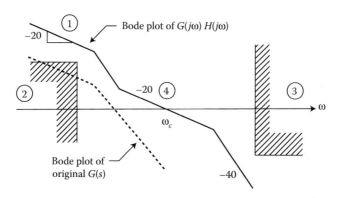

FIGURE 3.19
Control design using the Bode plot for $G(s)$ and $G(j\omega)H(j\omega)$.

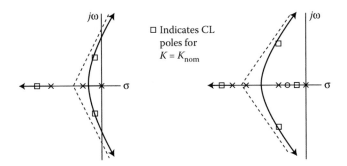

FIGURE 3.20
Root locus of CL poles of uncompensated and compensated systems.

Steps for the Bode plot approach:

i. Sketch the Bode plot of $G(s)$ and then add in gain plus poles and zeros of $H(s)$ to bend/shape $G(s)H(s)$ to meet specifications as required by the end user.

  • "Create a fair stretch of −20 dB/decade slope in the crossover region by choice of $H(s)$ with $\Phi_m \sim 45°$" "fair stretch" ~ ±1 octave $[\omega_c/2, 2\omega_c]$ or greater.

  • Must next evaluate CL poles, zeros, time response, and so on.

ii. Root locus approach:

  • Bend and shape RL of $G(s)$ by adding (real) poles and zeros so that the RL passes through "desirable" regions in the s-plane.

  • Then pick gain H to place poles. Consider mainly dominant poles. Figure 3.20 shows the pole placement of the uncompensated and compensated systems.

$$\square \text{ Indicates CL poles for } K = K_{nom}$$

  • Must next evaluate Φ_m, bandwidth, time response, and so on.

  • Useful approximation for second-order continuous system (Φ_m, in degree):

$$\zeta \sim \left(1 + \frac{\Phi_m}{190°}\right)\Phi_m/130°$$

3.5.1 Lag Compensator Design, $H(s)$

i. Lag network, magnitude, and phase are plotted in Figure 3.21.

$$H(s) = K\frac{1 + s/\alpha\omega_1}{1 + s/\omega_1} \quad \alpha > 1 \tag{3.26}$$

A lag network compensator is used to lower crossover frequency by reducing gain, but without changing very large frequency (VLF) and DC gain; $\omega_c - 10\,\alpha\omega_1$ as shown in Figure 3.21.

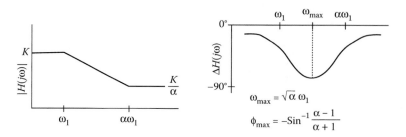

FIGURE 3.21
Lag network magnitude and phase angle plots for different ω and α values.

3.5.2 Lead Compensator Design

ii. Lead network, magnitude, and phase are plotted in Figure 3.22.

$$H(s) = K\frac{1 + s/\omega_2}{1 + s/\beta\omega_2} \quad \beta > 1 \tag{3.27}$$

A lead network compensator is used to add positive phase directly at the intended crossover frequency as shown in Figure 3.22.

$$\Rightarrow \omega_c = \sqrt{\beta}\,\omega_2, \quad \beta = \frac{1 + \sin(\Delta\Phi_{req})}{1 - \sin(\Delta\Phi_{req})} \tag{3.28}$$

iii. Lag–Lead combination.

$$H(s) = K\frac{(1 + (s/\alpha\omega_1))(1 + (s/\omega_2))}{(1 + (s/\omega_1))(1 + (s/\beta\omega_2))} \quad \alpha,\beta > 1 \tag{3.29}$$

Thus, finally, a generalization of proportional integral derivative (PID) is illustrated below:

$$H(s) = \left[K_0 + \frac{K_1}{\tau_1 s + 1} + \frac{K_2 s}{\tau_2 s + 1} \right] \tag{3.30}$$

iv. If an open-loop $G(s)$ does not contain a pole at $s = 0$ (DC), then must include one in $H(s)$ to have 0 ss error.

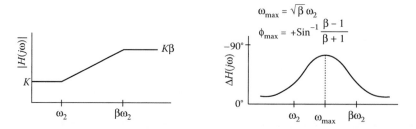

FIGURE 3.22
Lead network magnitude and phase angle plots for different ω and β values.

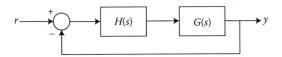

FIGURE 3.23
Lag compensator example block diagram.

3.5.3 Example of Lag Compensator Design

A block diagram for a lag-compensator is shown in Figure 3.23 for an easy reference. We have from Equation 3.26:

$$H(s) = K\frac{(1 + (s/\alpha\omega_1))}{(1 + (s/\omega_1))}; \ \forall \ \alpha > 1; \ \text{and} \ \ G(s) = \frac{5}{s(1 + (s/10))(1 + (s/50))}$$

Specifications:

 i. $Kv \geq 50$ (i.e., less than or equal to relative ss error to ramp)

 ii. $\Phi_m \sim 45°$ and no restriction on ω_c

To meet K_v requirement $\rightarrow$ choose $K = K_v/[sG(s)]|_{s=0} = 10$.
Sketching the Bode plot of $KG(s)$ as shown in Figure 3.24 for different values of α_1, ω_1 illustrates the following:

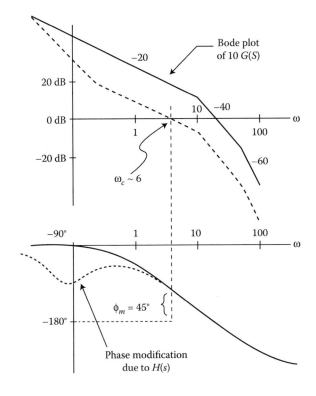

FIGURE 3.24
The Bode plots for lag compensator example.

1. With $K = 10$ system is unstable with $\Phi_m \sim 0$)
2. Would like to have $\omega_c \sim 6$, $[\angle G(j6) \sim 135°]$ to obtain desired Φ_m
3. Use lag compensation to lower gain at $\omega = 6$ so that $\omega_c \sim 6$ $(1/\alpha \sim -18.5\ dB) => \alpha \sim 8.3$
4. Select ω_1 with $\alpha\omega_1 \sim \omega_c/10 \rightarrow \omega_1 = 0.072$ and so $\alpha\omega_1 = 0.6$, which yields:

$$H(s) = 10\frac{1 + s/0.6}{1 + s/0.072}$$

Conclusions from the above example are

- If $\alpha\omega_1 \ll \omega_c$, then lag compensator will not affect Φ_m in region of intended crossover.
- Do not want ω_1 to be too small $\rightarrow$ too much mid-frequency attenuation.

3.5.4 Lag Compensation Design

i. Basically, lag compensation is trial and error with guidelines:

1. If $G(s)$ does not have a pole at $s \sim 0$, include one in $H(s)$:

$$H(s) = K\frac{(1 + (s/\alpha\omega_1))}{(1 + (s/\omega_1))} \quad \text{or} \quad \forall[(K_1/s) + H(s)]$$

2. Determine compensator gain K for suitable steady-state accuracy or mid-range accuracy
3. Sketch the Bode plot of $KG(s)$
4. Decide on the need for compensation
5. Select approximate frequency for intended ω_c where Φ_m is sufficient
6. Find approximate gain reduction α needed to have $|G(j\omega)H(j\omega)| = 1$ at desired ω_c
7. Select/adjust $\omega_1, \alpha, \omega_c$

ii. Lag-compensated designs are inherently sluggish due to the need to reduce ω_c (bandwidth)

3.5.5 Example of Lead Compensator Design

Thus as an example is for a lead compensator $H(s)$ as in Equation 3.27, a block diagram of which is shown in Figure 3.25 below we have:

$$H(s) = K\frac{1 + s/\omega_2}{1 + s/\beta\omega_2} \quad G(s) = \frac{5}{s(1 + (s/10))}$$

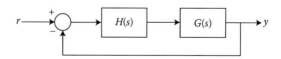

FIGURE 3.25
Lead compensator block diagram.

Specifications:

i. $K_v = 100$ (i.e., 1% to a ramp input)

ii. $|G(j\omega)\ H(j\omega)| \geq 50\ \omega < 1$ (i.e., < 2% relative error) to sinusoidal inputs; $20\log_{10} 50 = 34$ dB

iii. $\Phi_m \sim 45°$

To meet DC (K_v) requirement $\rightarrow K = 20$. Sketch the Bode plot.
Note: Lag compensator with $\Phi_m \sim 45°$ would violate mid-frequency specifications.

1. With $K = 20$, $\omega_c \sim 30$, $\Phi_m \sim 20°$
2. Introduction of lead network will increase ω_c slightly:

$$\rightarrow \Delta \Phi_{req} \sim 30° \text{ not } 25°$$

$$\beta = (1 + \sin 30°)/(1 - \sin 30°) = 3$$

3. Adjust ω_2 so that actual $\omega_c = \beta^{1/2} \omega_2 \rightarrow \omega_2 = 24$, $\omega_c = 41.5$, $\beta\omega_2 = 72$
4. Check Φ_m at ω_c:

$$\Phi_m = 180° + \text{angle of } (GH)|\omega_c = 44°$$

$$H(s) = 20\frac{1 + s/24}{1 + s/72}$$

- Desire ω_c at $\beta^{1/2} \omega_2 = \omega_{max}$ to get full benefit of lead
- Introduction of lead angle increases ω_c slightly $\rightarrow$ may require modification of β, ω_2
- Generally, $3 \leq \beta \leq 30$ corresponds to $30° \leq \Delta\Phi \leq 70$

3.5.6 Lead Compensation Design

i. Basically, trial and error with guidelines:
   1. If $G(s)$ does not have a pole at $s = 0$, include one in $H(s)$ (to meet DC error criteria)
   2. Determine compensator gain K for suitable steady-state accuracy or mid-range accuracy
   3. Sketch the Bode plot of $KG(s)$
   4. Decide on need for and type of compensation
   5. Pick $\Delta\Phi_{req}$ (plan for an increase in ω_c) $\rightarrow \beta$
   6. Obtain ω_2, ω_c (may require iteration with step 5)

Useful equations:

$$\omega_c = \sqrt{\beta}\,\omega_2 \tag{3.31}$$

$$\left| G(j\omega_c) \right| \cdot \left| H(j\omega_c) \right| = 1 \tag{3.32}$$

where

$$\left| H(j\omega_c) \right| = K \cdot \left(\frac{\omega_c}{\omega_2} \right) \quad \text{when } \omega_c \epsilon\, [\omega_2, \beta\omega_2]$$

- Lead-compensated designs have a high bandwidth (okay if not excessive noise in system), as shown in Figure 3.26
- Lead compensator ~ P-D design $K\left[1 + \dfrac{(\beta - 1)\tau_2 s}{\tau_2 s + 1} \right]$; $\tau_2 = \dfrac{1}{\beta\omega_2}$
- Root locus is shown in Figure 3.27
- Step response is shown in Figure 3.28

3.5.7 Critique of Continuous Time *H*(*s*) Design

- Classical design techniques are simple to use
  - Graphical techniques
  - Some trial and error
- Designs are easy to implement via analog circuitry

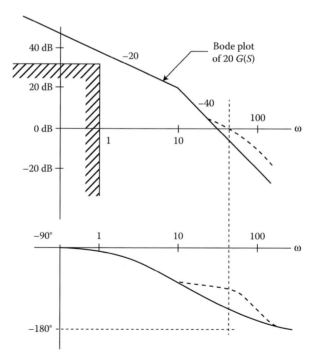

FIGURE 3.26
Root locus for a lead compensation design.

- Root locus:

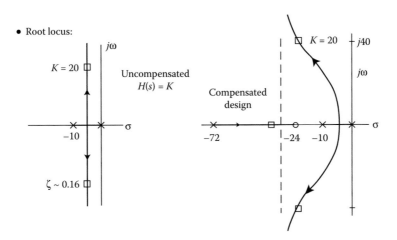

FIGURE 3.27
Root locus for uncompensated and compensated design respectively.

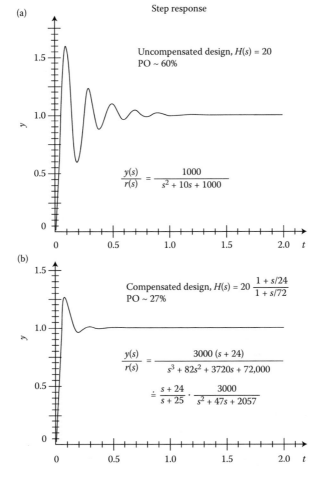

(a)

Step response

Uncompensated design, $H(s) = 20$
PO ~ 60%

$$\frac{y(s)}{r(s)} = \frac{1000}{s^2 + 10s + 1000}$$

(b)

Compensated design, $H(s) = 20 \dfrac{1 + s/24}{1 + s/72}$
PO ~ 27%

$$\frac{y(s)}{r(s)} = \frac{3000\,(s + 24)}{s^3 + 82s^2 + 3720s + 72{,}000}$$

$$\doteq \frac{s + 24}{s + 25} \cdot \frac{3000}{s^2 + 47s + 2057}$$

FIGURE 3.28
Step response for a lead compensator design.

- Consider Lag–Lead compensator when neither alone will suffice

$$\rightarrow \text{pick } \omega_1, \alpha, \omega_2, \beta$$

- Most-used design technique:
  - There are many such compensators "out there"
  - Can they be modified for digital implementation?

$$H(s) \rightarrow \tilde{H}(z)$$

But there are limitations:

- Simple lag, lead, and others may not be sufficient
- High-order compensator design via the Bode, or RL, is a challenging process, especially for humans
- Compensation does not use all available information
  - Uses only $y(t)$, not states $X(t)$
- Difficult to extend procedure to multi-input, multi-output systems

Note: Generally, root locus is much more tedious than the Bode plot, which serves as the most popular approach.

PROBLEMS

P3.1   A scalar system is given by

$$\dot{x} = -2x + 3u; \quad \text{and} \quad y = x$$

It is controlled by

$$u(k) = K_r r(k) - x(k)$$

With time step $h = 0.2$ s.
  1. Find the discrete equivalent model
  2. Check stability of closed loop ($K = +1$)
  3. Obtain the Φ_m via the Bode plot

P3.2   A continuous system is given by

$$G(s) = \frac{10}{s(1 + (s/15))(1 + (s/100))}$$

A lag compensator is shown in Figure P3.2.

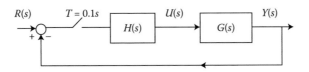

FIGURE P3.2
Lag compensator for Problem P3.2.

The specifications of the system are
1. $K_v \geq 100$
2. $\Phi_m \sim 45°$ and there is no restriction on ω_c
3. Choose $H(s) = 20(1 + s/0.6)/(1 + s/0.82)$

Verify answers using MATLAB program.

P3.3 A continuous system is given by

$$G(s) = \frac{10}{s(1 + (s/20))}$$

The lead compensator design is given in Figure P3.3.
Where $H(s)$ is given by

$$H(s) = 40\frac{1 + s/48}{1 + s/96}$$

The specifications are given by
1. $K_v = 100$
2. $|G(j\omega) H(j\omega)| \geq 75$ for $\omega < 1$
3. $\Phi_m \sim 45°$

Use MATLAB to verify results.

P3.4 A CL system is given in Figure P3.4.
$G(s)$ is given by

$$G(s) = \frac{5(s + 6)}{(s + 3)(s + 4)}$$

and

$$D(z) = 1$$

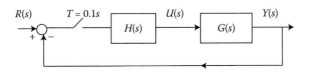

FIGURE P.3.3
Lead compensator for Problem P3.3.

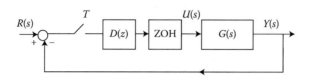

FIGURE P3.4
CL system for Problem P3.4.

With $T = 0.5\ s$.

Use MATLAB to solve this problem.

P3.5   A CL system is given in Figure P3.5.

   $G(s)$ is given by

$$G(s) = \frac{5(s + 6)}{(s + 3)(s + 4)}$$

and

$$D(z) = 1 + \frac{0.1z}{z - 1}$$

With $T = 0.5\ s$.

Use MATLAB to solve this problem.

P3.6   A CL system is given in Figure P3.6.

   $G(s)$ is given by

$$G(s) = \frac{5(s + 7)}{s(s + 3)(s + 4)}$$

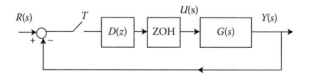

FIGURE P3.5
CL system for Problem P3.5.

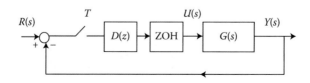

FIGURE P3.6
CL system for Problem P3.6.

and

$$D(z) = 1$$

With $T = 0.5\ s$.

Use MATLAB to solve this problem.

P3.7 For problems 4, 5, and 6, find the output response when the input to the system is a unit step, unit impulse, and unit ramp, respectively. The number of points of the response in each case should be >50. Assume zero initial conditions for all three problems and case of inputs.

P3.8 For a system in Figure P3.8, write a program to solve the transfer function using the Leverier algorithm. Write the code using MATLAB and other suitable syntax. Make the program interactive for future use.

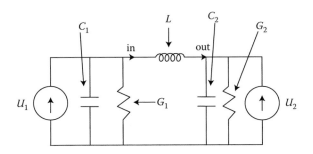

FIGURE P3.8
Two current sources and RLC combination π-circuit for Problem P3.8.

4

Compensator Design via Discrete Equivalent

4.1 Stability of Discrete Systems

We need a technique to ascertain the stability of the closed-loop (CL) system. Generally, this means whether the roots of CL characteristic polynomial (CP) $p(z)$ all lie within the unit circle. This is shown as

$$p(z) = \text{denominator of } T(z) = \frac{\check{G}(z)H(z)}{1 + \check{G}(z)H(z)} \tag{4.1}$$

$$\text{or } p(z) = |zI - \Phi + \Gamma K| \tag{4.2}$$

Equation 4.2 is a polynomial in z given by

$$p(z) = a_0 z^n + a_1 z^{n-1} + \cdots + a_n \qquad (\text{generally } a_0 = 1)$$

This technique should be simple and involve $\{a_i\}$ only and should be applicable to any polynomial in z.

In continuous time systems, stability analysis is achieved using the Routh–Hurwitz test to determine whether a polynomial $p(s)$ has its roots in the left half-plane (LHP) by analyzing the CP in s given by

$$p(s) = a_0 s^n + a_1 s^{n-1} + \cdots + a_n$$

A way to use the Routh–Hurwitz test in the z-domain is to:

1. Map a unit circle into LHP by replacing z with some suitable function $(z = e^{sh})$, but this will not work since the resulting $p(s)$ will not be a polynomial.

2. Use:

$$z = \frac{1 + wh/2}{1 - wh/2} \tag{4.3}$$

3. Substitute for z in $p(z)$; multiply by $(1 - wh/2)^n$ to obtain:

$$\tilde{p}(w) = n\text{th order polynomial in } w$$

4. Apply the Routh–Hurwitz test to $\tilde{p}(w)$.

4.1.1 Jury Test

The Jury test for $p(z) = a_0 z^n + a_1 z^{n-1} + \cdots + a_n$; is the characteristic equation in general. Now set up the Jury array as follows:

$$
\begin{array}{llllllll}
 & 1: & a_0 & a_1 & a_2 & \cdots & a_{n-1} & a_n \\
(n) & & & & & & & \\
 & 2: & a_n & a_{n-1} & a_{n-2} & \cdots & a_1 & a_0
\end{array}
\qquad \text{Let } r_n = \frac{a_n}{a_0}
$$

$$
\begin{array}{llllll}
 & 3: & a_0^{(n-1)} & a_1^{(n-1)} & a_2^{(n-1)} & \cdots & a_{n-1}^{(n-1)} \\
(n-1) & & & & & & \\
 & 4: & a_{n-1}^{(n-1)} & a_{n-2}^{(n-1)} & & \cdots & a_0^{(n-1)}
\end{array}
\qquad \text{Let } r_{n-1} = \frac{a_{n-1}^{(n-1)}}{a_0^{(n-1)}}
$$

$$
(0) \qquad 2n+1: \quad a_0^{(0)}
$$

where

$$
r_k = \frac{a_k^{(k)}}{a_0^{(k)}} \quad k = n, n-1, \ldots, 1 \tag{4.4}
$$

$$
a_i^{(k-1)} = a_i^{(k)} - r_k a_{k-i}^{(k)} \quad i = 0, 1, \ldots, k-1 \tag{4.5}
$$

where initially $a_i^{(n)} = a_i$.

In "English," the mechanism is as follows:

- Each odd row = previous odd row—r_k previous even row.
- Each even row = preceding odd row in reverse order.
- First row has coefficients of $p(z)$.
- Last row has one element.

Criteria:

1. If $a_0 > 0$, then all roots of $p(z)$ lie in the unit circle if and only if (iff)

$$
a_0^{(k)} {}_0^{(k)} > 0, \quad k = n-1, \ n-2, \ldots, 0 \tag{4.6}
$$

2. The number of negative $a_0^{(k)}$ = number of roots of $p(z)$ outside the unit circle.

4.1.1.1 *Applications of the Jury Test*

For every CP – $p(z)$:

- Test if the first entry in each odd row is greater than 0.
- If any $a_0^{(k)} = < 0$ is obtained, stop; $p(z)$ has roots $|\lambda| > = 1$.
- Create a simple computer program; need 2 scratch vectors.

EXAMPLE 4.1: $p(z) = z^2 - z + 0.5$

	1:	(1.0)	−1	0.5	$r = 0.5$
(2)	2:	0.5	−1	1.0	
	3:	$1 - 0.25 = (0.75)$	$-1 + 0.5 = -0.5$		$r = -0.5/0.75 = -0.67$
(1)	4:	−0.5	0.75		
(0)	5:	$0.75 - 0.33 = (0.42)$			

All $a_0^{(k)} > 0 \Rightarrow$ system is stable (all roots in unit circle).

EXAMPLE 4.2: $p(z) = z^2 - z + 2$

	1:	(1.0)	−1	2	$r = 2$
(2)	2:	2	−1	1.0	$r = 2$
	3:	$1 - 4 = (-3)$	$-1 + 2 = 1$		$a_0^{(1)} < 0 \Rightarrow$ system is unstable
(1)	4:	1	−3		
(0)	5:	$-3 - (-0.33) = (-8/3)$			$r = -\dfrac{1}{3} \Rightarrow$ 2 roots outside unit circle

EXAMPLE 4.3: $p(z) = z^3 - 0.15z^2 - 0.59$

	1:	(1.0)	−0.15	0.00	−0.59	$r = -0.59$
(3)	2:	−0.59	0.00	−0.15	1.0	
	3:	(0.65)	−0.15	−0.09		
(2)	4:	−0.09	−0.15	0.65		$r = -0.14$
	5:	(0.64)	−0.13			
(1)	6:	−0.13	0.64			$r = -0.20$
(0)	7:	(0.61)				

All $a_0^{(k)} > 0 \Rightarrow$ system are stable.

4.1.1.2 Application to State Variable Feedback (SVFB) Example

The equivalent discrete system is shown in

$$X(k + 1) = \begin{bmatrix} 1 & 1 \\ 0 & 1 \end{bmatrix} X(k) + \begin{bmatrix} 0 \\ 1 \end{bmatrix} u(k) \tag{4.7}$$

This system is to be controlled using the algorithm as in

$$u(k) = r(k) - \begin{bmatrix} 1 & 3 \end{bmatrix} X(k), \text{ where } K = \begin{bmatrix} 1 & 3 \end{bmatrix} \tag{4.8}$$

We now check if the CL system is stable:

- CL system matrix: $\bar{\Phi} = \Phi - \Gamma K$

$$\bar{\Phi} = \begin{bmatrix} 1 & 1 \\ 0 & 1 \end{bmatrix} - \begin{bmatrix} 0 \\ 1 \end{bmatrix} \begin{bmatrix} 1 & 3 \end{bmatrix} = \begin{bmatrix} 1 & 1 \\ 0 & 1 \end{bmatrix} - \begin{bmatrix} 0 & 0 \\ 1 & 3 \end{bmatrix} = \begin{bmatrix} 1 & 1 \\ -1 & -2 \end{bmatrix}$$

- CL CP

$$p(z) = |zI - \bar{\Phi}| = \begin{vmatrix} z-1 & -1 \\ 1 & z+2 \end{vmatrix} = z^2 + z - 1$$

- The Jury test is now applied to the $p(z)$ as follows:

(2)	1:	1.0	1	−1	$r = -1$
	2:	−1	1	1	
(1)	3:	0	Stop1		

- The CL system is unstable, but the roots are not on the unit circle.
  The roots of $p(z)$ are $z_1 = 0.618$, $z_2 = 1.618$, so $a_0^{(k)} = 0$ does not necessarily imply roots on unit circle.
  Note $|z_1 z_2| = 1$ here, corresponding to roots λ and $1/\lambda$.
- If some $a_0^{(k)} = 0$, can we replace $0 \to +\varepsilon$ and continue further, for example, as in the Routh–Hurwitz test, for continuous time systems.

4.1.2 Stability with Respect to a Parameter β

If the system (or controller) has a free parameter, β, we want to determine the range of values to find which system is stable.

EXAMPLE 4.4

A system $G(s) = a/(s + a)$, $a = 1$, is to be controlled using series compensation with control algorithm:

$$u(k) = Ke(k) + u(k - 1)$$

and time step $h = 0.69$ s. For what range of K is the system stable?
Applying techniques described earlier, we have:

1. Using the expression of the discrete equivalent yields

$$\tilde{G}(z) = \left. \frac{1 - e^{-ah}}{z - e^{-ah}} \right|_{ah=0.69} = \frac{0.5}{z - 0.5}$$

Input in the z-domain is

$$u(z) = Ke(z) + z^{-1}u(z) \to \frac{u(z)}{e(z)} = H(z) = \frac{K}{1 - z^{-1}} = \frac{Kz}{z - 1}$$

TABLE 4.1

The Jury Test for Example 4.4

(2)	1		$(K-3)/2$	$1/2$	$r = 1/2$
	$1/2$		$(K-3)/2$	1	
(1)	$3/4$		$(K-3)/4$		$r = (K-3)/3$
	$(K-3)/4$		$3/4$		
(0)	$3/4-(K-3)^2/12$				

$$1 + \tilde{G}(z)H(z) = \frac{Kz/2}{(z-(1/2))(z-1)} + 1$$

2. The CP is

$$p(z) = (z - 1/2)(z - 1) + Kz/2 = z^2 + [(K-3)/2]z + \tfrac{1}{2}$$

3. When applying the Jury test to $p(z)$, it yields the Jury criterion as in Table 4.1.

Applying the Jury criterion yields $\rightarrow 3/4 > (K-3)2/12 \rightarrow (K-3)^2 < 9$. This results in

$$-3 < K - 3 < -3 \rightarrow 0 < K < 6$$

Thus, reconciling with the root locus (RL) gives

$$1 + \tilde{G}(z)H(z) = 1 + \frac{K}{2}\frac{z}{(z-(1/2))(z-1)}$$

As shown in Figure 4.1, the RL for this example is illustrated for different values of $K > 0$, $K < 0$, and $K = 6$.

4.1.3 Stability with Respect to Multiple Parameters: α, β

If there is more than one parameter whose variation can affect the performance of a system, then one can determine the constraints that must be satisfied among a set of parameters. This is illustrated in Example 4.5.

EXAMPLE 4.5

Determine the region in the a_1a_2 plane for which $p(z) = z^2 + a_1 z + a_2$ has its roots in the unit circle, as shown in Figure 4.2.

Now generate the Jury array.

Applying the Jury criteria to Table 4.2 yields:

$$1 - a_2^2 > 0 \rightarrow -1 < a_2 < 1$$

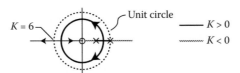

FIGURE 4.1
Root locus (RL) for Example 4.4.

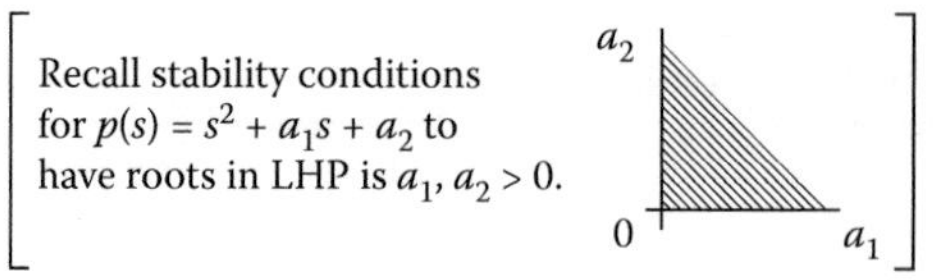

Recall stability conditions
for $p(s) = s^2 + a_1 s + a_2$ to
have roots in LHP is $a_1, a_2 > 0$.
a_2
0
a_1

FIGURE 4.2
Region of stability in the specified plane for multiple parameters.

TABLE 4.2
The Jury Test for Example 4.5

(2)	1	a_1	a_2	$r = a_2$
	a_2	a_1	1	
(1)	$1 - a_2^2$	$a_1(1 - a_2)$		$r = a_1/(1 + a_2)$
	$a_1(1 - a_2)$	$1 - a_2^2$		
(0)	$1 - a_2^2 - \frac{a_1^2(1 - a_2)}{1 + a_2}$			

and

$$1 - a_2^2 - \frac{a_1^2(1 - a_2)}{1 + a_2} > 0 \rightarrow (1 + a_2)^2 - a_1^2 > 0$$

since $1 - a_2 > 0$ and $1 + a_2 > 0 \rightarrow -(1 + a_2) < a_1 < 1 + a_2$. This is illustrated in Figure 4.3. The second-order $p(z)$ stability region is also accordingly illustrated in Figure 4.3.

4.1.4 A More Complicated, State-Space Example

An open-loop unstable continuous system is defined by

$$\dot{\mathbf{X}}(t) = \begin{bmatrix} 0 & 1 & -1 \\ 3 & -2 & 1 \\ 0 & 2 & -1 \end{bmatrix} \mathbf{X}(t) + \begin{bmatrix} 1 \\ 1 \\ 0 \end{bmatrix} u(t); \quad y(t) = \begin{bmatrix} 1 & 0 & 2 \end{bmatrix} \mathbf{X}(t)$$

This system is to be controlled using a digital computer with $h = 0.05$. We investigate CL stability using the SVFB algorithm as follows:

$$\begin{aligned} u(k) &= r(k) - 0.5\dot{\mathbf{X}}_1(k) - 2\mathbf{X}_2(k) - \mathbf{X}_3(k) \\ &= r(k) - \begin{bmatrix} 0.5 & 2 & 1 \end{bmatrix} \mathbf{X}(k); \quad K = \begin{bmatrix} 0.5 & 2 & 1 \end{bmatrix}; \quad (K_r = 1) \end{aligned}$$

a_2
1
a_1
−1
1
−1
$\Rightarrow -(1 + a_2) < a_1 < 1 + a_2$
2nd-order $p(z)$ stability region

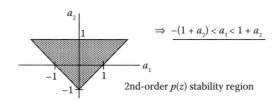

FIGURE 4.3
Second-order $p(z)$ stability.

1. Obtain the equivalent discrete system:

$$X(k + 1) = \Phi X(k) + \Gamma u(k)$$

using "Dscrt," algorithm as described in Chapter 3; yields:

$$\Phi = \begin{bmatrix} 1.0035 & 0.0453 & -0.0477 \\ 0.1430 & 0.9105 & 0.0429 \\ 0.0071 & 0.0930 & 0.9535 \end{bmatrix} ; \quad \Gamma = \begin{bmatrix} 0.0512 \\ 0.0513 \\ 0.0025 \end{bmatrix}$$

2. Form the CL system matrix:

$$\bar{\Phi} = \Phi - \Gamma K$$

Then use Leverier to obtain the CL transfer function:

$$T(z) = C(zI - \bar{\Phi})^{-1}\Gamma$$

Here one can notice that we need only obtain:

$$p(z) = |\, zI - \bar{\Phi}\,|$$

for a CL stability test; thus, we have a CP as

$$p(z) = z^3 - 2.733z^2 + 2.497z - 0.758$$

3. Applying the Jury test → shows $p(z)$ has all the roots in the unit circle → CL system stability.
4. The phase margin can be evaluated by using Leverier to obtain $K(zI - \Phi)^{-1}\,\Gamma$, and then using Bode (option 2 as in Chapter 3) to plot $LG(z)\,|_{z=e}^{jwh}$ → obtain $\omega_c \sim 2.8$ rad/s, $\Phi_m \sim 410$.

4.1.5 Example State-Space Example Plots

Figure 4.4 shows the loop gain versus ω and the state time response with $X(0) = [1\ 0\ 0]'$, $r(t) = 0$.

4.2 Fundamentals of Digital Compensator Design

Given a $G(s)$, or $\tilde{G}(z)$, design a series compensator $H(z)$ so that the CL system meets all end-user specifications.

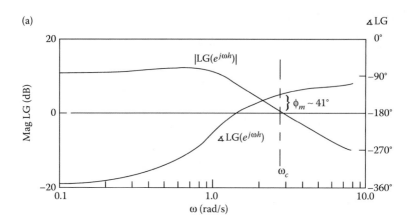

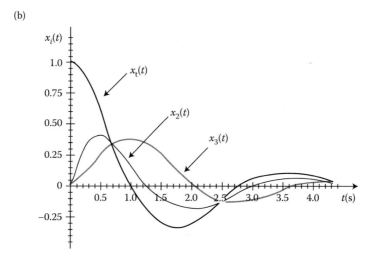

FIGURE 4.4
(a) Loop gain versus ω and (b) state time response with $x(0) = [1\ 0\ 0]'$, $r(t) = 0$.

Design approaches:

1. *H(z)* design via discrete:
   a. The idea is to use continuous time design methods to construct $H(s)$ given $G(s)$, then obtain from $H(s)$ a suitable discrete compensator: $\tilde{H}(z)$
   b. The scheme might be useful provided:

$$\tilde{G}(z)\,|_{z=e^{j\omega h}} \approx G(j\omega) \Rightarrow h \sim \text{small}$$

   c. Alternately, an analog $H(s)$ compensator often exists and we desire to replace the "older" analog system with a digital μ-processor controller. Generally, the designer faces a common problem such as Problem: Given $H(s)$, how do we obtain an $\tilde{H}(z)$?
   d. Direct design of $H(z)$ given $\tilde{G}(z)$.

The following evaluation tools can be used:

1. Stability tests
2. Loop gain analysis
3. RL
4. Simulation
5. Others

4.2.1 $H(z)$ Design via Discrete Equivalent: $H(s) - \tilde{H}(z)$

The following goals are pursued for such an approach:

1. Simplicity:

   Hold equivalence methods [viz., $G(s) \to \tilde{G}(z)$] and impulse transformation methods [$Z\{H(s)\}$] are not simple.

   $$\tilde{H}(z) = \frac{A(z)}{B(z)}; \quad \text{with} \quad A(z), B(z) = \text{polynomials}$$

   (Thus the "obvious" inverse relation $s = 1/h \, \log(z)$ is no good [NG].)

2. If $H(s) = m$th-order transfer function, then $\tilde{H}(z) = m$th-order transfer function is typically given by

   $$H(s) = \frac{b_0 s^m + b_1 s x^{m-1} + \cdots + b_m}{s^m + a_1 s^{m-1} + \cdots + a_m} \quad b_0 \neq 0 \tag{4.9}$$

   that is, $H(s)$ will invariably contain a pure gain (and state-variable model of $H(s)$ will have $d \neq 0$), which requires

   $$\tilde{H}(z) = \frac{\beta_0 z^m + \beta_1 z^{m-1} + \cdots + \beta_m}{z^m + \alpha_1 z^{m-1} + \cdots + \alpha_m} \quad \beta_0 \neq 0$$

3. Accuracy:

   $$\text{Desire } \tilde{H}(z) \left.\right|_{z=e^{j\omega h}} \approx H(j\omega)$$

   over the frequency range of interest/importance as specified by the end user.

An idea to ponder for future designs of $H(z)$ via discrete equivalent is to replace a given polynomial in "s" with some suitable rational $F(z)$.

- A given $H(s)$ can be synthesized, as an interconnection of integrators $1/s$ elements (recall elementary signal flow diagram using Mason's formula [Fletcher Powell]) $\to$ replace $1/s$ = continuous integrator; by $F(z)$ = transfer function of a discrete integrator.

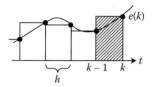

FIGURE 4.5
Discrete integrator form $e(k)$ via $F(z)$ to $g(k)$.

4.2.2 Forms of Discrete Integration

A discrete integrator form is used for $e(k)$ via $F(z)$ as under:
A typical discrete integration form is shown in Figure 4.5.

$$g(k-1) = \text{approximate value of} \int_{-\infty}^{(k-1)h} e(t)\, dt$$

$$g(k) = \text{approximate value of} \int_{-\infty}^{kh} e(t)\, dt$$

1. Forward integration: A representation of a forward integration scheme is shown in Figure 4.6.

$$g(k) = g(k-1) + he(k-1)$$

$$g(z) = z^{-1}g(z) + z^{-1}he(z)$$

$$\Rightarrow F(z) = \frac{h}{z-1} \sim \left.\frac{1}{s}\right|_{s \to \frac{z-1}{h}} \tag{4.10}$$

2. Backward integration:

$$g(k) = g(k-1) + he(k)$$

$$g(z) = z^{-1}\, g(z)\, he(z)$$

Backward integration is illustrated in Figure 4.7. Note the finer nuances between the forward and backward integration methods as shown in Equation 4.11:

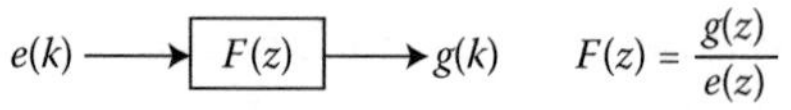

$$e(k) \longrightarrow \boxed{F(z)} \longrightarrow g(k) \qquad F(z) = \frac{g(z)}{e(z)}$$

FIGURE 4.6
Forward integration.

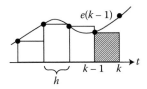

FIGURE 4.7
Backward integration.

$$\Rightarrow F(z) = \frac{h}{1 - z^{-1}} = \frac{zh}{z - 1} \sim \frac{1}{s}\bigg|_{s \to (z-1/zh)} \tag{4.11}$$

3. A trapezoidal, or Tustin, integration is described in the discrete time domain as

$$g(k) = g(k - 1) + h/2[e(k) + e(k - 1)]$$

and is shown in Figure 4.8.

The equivalent z-domain function of $g(k)$ is given by

$$g(z) = z^{-1}g(z) + h/1(1 + z^{-1})e(z)$$

Then the corresponding $F(z)$ is given by

$$\Rightarrow F(z) = \frac{h(1 + z^{-1})}{2(1 - z^{-1})} = \frac{h(z + 1)}{2(z - 1)} \sim \frac{1}{s}\bigg|_{s \to (2(z-1)/h(z+1))} \tag{4.12}$$

4.2.2.1 Relationship to True $s \to z$ Map

There are three distinct methods to find this relationship. Each method corresponds to a different rational approximation of e^{sh}:

Forward integration: $z = e^{sh} = 1 + sh$ gives $s = (z - 1)/h$ $\qquad$ (4.13a)

Backward integration: $z = 1/(e^{-sh}) = 1/(1 - sh)$ gives $s = (z - 1)/zh$ $\qquad$ (4.13b)

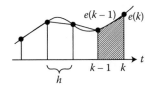

FIGURE 4.8
Signal form for trapezoidal or Tustin integration.

$$\text{Tustin integration: } z = \frac{e^{sh/2}}{e^{-sh/2}} \doteq \frac{1 + sh/2}{1 - sh/2} \quad \text{gives } s = \frac{2}{h} \frac{z - 1}{zh} \tag{4.13c}$$

Note: The above replacements maintain transfer function order.

$$\text{If } H(s) = \frac{b_0 s^m + b_1 s x^{m-1} + \cdots + b_m}{s^m + a_1 s^{m-1} + \cdots + a_m} \rightarrow \tilde{H}(z) = \frac{b_0 (z - 1)^m + \cdots}{(z - 1)^m + \cdots}$$

Use the forward integration $\leftrightarrow$ Euler method to predict $g(k)$.

$$\dot{g}(t) = g(t)$$

Correspondingly, in a digital domain:

$$g(k) - g(k - 1) \Rightarrow g(k) = g(k - 1) + he(k - 1)$$

$$\text{Even if } H(s) = \frac{r\text{th order}}{m\text{th order}}, \quad \tilde{H}(z) = \frac{m\text{th order}}{m\text{th order}}$$

This methodology looks okay since $H(s)$ is almost always mth order/mth order. In the case of the Tustin method, a first-order Padé approximation to z^{-1} can be successfully used.

Also mapping of the LHP to the unit circle yields some valid conclusions:

1. It is useful as a criterion for selecting integration scheme:
   a. Forward integration as shown in Figure 4.9 is now mapped into the z-plane.
   A stable $H(s)$ can yield an unstable $\tilde{H}(z)$! This is not generally good.
2. The backward integration process is shown again in Figure 4.10 in relation to this mapping from LHP to a unit circle.
   A stable $H(s)$ yields stable $\tilde{H}(z)$; some unstable $H(s)$ can also yield stable $\tilde{H}(z)$.
3. The Tustin integration mechanism for mapping from LHP to a unit circle is shown in Figure 4.11.
   This is a preferable map, since stability is mapped 1:1.

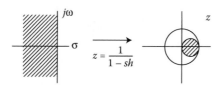

FIGURE 4.9
Mapping of LHP to unit circle for forward integration.

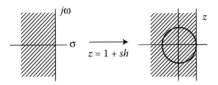

FIGURE 4.10
Mapping of LHP to unit circle for backward integration.

4.2.2.2 Computing $\tilde{H}(z)$ via Tustin Equivalent

Since any $H(s)$ can be decomposed (via PF expansion) into either a cascade or a sum of first- and second-order terms, equivalence can be done on a term-by-term basis as shown below:

1. Simple lag:

$$H(s) = K \frac{1}{\tau s + 1} \left(\text{or } K \frac{a_1}{s + a_1} \text{ with } a_1 = \tau^{-1} \right)$$

$$\tilde{H}(z) = K \left[\frac{1}{(2\tau/h)(z - 1/z + 1) + 1} \right] = \frac{(Kh/2\tau)}{1 + (h/2\tau)} \left[\frac{z + 1}{z - (1 - (h/2\tau)/1 + (h/2\tau))} \right];$$

where

$$\tilde{K} = 1 + (h/2\tau); \quad \alpha_1 = \frac{1 - (h/2\tau)}{1 + (h/2\tau)} \sim e^{-h/\tau}$$

2. General first-order factor:

$$H(s) = K \frac{b_0 s + b_1}{s + a_1} \rightarrow \tilde{H}(z) = \frac{z - \beta_1}{z - \alpha_1} \tag{4.14a}$$

where

$$\frac{b_1}{b_0} < a_1 \Rightarrow \text{lead}; \quad \frac{b_1}{b_0} > a_1 \Rightarrow \text{lag}$$

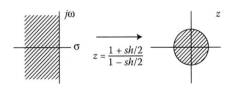

FIGURE 4.11
Mapping of LHP to unit circle for Tustin integration.

and

$$\beta_1 = \frac{b_0 - b_1 h/2}{b_0 + b_1 h/2}, \quad \alpha_1 = \frac{1 - a_1 h/2}{1 + a_1 h/2}, \quad \tilde{K} = K \frac{b_0 + b_1 h/2}{1 + a_1 h/2} \qquad (4.14b)$$

3. General second-order factor:

$$H(s) = K \frac{b_0 s^2 + b_1 s + b_2}{s^2 + a_1 s + a_2} \rightarrow \tilde{H}(z) = \frac{z^2 - \beta_1 z + \beta_2}{z^2 - \alpha_1 z + \alpha_2} \qquad (4.14c)$$

where

$$\alpha_2 = \frac{1 - (a_1 h/2) + (a_2 h^2/4)}{1 + (a_1 h/2) + (a_2 h^2/4)}, \quad \alpha_1 = \frac{2 - (a_2 h^2/2)}{1 + (a_1 h/2) + (a_2 h^2/4)}$$

and

$$\beta_2 = \frac{1 - (b_1 h/2) + (b_2 h^2/4)}{1 + (b_1 h/2) + (b_2 h^2/4)}, \quad \beta_1 = \frac{2b_0 - (b_2 h^2/2)}{b_0 + (b_1 h/2) + (b_2 h^2/4)}$$

with

$$\tilde{K} = K \frac{b_0 + (b_1 h/2) + (b_2 h^2/4)}{1 + (a_1 h/2) + (a_2 h^2/4)}$$

4.2.3 General Algorithm for Tustin Transformation

Generally,

$$H(s) = K \frac{b_0 s^m + b_1 s^{m-1} + \cdots + b_m}{s^m + a_1 s^{m-1} + \cdots + a_m} = \frac{u(s)}{e(s)}$$

1. Write a state variable model for $H(s)$ in SOF with $K = 1$ and $m = n$:

$$\text{Satet mdel: } \dot{X}(t) = AX(t) + Be(t),$$

$$\text{Measurement model: } y(t) = CX(t) + de(t)$$

where

$$A = \begin{bmatrix} -a_1 & 1 & 0 & & 0 \\ -a_2 & 0 & 1 & \cdots & 0 \\ & & & & 0 \\ \vdots & & & \ddots & \vdots \\ & & & & 1 \\ -a_n & 0 & & \cdots & 0 \end{bmatrix}; \quad B = \begin{bmatrix} \tilde{b}_1 \\ \tilde{b}_2 \\ \vdots \\ \tilde{b}_n \end{bmatrix}; \quad \tilde{b}_i = b_i - a_i b_0$$

$$C = \begin{bmatrix} 1 & 0 & \ldots & 0 \end{bmatrix} d = b_0$$

2. Take £ of the state model to yield: $sX(s) = AX(s) + Be(s)$ and replace:

$$s = \frac{2}{h}\left(\frac{z-1}{z+1}\right)$$

$$\frac{2}{h}\left(\frac{z-1}{z+1}\right)X(z) = AX(z) + Be(z)$$

3. Solve above for $X(z)$ and form: $u(z) = CX(z) + de(z)$:

$$u(z) = \left\{C(zI - \tilde{A})^{-1}\tilde{B}(z+1) + d\right\}e(z); \quad \tilde{A} = (I - h/2A)^{-1}\left(I + \frac{h}{2A}\right);$$

$$\tilde{B} = \left(I - \frac{h}{2}A\right)^{-1}B\frac{h}{2}; \quad \tilde{H}(z) == \left\{C(zI - \tilde{A})^{-1}\tilde{B}(z+1) + d\right\}$$

4. Use the Leverier algorithm to obtain $\bar{a}_i, \bar{b}_i$, of denominator and numerator of $C(zI - \tilde{A})^{-1}\tilde{B}$

5. This yields a final form as

$$\tilde{H}(z) = K\frac{\beta_0 z^m + \beta_1 z^{m-1} + \cdots + \beta_m}{z^m + \alpha_1 z^{m-1} + \cdots + \alpha_m}$$

where

$$\beta_i = \bar{b}_i + \bar{b}_{i+1} + d\bar{a}_i; \quad i = 0, 1, 2, \ldots, m-1$$

We can now make some Bode plot comparisons.

4.2.3.1 *Bode Plot Comparisons*

The above process is compared with the Bode plot performance. Usually,

$$\tilde{H}(z)\big|_{z=e^{j\omega h}} \equiv H(s)\big|_{s=2/h(e^{j\omega h}-1/e^{j\omega h}+1)} \approx H(j\omega) \quad \text{for Tustin equivalence}$$

(Include option 3 in the Bode plot program with $x = (z-1)/zh$, and option 4, with $x = 2/h\{(z-1)/(z+1)\}$, where $z = e^{j\omega h}$).

EXAMPLE 4.6

$$H(s) = \frac{2s^2 + 3s + 4}{s^2 + 2s + 6} \xrightarrow[h = 0.5]{\text{Tustin}} \tilde{H}(z) = \frac{1.6z^2 - 1.867z + 0.8}{z^2 - 0.667z + 0.467}$$

The results for the Bode plot of this Tustin equivalent system for $H(s)$ is shown in Figure 4.12.

- Tustin equivalence is usually superior to the backward difference equivalent when comparing:

$$\tilde{H}(z) \big|_{z=e^{j\omega h}} \text{ to } H(j\omega)$$

We will discuss this Tustin equivalent with frequency prewarping further in the sequel.

4.2.4 Tustin Equivalence with Frequency Prewarping

The following aspects have to be considered in this analysis:

- Is it possible to improve the match between Tustin equivalent

$$\tilde{H}(z) \big|_{z=e^{j\omega h}}$$

and original $H(j\omega)$?

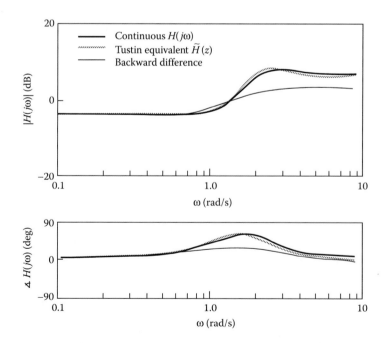

FIGURE 4.12
The Bode plot for Example 4.6 with Tustin equivalence.

- At which frequencies ω, does the equality hold?

  We can see: Tustin $\tilde{H}(z)\big|_{z=e^{j\omega h}} = H(s)\big|_{s=j\omega}$ if and only if (iff):

  $$\frac{2}{h}\left(\frac{e^{j\omega h}-1}{e^{j\omega h}+1}\right) = j\omega \quad \text{or} \quad \tan\left(\frac{\omega h}{2}\right) = \frac{\omega h}{2}$$

  Thus, for $0 \le \omega < \pi/h$ equality holds only at $\omega = 0$.

- Can one obtain equality at one other $\omega \ne 0$, if we have

  $$\tan\left(\frac{\omega h}{2}\right) = a\frac{\omega h}{2}; \quad a > 1?$$

  This corresponds to replacement

  $$s \to \frac{2}{ah}\left(\frac{z-1}{z+1}\right) \tag{4.15}$$

- For equality at $\omega = \omega_1$, usually some important frequency for evaluation is given by

  $$a = \frac{\tan(\omega_1 h/2)}{(\omega_1 h/2)} \tag{4.16}$$

- Tustin with prewarp (include as option 5 in the Bode plot)

  $$s \to \frac{2}{h}\frac{(\omega_1 h/2)}{\tan(\omega_1 h/2)} h\left(\frac{z-1}{z+1}\right) \tag{4.17}$$

(Like a "modified" $h \to ah$).

This is illustrated in Figure 4.13 as $\tan(\omega h/2)$ versus $\omega h/2$.

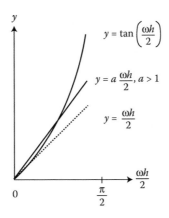

FIGURE 4.13
Plot of $\tan(\omega h/2)$ versus $\omega h/2$.

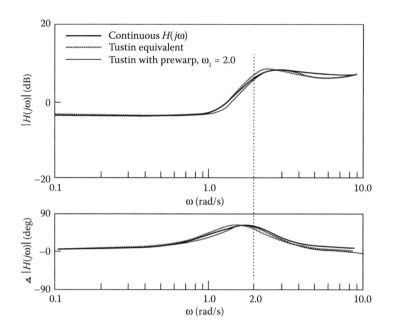

FIGURE 4.14
The Bode plot for Example 4.7 with Tustin equivalence with prewarping.

EXAMPLE 4.7: TUSTIN EQUIVALENCE WITH PREWARPING

Given

$$H(s) = \frac{2s^2 + 3s + 4}{s^2 + 2s + 6}; \quad h = 0.5$$

we require

$$\tilde{H}(z)\big|_{z=e^{j\omega h}} = H(s)\big|_{s=j\omega} \quad \text{at } \omega = 2$$

(corresponds approximately to where $\angle H(j\omega)$ is max).
Then,

$$a = \frac{\tan 0.5}{0.5} = 1.093; \quad \tilde{H}(z) = \frac{1.563z^2 - 1.706z + 0.742}{z^2 - 0.553z + 0.452}$$

The Bode plot is illustrated in Figure 4.14.

- Results show that this method gives a better match in the region:

$$\omega \sim [1.2, 3]$$

EXAMPLE 4.8: TUSTIN EQUIVALENCE WITH PRE-WARPING

In this example, we have:

$$H(s) = \frac{2s^2 + 3s + 4}{s^2 + 2s + 6}; \quad h = 0.5$$

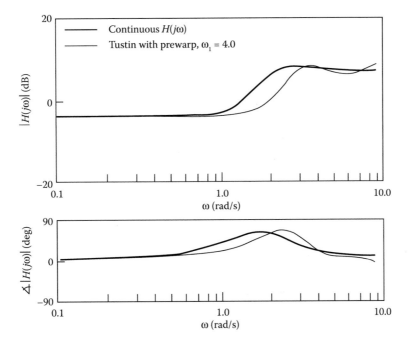

FIGURE 4.15
The Bode plot for Example 4.8 with Tustin equivalence with prewarping and with poor choice of ω.

- A poor choice of ω_1 can result in substantial $H(j\omega)$ versus $\tilde{H}(z)\big|_{z=e^{j\omega h}}$ mismatch for $\omega \neq \omega_1$.
- For example, $\omega_1 = 4$, $a = \tan(0.5)/0.5 = 1.558$.

$\rightarrow$ To avoid problems keep $\omega_1 \leq 1/h < \pi/h$ and examine the Bode plot.

Comparisons of $\tilde{H}(z)\big|_{z=e^{j\omega h}}$ versus $H(s)\big|_{s=j\omega}$ are illustrated below in Section 4.3.

The Bode plot results are illustrated in Figure 4.15, which shows the difference when compared to the previous Figure 4.14.

4.3 Discrete Equivalent Designs

Other techniques for $H(s) \rightarrow \tilde{H}(z)$ equivalences:

- Pole-zero mapping from the s- to z-plane:

$$H(s) = K \frac{\prod_{i=1}^{p}(s - \delta_i)}{\prod_{i=1}^{m}(s - \lambda_i)} \rightarrow \tilde{H}(z) = \tilde{K} \frac{\prod_{i=1}^{m}(z - \tilde{\delta}_i)}{\prod_{i=1}^{m}(z - \tilde{\lambda}_i)} \tag{4.18}$$

where

- If $H(s)$ has a pole at $s = \lambda_i$, then $\tilde{H}(z)$ has a pole at $z = \tilde{\lambda}_i = e^{\lambda_i h}$.

- If $H(s)$ has a zero at $s = \delta_i$, then $\tilde{H}(z)$ has a zero at $z = \tilde{\delta}_i = e^{\delta_i h}$.

- If $H(s)$ has a zero at $s = \infty$, then $\tilde{H}(z)$ has a zero at $z = -1$. (There will be m-p zeros, but usually $p = m$.)

- Pick $\tilde{K}$ such that $H(s)|s=0 = \tilde{H}(z)\,|_{z=1}$; (use $s = 2\pi/1000\, h$ if $H(0) = 0$).

- *Zero-order hold:* The state model (SOF) for $H(s)$ is given, then

$$\tilde{H}(z) = C(zI - \Phi)^{-1}\Gamma + d$$

(Has "effective" $h/2$ s delay due to hold equivalence)

- Higher-order polynomial approximations to $1/s$
  Tustin ~ first-order polynomial through $e(k-2)$, $e(k-1)$, $e(k)$

  Simpson ~ second-order polynomial through $e(k-2)$, $e(k-1)$, $e(k)$

where

$$\frac{1}{s} \rightarrow \frac{h(z^2 + 4z + 1)}{3(z^2 - 1)} \tag{4.19}$$

This method gives a better equivalence! in $\tilde{H}(e^{j\omega h})$ versus $H(j\omega)$, but in the order of $\tilde{H}(z)$ is 2 m.

4.3.1 Summary of Discrete Equivalence Methods

A summary of discrete equivalence methods is shown in Figure 4.16. Any reasonable approximation to $H(s)$

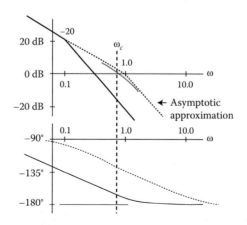

FIGURE 4.16
Discrete equivalent methods.

- Tustin equivalence, using,

$$s \to \frac{2}{h}\left(\frac{z-1}{z+1}\right)$$

gives a good approximation with minimum effort. This is the most commonly used method.

- Consider the use of prewarping if there is a frequency ω_1 or frequency region about ω_1, where it is important that $\tilde{H}(e^{j\omega h}) \approx H(j\omega)$ for example, in the vicinity of ω_{max} of lead network, or around crossover frequency ω_c.

- Pole-zero mapping is frequently used (very similar in results to Tustin), but does not permit frequency prewarping.

- $H(s) \to \tilde{H}(z)$ equivalent transformations are very frequently used in digital filtering and digital filter design.

4.3.2 Example of a Discrete Equivalent Design

- Radar positioning system: A Radar positioning system is shown in Figure 4.17.
- CL requirements are illustrated in Figure 4.18 with a desired end-user specifications as in Franklin and Powell (1980) [1].

4.3.2.1 End-User Specifications

Desire a 15% overshoot (OS) to a step command input ($\zeta \to \sim 0.5$) and settling time (TS) $t_s|1\% \sim 0.5$ with a phase margin (PM) of $\Phi_m \geq 50°$.

- "Solution," $H(s)$ = lead network (NW)

$$\text{Lead } NW = \frac{10s + 1}{s + 1} \quad (\omega_2 = 0.1, \quad \beta = 10, \quad K = 1)$$

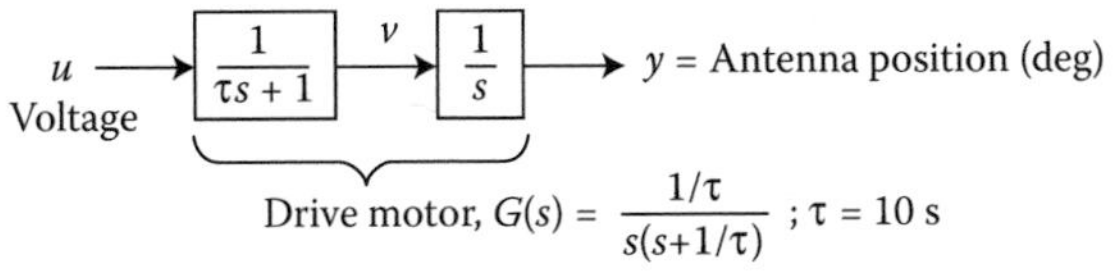

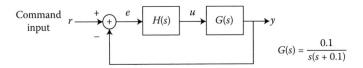

FIGURE 4.17
Radar positioning system.

FIGURE 4.18
CL requirements of Radar positioning system.

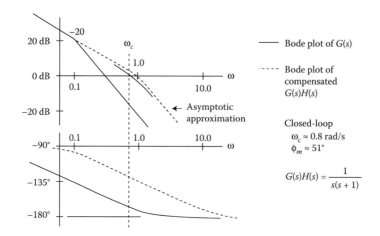

FIGURE 4.19
The Bode plot for CL system per end-user specifications.

- This is not a good CL design; there is not as large a region of −20 dB slope around the crossover as there is under, as seen from Figure 4.19.

$$\omega_c \neq \omega_2\sqrt{\beta}$$

RL and Time domain response of continuous design:

- The RL of uncompensated and compensated systems is shown in Figure 4.20.
- CL step response is shown in Figure 4.21.

4.3.3 Discrete Equivalent Computations

- Select time step $h = 1.0$ s.

  Note: The state model of a system with

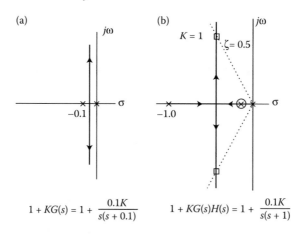

FIGURE 4.20
Root locus (RL) of uncompensated and compensated system for Radar positioning system. (a) Root locus of uncompensated system $1 + KG(s) = 1 + ((0.1K)/s(s + 0.1))$. (b) Root locus of compensated system $1 + KG(s)H(s) = 1 + ((K)/s(s + 1))$.

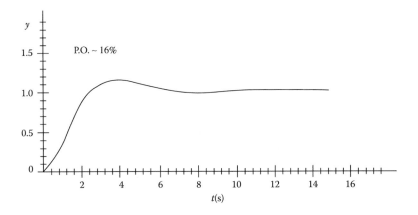

FIGURE 4.21
CL step response compensated system for Radar positioning system.

$$x_1 = v, \quad x_2 = y$$

$$\dot{X}(t) = \begin{bmatrix} -0.1 & 0 \\ 1.0 & 0 \end{bmatrix} X(t) + \begin{bmatrix} 0.1 \\ 0 \end{bmatrix} u(t); \quad y(t) = \begin{bmatrix} 0 & 1 \end{bmatrix} X(t)$$

$$\|A\| = \sqrt{1.01/2} \doteq 0.7; \quad |\lambda_{max}(A)| = 0.1.$$

so $h = 1.0$ is compatible with criterion $h < (0.5 \rightarrow 1.0)/|(\|A\|)|$.

- Zero-order hold equivalent:

$$\tilde{G}(z) = \frac{z + 0.967}{(z - 1)(z - 0.905)}$$

- Tustin equivalent:

$$\tilde{H}(z) = H(s)\big|_{s=2(z-1/z+1)}$$

$$\tilde{H}(z) = 7\left(\frac{z - 0.905}{z - 0.333}\right) = 7\left(\frac{1 - 0.905z^{-1}}{1 - 0.333z^{-1}}\right) = \frac{u(z)}{e(z)}$$

Algorithm:

$$u(k) = 7e(k) - 6.335e(k - 1) + 0.333u(k - 1)$$

This discrete equivalent system is shown in Figure 4.22.

We need to examine the CL step response to LG (z), for a discrete system.

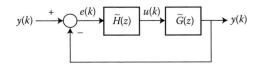

FIGURE 4.22
Discrete equivalent CL system for radar positioning system.

4.3.4 Evaluation of Digital Control Performance

- Step response, $r(t) = 1$: Step response is illustrated in Figure 4.23 with % OS of 50% ($y_{max} \sim 1.5$). This corresponds to $\zeta \sim 0.22$, while a continuous design had $\zeta \sim 0.5$.
- We need to find what happened here:
    - Clearly, there has been a decrease in Φ_m.
    - $\tilde{H}(e^{j\omega h}) \approx H(j\omega)$, at least in the crossover region.
    - Clearly here the problem is that in the related crossover region

$$\tilde{G}(e^{j\omega h}) \neq G(j\omega)$$

- Heuristic analysis:
    - The first (crude) approximation $\tilde{G}(e^{j\omega h}) \approx e^{-j\omega h/2}G(j\omega)$ that is, sampling, introduces a delay of $h/2$ s.
    - At ω_c we get a decrease in Φ_m of 57.3 $\omega_c h/2$ deg. $\rightarrow 23°$ loss of phase margin here!
    - Φ_m of discrete system $\sim 51° - 23° = 28°$ corresponds to

$$\zeta \sim 0.25 \text{ (for a second-order continuous system)}$$

4.3.5 Continuous versus Discrete System Loop Gain

Figure 4.24 shows aliasing properties of discrete LG for $\omega > \pi/h = 3.14$.
  Repetition for $\omega > 2\pi/h$; LG(z) at $\omega = 2\pi/h$ ($z = 1$).

4.3.6 Methods to Improve Discrete CL Performance

- Pick a smaller time step, h:

$$\Delta\Phi_m = 57.3 \left(\frac{\omega_c h}{2} \right) \text{deg.} < 5 - 10°$$

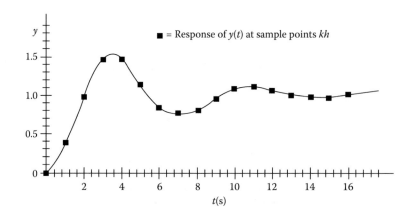

FIGURE 4.23
Digital equivalent CL system performance for radar positioning system.

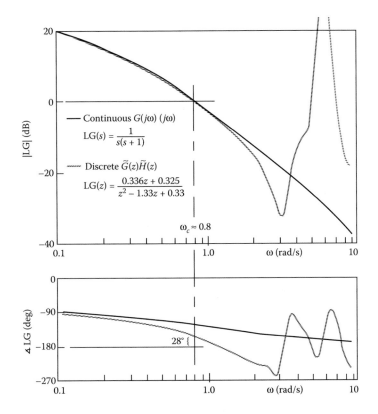

FIGURE 4.24
Continuous versus discrete system loop gain comparison.

Choosing in this manner will give a smaller h than does the criterion $h \leq (0.5 \rightarrow 1.0)/||A||$. But note that very small h may cause CPU timing and other problems.

- Use Tustin with prewarp.
  Not particularly useful, here but could be used to assure $\tilde{H}(z)$ gives little or no magnitude and/or phase distortion in the crossover region.
- Redesign $H(s)$ to give additional positive phase.
  - Precompensate for several phase decrease in $\tilde{G}(z)$.
  - For a given $h = 1.0$, we need a continuous system phase margin of ~70°!; (An unreasonable $H(s)$ design.
  - Good approach if $\Delta\Phi_m < 15°$.
- Design $H(z)$ directly in the z-plane.
  - $\tilde{G}(z)$ is fundamentally different than $G(s)$.
  - Avoids small time step constraints needed to make Tustin equivalent $\tilde{H}(z)$ perform satisfactorily.
  - Less guesswork to modify design.
  - May be possible to use $\tilde{H}(z)$ as a starting point.
  - $\rightarrow$ Use *Tustin* if $\omega_c h$ is small, otherwise consider direct design of $H(z)$.

PROBLEMS

Compensator Design via Discrete Equivalent

P4.1   A system is given by

$$G(s) = \frac{2}{s + 2}$$

This system is managed by a controller given by

$$u(k) = Ke(k) + u(k + 1)$$

With $h = 0.72$ s, find the range of K if the CL system is stable using the Jury test.

P4.2   Determine the region in the $a_1 - a_2$ plane for which

$$p(s) = s^2 + a_1 s + a_2$$

has its roots in the unit circle for the following values of a_1 and a_2.

1. $a_1 = 1.9; a_2 = 0.9$
2. $a_1 = 0.1; a_2 = 0.9$
3. $a_1 = 1.9; a_2 = 0.9$
4. $a_1 = 1.1; a_2 = 0.1$

P4.3   The open-loop unstable continuous system is defined by

$$\dot{X}(t) = \begin{bmatrix} 0 & 2 & -2 \\ 6 & -4 & 2 \\ 0 & 4 & -2 \end{bmatrix} X(t) + \begin{bmatrix} 2 \\ 2 \\ 0 \end{bmatrix} u(t);$$

$$y(t) = \begin{bmatrix} 2 & 0 & 4 \end{bmatrix} X(t)$$

This system is to be controlled by using a digital computer with $h = 0.1$ s.

1. Investigate the CL stability using the SVFB algorithm

$$u(k) = r(k) - 1.0x_1(k) - 4x_2(k) - 2x_3(k)$$

2. Find the phase margin using Leverier algorithm.
3. Plot the Bode plots of the loop gain (LG).
4. Plot the state response with $X(0) = [2\ 0\ 0]'$, $r(t) = 0$.

P4.4   Using the Tustin equivalence method, find the discrete equivalent of the following $H(s)$:

$$H(s) = \frac{2s^2 + 5s + 7}{s^2 + 3s + 8}$$

With $h = 0.4$, conduct a Bode plot comparison of continuous, Tustin equivalent, and backward differences.

P4.5 Using the Tustin equivalence with the prewarping method, find the discrete equivalent of the following $H(s)$:

$$H(s) = \frac{2s^2 + 5s + 7}{s^2 + 3s + 8}$$

With $h = 0.5$, conduct a Bode plot comparison of continuous, Tustin equivalent, and backward differences.

P4.6 Using the pole-zero mapping method, find the discrete equivalent of the following $H(s)$:

$$H(s) = \frac{2s^2 + 5s + 7}{s^2 + 3s + 8}$$

With $h = 0.4$, conduct a Bode plot comparison of continuous, zero mapping, and backward differences.

P4.7 An open-loop unstable continuous system is defined by

$$\dot{X}(t) = \begin{bmatrix} 0 & 1 & -1 \\ 3 & -2 & 1 \\ 0 & 2 & -1 \end{bmatrix} X(t) + \begin{bmatrix} 1 \\ 1 \\ 0 \end{bmatrix} u(t)$$

$$y(t) = \begin{bmatrix} 1 & 0 & 2 \end{bmatrix} X$$

and is controlled using a digital computer with $h = 0.5$. Investigate CL stability using the SVFB algorithm:

$$u(k) = r(k) - 0.5x_1(k) - 2x_2(k) - x_3(k); \quad \forall K_r = 1$$

P4.8 An antenna positioning system control has a plant transfer function:

$$G(s) = \frac{0.1}{s^2(s + 0.1)}$$

To the following specifications:
1. PO to step input ~15%.
2. $t_s|_{1\%} \sim 10$ s.
3. $\Phi_m \geq 50°$.
4. Design a discrete compensator $H(z)$ using w to z (backward) transformation with $h = 1.0$. Show step response and the corresponding control "u" in the time domain.
5. Conduct a frequency domain evaluation.

Use all routines like, "Discrete," "Leverier," etc.

P4.9 The system

$$\dot{x}(t) = 0.5x(t) + 0.96u(t)$$

is controlled digitally using the following algorithm:

$$u(k) = K_r r(k) - x(k)$$

With a time step of $h = 0.4$ s,

1. Find the discrete equivalent model.
2. Check stability of CL system for $K = +1$.
3. Obtain the phase margin via the Bode plot.

P4.10 A discrete system block diagram is shown in Figure P4.10. Using the discrete decomposition, the state diagram of the controlled process $G(s)$ is drawn as shown in Figure P4.11.

The state diagram of the overall system was constructed by connecting the state diagram of Figure P4.11 so that the zero-order hold and using relationship:

$$e(kT) = r(kT) - c(kT)$$
$$= r(kT) - x_1(kT)$$

also letting $t_0 = kT$, and

$$h(kT^+) = h(t) = e(kT) \forall kT \leq t < (k+1)T$$

The complete state diagram is shown in Figure P4.12.

1. Write the state equations using $T = 1$ s and $r(t)$ is a unit-step input.
2. Find $\Phi(T)$ and $\Gamma(T)$ matrices.

The state diagram between the error node and the input node is shown in Figure P4.13

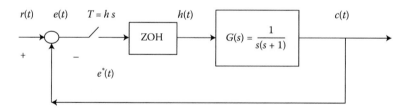

FIGURE P4.10
Discrete system block diagram for Problem P4.10.

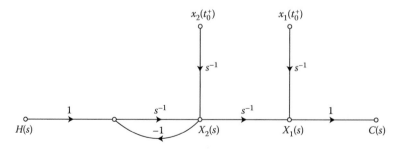

FIGURE P4.11
Signal flow graph from $H(s)$ to $C(s)$ to match block diagram from $h(t)$ to $c(t)$.

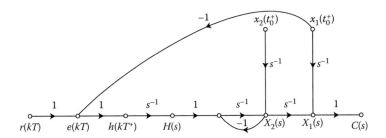

FIGURE P4.12
Signal flow graph for digitized $r(kT)$ to $t(s)$ respectively.

3. The digital controller is described by

$$D(z) = \frac{b_1 + b_0 z^{-1}}{1 + a_0 z^{-1}}$$

4. Write the dynamic equations of $D(z)$, that is, it's state and output equations, as shown in Figure P4.14.
5. Draw the state diagram of the sampled data system.
6. Using the state-variable analysis of response between sampling instants, $T = 1$ s, and increment $\Delta = 0.5$ s, plot t versus $x_1(t)$ and $x_2(t)$.

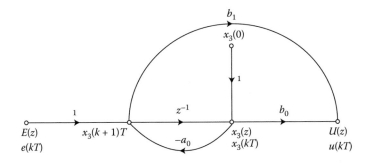

FIGURE P4.13
Signal flow graph from $e(kT)$ to $u(kT)$.

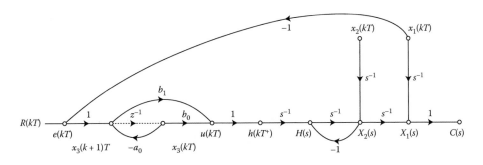

FIGURE P4.14
Signal flow graph from $R(kT)$ to $C(s)$.

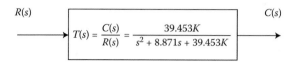

FIGURE P4.15
Transfer function $T(s)$ for Problem P4.15.

7. Does your plot indicate that the maximum OS of the system is closer to 1.448 rather than 1.399, as presented by the results only at the sampling instants?

8. Show all your plots clearly. Indicate all labels, titles, and variables. Use appropriate legend.

P4.11 Consider the following CE of a discrete data system:

$$z^3 + 3.3z^2 + 4z + 0.8 = 0$$

Carry out the Jury test to assess the stability of the system.

P4.12 A space vehicle control system is shown in Figure P4.15.

The CL transfer function is given by

$$\frac{C(s)}{R(s)} = \frac{39.453K}{s^2 + 8.871s + 39.453K}$$

for

$$\frac{C(s)}{T_m(s)} = \frac{1}{J_v s^2}$$

$K_p = 1.65 \times 10^6$ position sensor gain

$K_R = 3.71 \times 10^5$ rate sensor gain

K = variable—amplifier gain

$J_v = 418{,}222$—moment of inertia

1. Find ω_n—natural angular velocity.

2. ζ = damping coefficient of the system.

3. Find the range of K for the system to be stable.

4. The continuous data system in Figure P4.16 is subject to a sampled-data control with the position feedback and the input processed by a sample-and-hold (zoh) device. Draw the system block diagram of this digital space vehicle control system.

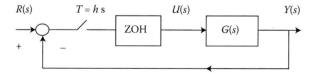

FIGURE P4.16
Block diagram for *CL* space vehicle system.

5

Compensator Design via Direct Methods

5.1 Direct Design Compensation Methods

The following are the different methods for direct design compensation schemes for a system model as illustrated in Figure 5.1. These schemes work directly with $\tilde{G}(z)$ to design $H(z)$ and so are not limited by the requirement that h be small.

1. Root locus (RL) design methods: Compensator design in z-plane using standard RL design procedures to move CL poles.

2. w-plane design methods: This method is similar to classical frequency (ω) domain design procedures where w is a rational approximation to $(1/h)\ln(z)$ (where $\ln(.) \to$ natural log).

3. Fixed-form parametric design: This method assumes a structural form for $H(z)$, for example, proportional integral derivative (PID) design, and adjusts free parameters.

4. Miscellaneous approaches: [2].

5. Closed Loop (CL) Transfer function:

$$T(z) = \frac{\tilde{G}(z)H(z)}{1 + \tilde{G}(z)H(z)} = \frac{y(z)}{r(z)}$$

6. Zeros of $T(z)$ are the zeros of $\tilde{G}(z)H(z)$ = zeros of $\tilde{G}(z)$ plus those added by $H(z)$.

7. Poles of $T(z)$ are the roots of

$$1 + \tilde{G}(z)H(z)$$

5.1.1 RL Design of H(z)

$$H(z) = K\frac{\prod_{i=1}^{m}(z + \delta_i)}{\prod_{i=1}^{m}(z + \lambda_i)} = KH_0(z) \tag{5.1}$$

- In this method with $H(z)$ as defined in Equation 5.1, the designer should pick poles and zeros of $H(z)$ so that the RL of

$$1 + K\tilde{G}(z)H_0(z)$$

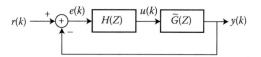

FIGURE 5.1
System model for direct design.

with respect to gain K passes through the region in the z-plane where damping, ζ, and natural frequency, ω_n, are suitable.

- One can plot on a z-plane with ζ, ω_n overlay.
- Pick δ_i, λ_i real, generally with $|\lambda_i| \leq 1$.
- Any added zeros δ_i must have an associated pole (no free zeros).
- Generally a first- or second-order $H(z)$ suffices, for example,

$$H(z) = \frac{K(z + \delta_1)}{(z + \alpha_1)} = KH_0(z) \qquad (5.2)$$

- If $\lambda_1 < \delta_1 \Rightarrow$ lead compensation
- If $\lambda_1 > \delta_1 \Rightarrow$ lag compensation
- Then pick K so that (dominant) CL poles are at some desired location on the RL and specifications are met.

$$K = \left.\frac{1}{\tilde{G}(z)H_0(z)}\right|_{z=z_{des}} \qquad (5.3)$$

- Next evaluate time response, loop gain, $K\tilde{G}(z)H_0(z)@z = e^{j\omega h}$, and so on.
- Adjust δ_i, λ_i, and K until the system meets specifications.

$\rightarrow$ This further emphasizes the trial-and-error design.

5.1.1.1 Some Helpful Hints for RL Design

Recall—RL bends toward zeros, away from poles.

EXAMPLE 5.1

As shown in Figure 5.2, this phenomenon is illustrated for an added zero and an added pole.

- If a zero steady-state (ss) error to a constant input is required, $\tilde{G}(z)H_0(z)$ must have a pole at $z = 1$.
- Try not to have CL poles on $-1 < z < 0$. If there is a pole at $z = -a$, then $y(k)$ or $u(k)$ has a term of the form $(-a)^k \rightarrow 1, -a, +a^2, -a^3, \ldots$ point-to-point oscillation.
  - If one cannot avoid this then try to keep $|a|$ small. This is illustrated in Figure 5.3.
  - A "nice" region in the z-plane, especially for dominant pair, is illustrated in Figure 5.4.

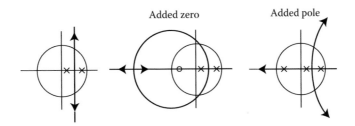

FIGURE 5.2
RL-shape change with an added zero and an added pole, respectively.

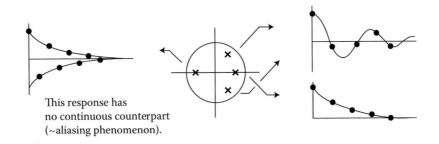

FIGURE 5.3
Relation of time response to location of poles in the unit circle.

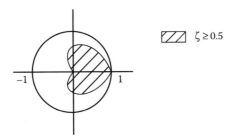

FIGURE 5.4
Nice region in z-plane.

5.1.2 Example of Design Approach: Antenna Positioning Control

The s-domain process transfer function for an antenna positioning system is shown below. The corresponding discrete equivalent is also shown for $h = 1$:

$$G(s) = \frac{0.1}{s(s + 0.1)} \xrightarrow{h = 1} \tilde{G}(z) = 0.048 \frac{z + 0.97}{(z - 1)(z - 0.905)}$$

$H(Z) = K$

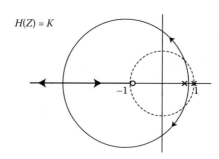

FIGURE 5.5
Uncompensated RL for Example 5.1.2.

End-user specifications are as follows:

1. PO to step input = 15%
2. Settling time $TS|_{1\%} = 10$ s
3. Phase margin PM; $\Phi_m \geq 50°$

The uncompensated RL is shown in Figure 5.5. The uncompensated RL shows a pole already at $z = 1$ via $\tilde{G}(z)$. The RL is not very encouraging and not very good.

- Need a zero on [0, 1] to bend RL inward more.
- Place associated pole on [−1, 0] away from added zero.

To have $t_{s|1\%} = 5/\zeta\omega_n \sim 10 \rightarrow \zeta\omega_n \sim 0.5$ for dominant CL poles

$\rightarrow$ need $|z| = e - \zeta^{\omega}_n \sim 0.6$, with $\zeta = 0.5$ to obtain PO $\sim 15\%$

- First trial design of the compensated system yields

$$H(z) = K(z - 0.5)/(z + 0.6)$$

The compensated RL is shown in Figure 5.6.

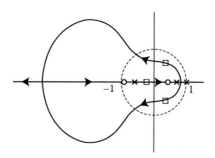

FIGURE 5.6
RL of compensated system with $H(z)$.

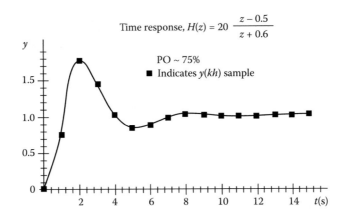

$$\text{Time response, } H(z) = 20 \ \frac{z - 0.5}{z + 0.6}$$

PO ~ 75%
■ Indicates $y(kh)$ sample

FIGURE 5.7
Time response of compensated system.

This yields a RL that is not too bad, with $K \sim 20$ to obtain a dominant CL pair. With $\zeta = 0.5$ one also gets a CL pole at $z = -0.2$. Will this give rise to some ensuring problems?

- Examine CL response via simulation:

$$u(k) = 20e(k) - 10e(k-1) - 0.6\,u(k-1)$$

The time response is shown in Figure 5.7.
Where the PO is 75%, $y(kh)$ samples are shown by the bolded squares in Figure 5.7.
The corresponding discrete input $u(k)$ is plotted in Figure 5.8.

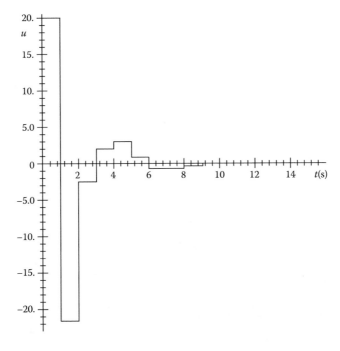

FIGURE 5.8
Discrete input $u(k)$.

It is clear from the results that we need to reduce gain, and move zero at 0.5 closer to the pole at $z = -0.905$, which needs the movement of the pole at -0.6 closer to $z = 0$.

5.1.3 RL Redesign (After Much Trial and Error)

Now, use zero to cancel the pole at $z = 0.905$, and place the pole so that the RL goes through the "nice" region ($|zl \le 0.6$, $\zeta \sim 0.5$), as described earlier.
    For this trial we choose

$$H(z) = K\frac{z - 0.905}{z + 0.2}$$

The new RL is shown in Figure 5.9. $K \sim 9$ gives poles at $0.18 \pm j\,0.44$.
    A step response is shown in Figure 5.10 for the discrete input shown in Figure 5.11. This redesign is good, but K_v has gone from 1.0 (continuous design) to 0.71,

$$\text{Since } \frac{1}{h}(1 - z^{-1})\tilde{G}(z)H(z)\Big|_{z=1} = K_v = 0.71$$

The Bode plot of

$$LG(z) = \frac{0.432z + 0.418}{(z - 1)(z + 0.2)} \Rightarrow \omega_c \sim 0.70\frac{\text{rad}}{\text{s}}, \ \Phi_m \sim 55°$$

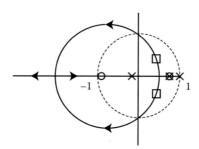

FIGURE 5.9
RL for redesign.

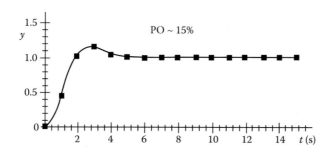

FIGURE 5.10
Step response for redesign.

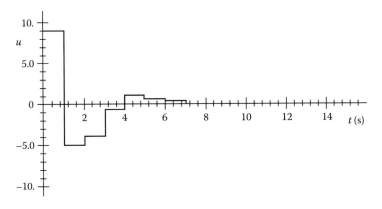

FIGURE 5.11
Discrete input for redesign.

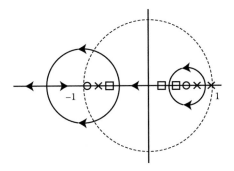

FIGURE 5.12
RL of a poor design.

5.1.4 An Example of a Poor Design Choice

To reduce the PO further, we can move the zero of $H(z)$ closer to the pole at $z = 0.905$ and move the pole of $H(z)$ further out toward $z = -1$.

The corresponding $H(z)$ is shown below for which the RL is shown in Figure 5.12. The output response is shown in Figure 5.13 and the discrete input is shown in Figure 5.14. Intersample "ripples" in $y(t)$ and oscillatory $u(k)$ are indicative of CL poles on negative real axis.

5.2 w-Plane Design of $H(z)$

In this design method, the below steps are followed:

- Attempt to use the Bode design techniques to obtain $H(z)$ starting with $\tilde{G}(z)$
- Cannot go into the s-plane to design $H(s)$ and then get $H(z)$
  - Map from $z \rightarrow s$ plane is not a very rational way to follow
  - Need a rational approximation to $z = e^{sh}$

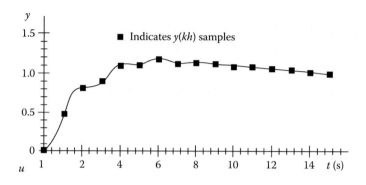

FIGURE 5.13
Step response of a poor compensated design.

- Define the "w-plane" with $w \sim s$; as

$$z = \frac{1 + wh/2}{1 - wh/2} \tag{5.4}$$

which yields

$$w = \frac{2}{h}\left(\frac{z-1}{z+1}\right) = \mu + j\upsilon \tag{5.5}$$

- On the unit circle we then have:

$$\upsilon = \frac{2}{h}\tan\left(\frac{\omega h}{2}\right) \approx \omega \quad \text{when } \omega h < 1$$

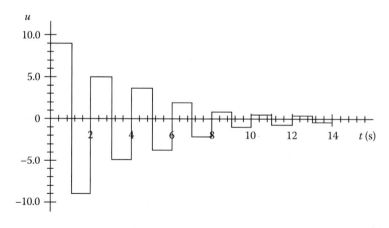

FIGURE 5.14
Discrete input to the poor design.

- Rational mapping can then be found as follows:

$$\tilde{G}(z) = \frac{b_0 z^n + b_1 z^{n-1} + \cdots + b_n}{z^n + a_1 z^{n-1} + \cdots + a_n} \rightarrow \tilde{G}(w) = \frac{c_0 w^n + c_1 w^{n-1} + \cdots + c_n}{w^n + d_1 z^{n-1} + \cdots + d_n}$$

- $\tilde{G}(w)$ will always be the n-th order/n-th-order polynomial in "w."
- Unit disk $|z| \leq 1$ is mapped into LHP $\mathrm{Re}(w) \leq 0$:

$$\left. \cdot \tilde{G}(w) \right|_{w=j\upsilon} \approx \left. \tilde{G}(z) \right|_{z=e^{j\omega h}} \quad \text{if } \omega h < 1 \tag{5.6}$$

- To the first approximation, ($\omega \ll \pi/h$) yields

$$\left. \tilde{G}(w) \right|_{w=j\upsilon} \doteq \left. G(s)e^{-(sh/2)} \right|_{s=j\omega} \tag{5.7}$$

- Equation 5.7 be included as an additional option in the Bode plot subroutine that we evaluated in Chapter 3.

5.2.1 Design Approach

The w-plane design approach is shown in Figure 5.15.

- $z \rightarrow w$ transformation can be illustrated as follows:
  - If

$$\tilde{G}(z) = K \frac{\prod_{i=1}^{m}(z + \delta_i)}{(z-1)^k \prod_{j=1}^{n-k}(z + \lambda_j)} \tag{5.8}$$

then, using the above transformation as shown in Figure 5.15,

$$\tilde{G}(w) = \frac{K\left[\prod_{i=1}^{m}(1 + \delta_j)\right]\left(1 - (w/(2/h))\right)^{n-m} \prod_{i=1}^{m}\left(1 + (w/(2/h)((1 + \delta_i)/(1 - \delta_i)))\right)}{\left[\prod_{j=1}^{n-k}(1 + \lambda_j)\right]h^k w^k \prod_{j=1}^{n-k}\left(1 + (w/(2/h)((1 + \lambda_j)/(1 - \lambda_j)))\right)} \tag{5.9}$$

$$\tilde{G}(w) \rightarrow \quad \rightarrow H(w)$$

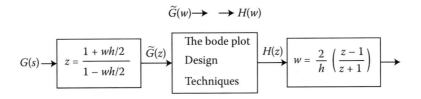

FIGURE 5.15
Design approach from $G(s)$ to $H(z)$ via w-plane.

- – Useful formula when $\tilde{G}(z)$ has only real poles and zeros
- • State-space approach in a general case
  - – Need a general technique that is computer-oriented
- • w-to-z transformation
  - • Identical to Tustin transform on $H(w)$

5.2.2 General $z \to w$ Plane Mapping

Given a system to be controlled as with state and measurement model by

$$\text{State Model: } \dot{X}(t) = AX(t) + Bu(t),$$

$$\text{Measurement Model: } y(t) = CX(t)$$

Here, all bold upper-case variables are $n \times 1$ vectors and upper-case letters represent matrices like $A = n \times n$, $B = n \times 1$, and $C = 1 \times n$; lower-case variables are scalars. Thus in the above model the system is a single-input, single-output (SISO) system.
Determine $\tilde{G}(w)$.

1. Obtain the equivalent discrete system in the usual manner:

$$X(k + 1) = \Phi X(k) + \Gamma u(k)$$

$$y(k) = CX(k)$$

This yields

$$\tilde{G}(z) = C(zI - \Phi)^{-1}\Gamma$$

2. z-transform: $zX(k) = \Phi X(z) + \Gamma u(z)$.
3. Let

$$z = \frac{1 + wh/2}{1 - wh/2}$$

which yields

$$\left(1 + \frac{wh}{2}\right)X(w) = \left(1 - \frac{wh}{2}\right)\Phi X(w) + \left(1 - \frac{wh}{2}\right)\Gamma u(w)$$

and

$$y(w) = CX(w)$$

4. Solve for $y(w)$.

$$y(w) = C\left[wI - 2/h(\Phi + I)^{-1}(\Phi - I)^{-1}(\Phi + I)^{-1}\Gamma\left(\frac{2}{h} - w\right)\right]u(w) \tag{5.10}$$

where

$$\tilde{\Phi} = 2/h(\Phi + I)^{-1}(\Phi - I); \quad \tilde{\Gamma} = (\Phi + I)^{-1}\Gamma;$$

$$\tilde{G}(w) = C\left[wI - 2/h(\Phi + I)^{-1}(\Phi - I)^{-1}(\Phi + I)^{-1}\Gamma\left(\frac{2}{h} - w\right)\right]$$

- Use Leverier with $\tilde{\Phi}$ and $\tilde{\Gamma}$ to obtain $C(wI - \tilde{\Phi})^{-1}\tilde{\Gamma}$, then include the $(2/h - w)$ factor.
- Note the non-minimum phase zero at $w = 2/h$.

Follow the Tustin state-space approach for w-to-z plane $H(w) \to H(z)$.

5.2.3 Example of Design Approach

We consider again the earlier example of the antenna-positioning controller with

$$G(s) = 0.1/s(s + 0.1)$$

Then,

$$\tilde{G}(z) = 0.048\frac{z + 0.967}{(z - 1)(z - 0.905)} \quad \to \quad \tilde{G}(w) = \frac{(1 + (w/120))(1 - (w/2))}{w(1 + (w/0.1))} \qquad (5.11)$$
$$(H = 1.0)$$

The ensuing results are shown in Figures 5.16 and 5.17 for the Bode plot magnitude versus frequency and phase angle versus frequency, respectively, in the w-plane.

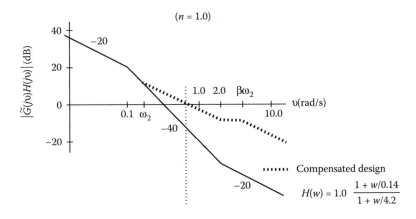

FIGURE 5.16
The Bode magnitude plot of an antenna-positioning controller with w-plane design.

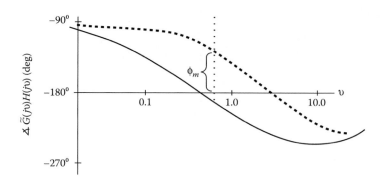

FIGURE 5.17
The Bode angle plot of an antenna positioning controller with w-plane design.

The plots in Figures 5.16 and 5.17 show both the compensated and uncompensated designs, respectively, and one can surmise that it is advisable to

- Use lead compensation to keep the bandwidth up
- Make ω_c as large as possible with a $\Phi_m \sim 55°$
- Use a limit value of $\beta = 30$ (corresponds to $\Delta\Phi_{max} \sim 69°$)
  - $\rightarrow \omega_c \sim 0.77$ (where angle of $G(jv) = -180° + \Phi_m - 69° = -194°$)
  - $\rightarrow \omega_2 = \omega_c/(\beta)^{1/2} = 0.14$, $\beta\omega_2 = 4.2$
- Pick K so that

$$K\left|\tilde{G}(w)H_0(w)\right|\big|_{w=j0.77} = 1 \Rightarrow K = 1.0$$

5.2.3.1 Example of w-to-z (Backward) Transformation

$$H(z) = 1.0\frac{1 + w/0.14}{1 + w/0.42}\bigg|_{w=(2/h)(z-1/z+1)} = 10.5\frac{z - 0.87}{z + 0.35} \tag{5.12}$$

- $H(z) \rightarrow$ No reduction in low-frequency gain

$$K_v = \left(\frac{z-1}{zh}\right)\tilde{G}(z)H(z)\bigg|_{z=1} = 1.0 \text{ (same as continuous design)}$$

The time response and discrete input to the system are shown in Figures 5.18 and 5.19, respectively, where one can see that:
- This scheme is very similar to the RL design:

$$H(z) = 9\frac{z - 0.905}{z + 0.2}\left(\text{a bit }\frac{\text{faster}}{\text{better}}\right)$$

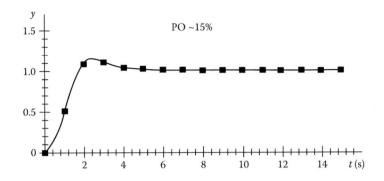

FIGURE 5.18
Time response for w–z backward transformation.

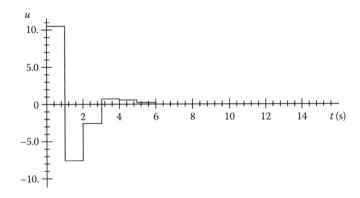

FIGURE 5.19
Discrete input for w–z backward transformation.

5.2.4 Frequency Domain Evaluation

- Examine actual loop gain $(LG)(z)\,|_{z=e^{i\omega h}}$ to find true ω_c, Φ_m:

$$LG(z) = \tilde{G}(z)H(z) = 0.504\,\frac{(z + 0.967)(z - 0.87)}{(z - 1)(z - 0.905)(z + 0.35)}$$

Thus, here

- Compare $\tilde{G}(z)H(z)$ with $w = jv$.

The frequency domain evaluation is shown in Figures 5.20 and 5.21. One can see that:

- Discrete loop gain is very similar to RL design, with ~3 dB higher very-low-frequency gain.
- w-plane design approximation is OK for $v \sim \omega < 1/h$.
- Actual $\Phi_m \sim 56°$, $\omega_c \sim 0.73$ (system will tolerate a maximum loop delay).

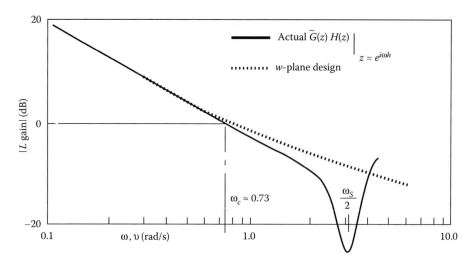

FIGURE 5.20
The Bode plot gain of LG.

and

$$\tau_{max} = \frac{\Phi_m}{\omega_c} = 1.34 \text{ s}$$

5.2.4.1 RL versus w-Plane Design Comparison

Either approach, used correctly, will give a good design.
   RL Design:

- The RL plot is more difficult to draw than the Bode plot.
- It is hard to determine where to place poles and zeros of $H(z)$ to properly shape RL as desired.
- The RL plot seems to require more trial and error than does the Bode approach.
- Needs overlay $\zeta - \omega_n$ contours on RL versus ω-plane design.
- Difficult to make engineering approximations.
- If $h \sim$ small, the RL tends to crowd into regions around $z = 1$.

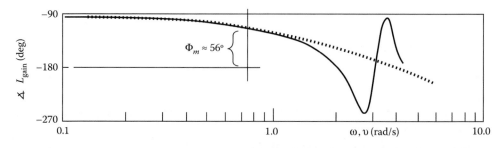

FIGURE 5.21
Angle of loop gain.

Bode/*w*-Plane Design:

- Easier to work with and to modify than the RL.
- Requires $z \to w$ mapping on $\tilde{G}(z)$ then reverse mapping on H.
- Still needs to evaluate frequency plot of LG in z-domain, since $w \neq s$.
- No guarantee that a good w-plane design will yield a good z-plane design (unless $v < 1/h$).
- Gives no explicit knowledge of CL pole locations.

5.3 PID Design

- Most common packaged form of digital controller
  - Very popular in the process control industry
- Continuous PID, $u(s) = H(s)\, e(s)$,

$$u(s) = K\left[1 + \frac{1}{T_1 s} + \frac{T_2 s}{1 + (T_2 s/N)}\right] e(s) \tag{5.13}$$

Proportional Integral Derivative
term   term   term

The three terms on the RHS of Equation 5.13 are respectively the proportional, integral, and derivative terms of the PID design scheme.

T_1 = integrator or reset time, which is generally a big number.
T_2 = derivative time.
$N \sim 2 \to 20$ and is usually a fixed number.

- Integral term is not necessary if there is an integrator (k/s) in the loop already.
- Equivalent to the lead compensator (PD part) + integral term.

$$\text{PD}: 1 + \frac{T_2 s}{1 + (T_2 s/N)} \iff \frac{1 + (s/\omega_2)}{1 + (s/\beta\omega_2)} \tag{5.14}$$

which yields

$$N = \beta - 1 \quad \text{or} \quad \beta = N + 1$$

$$T_2 = (\beta - 1)/(\beta\omega_2) \quad \text{or} \quad \omega_2 = N/[(N+1)T_2]$$

- There are various "tuning rules" that exist for T_1, T_2, and K as illustrated by Ziegler and Nichols [2, 1942].
- Not all parts in this scheme are always necessary for good control (i.e., Proportional (P), Proportional Integral (PI), Proportional Derivative (PD), Proportional Integral Derivative (PID), etc.).

5.3.1 Digital PID Controller

- The discrete equivalent obtained from backward difference (other methods may also be used), $s \to (z-1)/hz$:

$$u(z) = K\left[1 + \frac{hz}{T_1(z-1)} + \frac{T_2}{\left(h + \dfrac{T_2}{N}\right)} \cdot \frac{(z-1)}{\left(z - \dfrac{T_2}{Nh + T_2}\right)}\right]e(z)$$

- General parametric form is given by Equation 5.15:

$$u(z) = K\left[1 + \frac{h}{T_{1d}} \cdot \frac{1}{1-z^{-1}} + \frac{T_{2d}}{h} \cdot \frac{1-z^{-1}}{1-\gamma z^{-1}}\right]e(z) \tag{5.15}$$

which necessitates parameters to be determined: K, T_{1d}, T_{2d}, and possibly:

$$\gamma = \frac{T_{2d}}{Nh}$$

- Implementation is done as shown below [3,4]. Sum up the three parts separately as in Figure 5.22.

Thus, we have the three parts of the digital controller as

1. Integral: $UI(k) = (h/T_{1d})e(k) + UI(k-1)$; $UI(k)$ is the integral component of the controller
2. Proportional: $UP(k) = e(k)$; is the proportional component of the controller
3. Derivative: $UD(k) = (T_{2d}/h)[e(k) - e(k-1)] + \gamma\, UD(k-1)$; $UD(k)$ is the derivative component of the controller

Then we can obtain:

$$u(k) = K\,[UI(k) + UP(k) + UD(k)], \text{ as shown in Figure 5.22.}$$

5.3.1.1 PID Algorithm Implementation

- Algorithm at step k:

$e = r - y$

$UI = (h/T_{1d})e + UI$

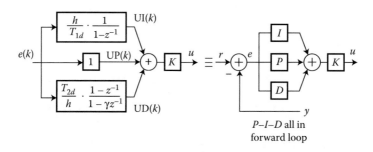

FIGURE 5.22
PID sum of three parts: P, I, and D-PID.

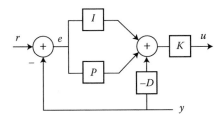

FIGURE 5.23
Derivative of output form of PID algorithm.

$$UP = e$$
$$UD = (T_{2d}/h)[e - e_{last}] + \gamma\, UD$$
$$e_{last} = e$$
$$u = K(UI + UP + UD)$$

- Derivative of output form:
  - If r suddenly changes from time $k - 1$ to time k, for example, a step change, then $e(k) - e(k - 1)$ may be large and UD will have a "spike" at step k. This is undesirable for proper control implementation.
  - Modify UD computation to use only $\Delta y = y(k) - y(k - 1)$, as shown in Figure 5.23.

$$UD(k) = -(T_{2d}/h)\,[y(k) - y(k - 1)] + \gamma\, UD(k - 1) \tag{5.16}$$

  - CL stability is unaffected (stability is not a function of r).
- "Set-point on I" structure.
  - Move P to act only on y also, $UP = -y(k)$.
  - Only integral compensation uses the error signal.
  - Popular structure in process control (keeps the control signal very smooth).

5.3.2 Integral Windup Modifications

- A problem that arises when u is limited, for example,

$$B^- \le u(k) \le B^+$$

(Symmetric limits are most common, $B^- = -B^+$)

- Limits are imposed by the system under control, for example, actuator constraints or the dynamic range of electronic components.
- Match these limits in controller software as

$$\text{If } (u \ge B^+) \text{ set } u = B^+, \text{flag} = +1$$

$$\text{If } (u \le B^-) \text{ set } u = B^-, \text{flag} = -1$$

$$\text{Else flag} = 0$$

- One can realize that the control is probably saturated because $e(k)$ was large.
  - Because u is limited, the error e will not be reduced to zero as fast (this results in a slower system).
  - This is not indicative of an ss error e and its nature.
    - Turn off/skip the integration of $e(k)$ in UI if the last control value was at a limit.

$$\text{if (flag = 0) UI = UI + } (h/T_{1d})\ e$$

$$\text{if (flag} \neq 0) \text{ UI = UI}$$

- Provide integral protection.
  - Value of UI does not change if/when u is saturated.
- Include PID structure in "Cntrl" subroutine, OPT = 4. "Cntrl" algorithm is described in Section 3.3.4 in Chapter 3 of the prequel.

5.3.3 Example

Lack of integral protection will often result in large overshoots in the system response, since long periods of $+e$ (or $-e$) will cause UI to build up large values, then e reverses ....

EXAMPLE. 5.2

A motor with transfer function $G(s) = 1/[s(s + 1)]$ is to be controlled using a digital PI controller:*

$$u(z) = K\left[1 + \frac{h}{T_{1d}} \cdot \frac{1}{1 - z^{-1}}\right] e(z)$$

(5.17)

With $K = 0.4$, $T_{1d} = 5$ s, and $h = 0.5$ s, examine the step response when $|u(k)| \leq 0.2$, with and without integral windup protection. The step response is shown in Figure 5.24.

5.3.4 Other PID Considerations

- Further integration term modifications:
  - UI removes the ss error, but introduces $-90°$ phase lag$\rightarrow T_{1d} \sim$ large
  - It is common to limit $|UI|$, for example, $|UI| < M$
  - Consider integrating only when e is small (pros and cons)
- Compute UI in double precision, UI = UI + $he/(T_{1d}) \rightarrow$ second term is small in relation to UI
  - Numerical roundoff can cause UI to "stall"
- Alternate implementation forms:
  - "Velocity" form: computes Δu
  - "Bumpless" transfer: for changing manual $\leftrightarrow$ auto mode

---

* The "I" part of the controller is not really needed here since $G(s)$ contains a 1/s. But it is only an example. The integrator at the origin helps in evaluating the final value easily; one can use the final value theorem and verify the ss value as

$$\rightarrow \lim_{s \to 0} sG(s)$$

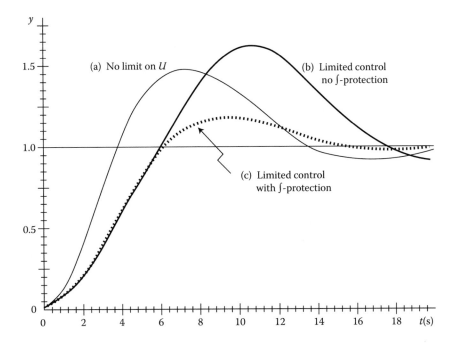

FIGURE 5.24
Step response of example from Example 5.2.

- Parameter selection: K, T_{1d}, T_{2d}, (γ):
  - Generally by trial and error and on-line tuning
  - Preliminary design approach:
    - Step 1: design lead compensator via RL, w-plane, ... $\rightarrow$ obtain γ, K, T_{2d}
    - Step 2: add in I term with T_{1d} ~ large (if needed; gives a pole-zero pair at z ~ 1)
    - Step 3: go back to step 1 for modifications
  - These are crude, heuristic initial tuning rules

5.3.5 PID Initial Tuning Rules

- Based on continuous $H(s) \rightarrow H(z)$ if h is small.
- Can be used directly on physical process [no need to model $G(s)$].

5.3.5.1 Transient Response Method

For different values of parameters, a step response is illustrated in Figure 5.25 and in Table 5.1.

5.3.5.2 Ultimate Sensitivity Method

1. Use a Proportional "P" controller ($u = Ke$) to stabilize system.
2. Slowly increase gain K until the system is on the stable Boundary $\Rightarrow K_{max}$.
3. Obtain the time period of oscillations, T_p.

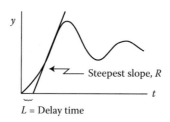

FIGURE 5.25
Unit step response of open-loop system.

TABLE 5.1

PID Tuning Rules

	K	T_{1d}	T_{2d}
P	1/RL	—	—
PI	0.9/RL	3 L	—
PID	1.2/RL	2 L	0.5 L

TABLE 5.2

Period of Oscillations T_p for Different Controller Configurations and Parameter Values

	K	T_{1d}	T_{2d}
P	$0.5\,K_{max}$	—	—
PI	$0.45\,K_{max}$	$T_p/1.2$	—
PID	$0.6\,K_{max}$	$T_p/2$	$T_p/8$

The parameter values are illustrated in Table 5.2 for different types of controllers for ultimate sensitivity of the controller:

- A guideline for time-step selection:

$$h \sim 0.01\ T_p \text{ to } 0.05\ T_p$$

This is quite conservative in general and additional analysis for further clarification and pondering is provided in the sequel.

5.3.6 Real-Time PID Control of an Inverted Pendulum Using FEEDBACK≪®: Page 26 of 33–936 S of FEEDBACK≪® Document

PID control covers three control areas as described in FEEDBACK≪®:

1. Inverted pendulum control
2. Swing-up control related to windup protection
3. Crane control—reverse of inverted pendulum

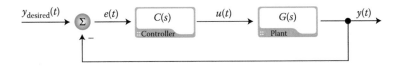

FIGURE 5.26
Simple CL control system for an inverted pendulum.

A simple CL plant control system has desired characteristics described by the transfer function as in

$$T_{CL} = \frac{C(s)G(s)}{1 + C(s)G(s)} \tag{5.18}$$

Where in general $G(s)$ is the plant transfer function and $C(s)$ is the controller in a series with $G(s)$ and a unity feedback as shown in Figure 5.26.

The three blocks of the PID controller are proportional, integral, and derivative as shown in Equation 5.19:

$$u(t) = P \cdot e(t) + I \cdot \int e(t)d(t) + D \cdot \frac{de(t)}{dt} \tag{5.19}$$

The error is defined in Equation 5.19 as

$$e(t) = y_{\text{desired}}(t) - y(t) \tag{5.20}$$

The transfer function of Equations 5.19 and 5.20 yields:

$$U(s) = \left(P + \frac{I}{s} + D \cdot s \right) \cdot E(s) \tag{5.21}$$

$$C(s) = \frac{U(s)}{E(s)} = \left(P + \frac{I}{s} + D \cdot s \right) = \frac{Ds^2 + Ps + I}{s} \tag{5.22}$$

Using real-time PID control of the inverted pendulum using (page 30 of 33–936 S of FEEDBACK≪® document).

Results of RL are illustrated in Figure 5.27 for $P = 27.84$, $I = 50$, and $D = 3.9$.

There are more results with variation in P. I and D as are illustrated in 33–936 S of FEEDBACK≪® document.

5.4 A Technique for System Control with Time Delay

$$\tau = Mh + \epsilon$$
$$\tilde{G}(z) = z^{-M}\tilde{G}_1(z), \quad M = \text{integer} \tag{5.23}$$

"Fractional" delay part of $0 \le \epsilon < h$ is embedded in $\tilde{G}_1$.

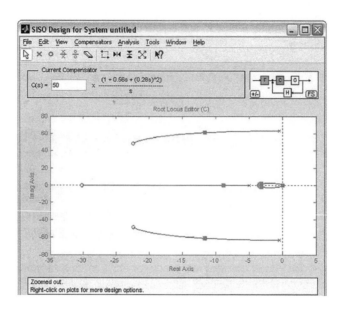

FIGURE 5.27
RL with PID control.

5.4.1 Smith Predictor/Compensator

Theoretically,

- Design $H(z)$ using $\tilde{G}_1^m(z)$ = "model" of $\tilde{G}_1(z)$ (usually $(\tilde{G}_1)^m \equiv \tilde{G}_1$)
- Implementation:

  Implementation of the Smith predictor/compensator is shown in Figure 5.28 $y_p(k)$ = "predicted" value of $y(k)$:

$$y_p(z) = z^{-M}(\tilde{G}_1)^m(z)u(z) \tag{5.24}$$

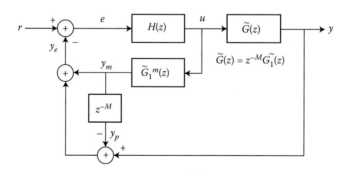

FIGURE 5.28
Smith predictor implementation.

- Nominally $y(k) - y_p(k)$ = prediction error should be small
- Control is based primarily on $r - y_m$, where

$$y_m(z) = (\tilde{G}_1)^m(z)u(z) \sim M - \text{step ahead prediction of } y$$

and

$$u(z) = H(z)\big\{r(z) - \big[y_m(z) + y(z) - y_p(z)\big]\big\} \qquad (5.25)$$

The effective output is

$$y_e = y_m(z) + y(z) - y_p(z) \sim \text{effective } output$$

- Basic idea is to build a control that approximates

$$u(z) = H(z)z^{+M}[r(z) - y(z)]$$

(Need to know/estimate future "r" if it is changing.)

5.4.1.1 Smith Compensator Application

- Model of system in feedback loop
  - Possible numerical problems if $\tilde{G}_1$ is unstable
  - Initialize $(\tilde{G}_1)^m$ to rest condition ($\equiv 0$)

Implementing the $z^{-M} = M$-step delay line by an $(M + 1)$—dimensional push-down stack is shown in Figure 5.29.

5.4.2 Example of Smith Predictor Motor-Positioning Example with τ 1 S, $h = 1$ S (i.e., $M = 1$)

$$H(z) = 10.5\frac{z - 0.87}{z + 0.35} \quad \text{(from } w\text{-plane design)}$$

Step response using Smith compensation is shown in Figure 5.30. One can realize that:

- Recalling $\Phi_m \sim 56°$ and $\omega_c \sim 0.73 \rightarrow 1.34$ s, one can expect poor performance with no delay compensation, as Φ_m would drop to ~14°.

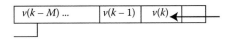

FIGURE 5.29
Push down stack. Initialize stack with $v(k - j) = y(k)$ for all j at $k = 0$.

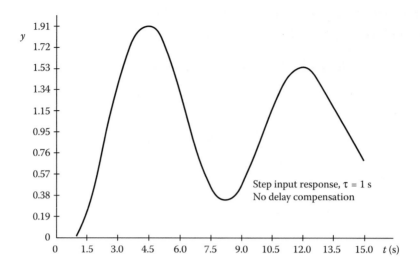

FIGURE 5.30
Step input response using Smith compensation.

5.4.2.1 Results with Delay Compensation

- $M = 1, (\tilde{G}_1)^m(z) = \tilde{G}_1(z) = \dfrac{(z + 0.97)}{(z-)(z - 0.905)}.$

System initially at rest, $r(k)$ = step input.

The step response with delay compensation is shown in Figure 5.31 for a system with and without delay.

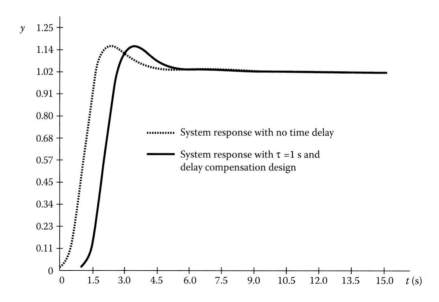

FIGURE 5.31
Step response of a system with and without delay.

- CL response is identical to the undelayed case, with a time-shift of M steps.
  - If the system is not initially at rest, output response would "drift" from the first M steps until the first control begins to affect response.
- As M increases, $(\tilde{G}_1)^m \sim G_1(z)$ becomes more critical.

EXAMPLE 5.3: ALTERNATE IMPLEMENTATION OF SMITH COMPENSATOR
- Consolidate FB loops as shown in Figure 5.32 for an alternate implementation of the Smith Predictor.

$$u(z) = H(z)e(z) - H(z)(\tilde{G}_1)^m(z)(1 - z^{-M})u(z) \tag{5.26}$$

- Also consolidate the inner loop, between e and u, as shown in Figure 5.33.

$$H^*(z) = \frac{H(z)}{1 + H(z)(\tilde{G}_1)^m(z)(1 - z^{-M})} \tag{5.27}$$

- Typically, $H^*(z)$ will be a high-order compensator.
  - $\gg$ 1–2 are usually associated with lag, lead, and PID.
  - Implementation methods are critical
  - Speed/timing for real-time
  - Accuracy

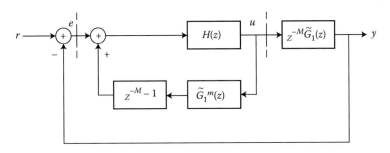

FIGURE 5.32
Alternate implementation of a Smith compensator.

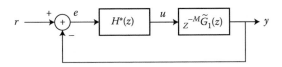

FIGURE 5.33
Consolidation of an inner loop between e and u.

5.5 Implementation of High-Order Digital Compensators

For high-order digital compensators, we use

$$H(z) = \frac{\beta_0 z^m + \beta_1 z^{m-1} + \cdots + \beta_m}{z^m + \alpha_1 z^{m-1} + \cdots + \alpha_m} \qquad (5.28)$$

- With direct form, the input takes form as

$$
\begin{aligned}
u(k) = \beta_0 e(k) + \left[\beta_1 e(k-1) + \cdots + \beta_m e(k-m)\right] \\
- \left[\alpha_1 u(k-1) + \cdots + \alpha_m u(k-m)\right]
\end{aligned} \qquad (5.29)
$$

where we define

$$\text{SE} \triangleq \left[\beta_1 e(k-1) + \cdots + \beta_m e(k-m)\right]$$

$$\text{SU} \triangleq \left[\alpha_1 u(k-1) + \cdots + \alpha_m u(k-m)\right]$$

- SE and SU for time k: computed at step $k-1$; where SE and SU are the first and second terms of Equation 5.29 of the right hand side (RHS) respectively.
- Needs storage of last m, $e(i)$, and $u(i)$.
- This method has very poor numerical properties like:
  - Small changes in α_i and β_i coefficients (especially α_m and β_m) can cause large changes in roots = poles and zeros of $H(z)$.
  - Errors in $e(k)$ and $u(k)$ "hangaround" formsteps.
- Decomposition approach:
- Decompose $H(z)$ into a sum of low-order subparts (e.g., as in PID) and then add up parts as follows:

$$H(z) = \beta_0 + \frac{\tilde{\beta}_1 z^{m-1} + \cdots + \tilde{\beta}_m z^{-m}}{z^m + \alpha_1 z^{m-1} + \cdots + \alpha_m}; \quad \tilde{\beta}_i = \beta_i - \beta_0 \alpha_i$$

PF expansion (assume no repeated roots):

$$H(z) = \beta_0 + \sum_{i=1}^{N_f} \frac{A_i}{z + k_i} + \sum_{i=1}^{N_s} \frac{A_{i1} z + A_{i2}}{z^2 + k_{i1} z + k_{i2}} \qquad (5.30)$$

N_f first-order factors; N_s second-order factors.

Thus, the implementation structure of $H(z)$ is given by

$$H(z) = \beta_0 + \sum_{i=1}^{N_f} \frac{A_i z^{-1}}{1 + k_i z^{-1}} + \sum_{i=1}^{N_s} \frac{(A_{i1} + A_{i2} z^{-1}) z^{-1}}{1 + k_{i1} z^{-1} + k_{i2} z^{-2}} \qquad (5.31)$$

Equation 5.26 is illustrated in Figure 5.34.

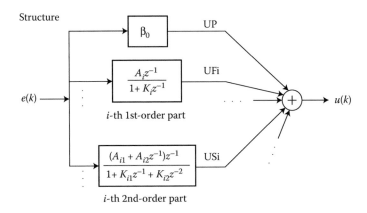

FIGURE 5.34
Structure of the higher-order $H(z)$ with constant, first-order, second-order, and so on.

Structure
Note the 1-step delay in all first- and second-order parts $\rightarrow$ can compute these at step $k-1$ for use at time k. Thus, the algorithm follows as

Algorithm (initialiaze R, e_1, USi$_1$, USi$_2$, UFi$_1$ = 0)

Obtain $e(k) = e = r(k) - y(k)$
$U = \beta_0 e + R$
Output $u(k) = U$

Do for each
Obtain next value of UF$_i$ $\left\{ \text{UFi} = A_i e - k_i \text{UFi}_1 \right\}$

Ist-order part $\qquad$ UFi$_1$ = UFi $\qquad$ save it for next time

Do for each $\qquad \left[\text{USi} = A_{i1} e + A_{i2} e_1 - k_{i1} \text{USi}_1 - k_{12} \text{USi}_2 \right.$
Obtain

2nd-order part $\qquad \left. \qquad\qquad \text{USi}_2 = \text{USi}_1 \right.$

$\qquad$ next USi
$\qquad\qquad\qquad\qquad \text{USi}_1 = \text{USi}$
$\qquad$ save last

This results in two values of USi.
Where

$$R = \sum_{i=1}^{N_f} \text{UFi} + \sum_{i=1}^{N_S} \text{USi}$$

$$e_1 = e$$

- Include in Cntrl subroutine, OPT = 3.

5.6 Summary of Compensator Design Methods

- Indirect design $H(s) \rightarrow \tilde{H}(z)$ by discrete equivalent:
  - Generally requires small h
  - Easy and straightforward
- Direct design methods:
  - RL w-plane, PID
  - Only have Nyquist restrictions on h
  Advantages:
  - Generally easy to design $H(z)$
  - A low-order design, easily realized and found
  - Higher-order dynamics in $G(s)$ accommodated with little extra effort
  - Universally used techniques, and time-tested
  Disadvantages:
  - Low-order compensator designs do not always work
  - Does not use all available information about system behavior (e.g., y instead of x)
  - Measures used are not 1:1 with time response (requires trial and error with CL simulation)
  - Limited by human insight
  - Extremely difficult for MIMO systems

PROBLEMS

P5.1 The CE of a discrete-data control system is given by

$$z^3 + 5.94z^2 + 7.7z - 0.368 = 0$$

Use the w-plane bilinear transformation and apply the Jury–Raible test to assess stability.

P5.2 A direct design $H(z)$ is given by

$$H(z) = 20 \frac{z - 0.6}{z + 0.7}$$

1. Find the time response of $y(k)$ for $T = 1$ s.
2. Find PO and the corresponding $u(k)$ needed.
3. Change $H(z)$ to obtain a PO of 15% and the corresponding time response of $y(k)$ and $u(k)$ needed.
4. Evaluate the corresponding K if

$$H(z) = K \frac{z - 0.6}{z + 0.6}$$

with a PO of 15%.

P5.3 A resistance, inductance, capacitance (RLC) system is shown in Figure P5.1.

The state of this system can be described in terms of a state vector consisting of capacitor voltage and inductor current as $x_1(t)$ and $x_2(t)$, respectively. For $R = 3$, $L = 1$, and $C = \frac{1}{2}$, the state equation is given by

$$\dot{X} = \begin{bmatrix} 0 & -\left(\dfrac{1}{C}\right) \\ \dfrac{1}{L} & -\left(\dfrac{R}{L}\right) \end{bmatrix} X + \begin{bmatrix} +\left(\dfrac{1}{C}\right) \\ 0 \end{bmatrix} u(t)$$

$$= \begin{bmatrix} 0 & -2 \\ 1 & -3 \end{bmatrix} X + \begin{bmatrix} +2 \\ 0 \end{bmatrix} u(t)$$

Using $T = 0.2$ s:

1. Find the discrete equivalent system.
2. Find the response of the system for $x_1(0) = x_2(0) = 1$ and $u(t) = 0$ from 0 to 10 s.
3. Find the error at $T = 0.1$ s and $T = 0.05$ s and show the reduction in the error to 1.5% of the initial value.

P5.4 The state variable representation of the spread of an epidemic disease is given by

$$\dot{X} = \begin{bmatrix} -1 & -1 & 0 \\ 1 & -1 & 0 \\ 1 & 1 & 0 \end{bmatrix} X + \begin{bmatrix} 1 & 0 \\ 0 & 1 \\ 0 & 0 \end{bmatrix} u$$

The rate of the new susceptible is zero, that is, $u_1 = 0$, and the rate of adding infectives is represented by $u_2(0) = 1$ and $u_2(k) = 0$ for $K \geq 1$. Using $T = 0.2$ s.

1. Find the discrete time equation.
2. Find the time response from 0 to 10 s.

P5.5 A CL system is shown in Figure P5.2 for the pitch control of an aircraft with a transfer function:

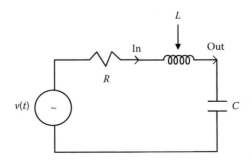

FIGURE P5.1
RLC circuit with a voltage.

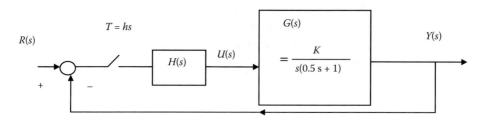

FIGURE P5.2
CL system block diagram for the pitch control of an aircraft.

$$G(s) = \frac{K}{s(0.5\,s + 1)}$$

Select gain K and sampling time T so that the OS is limited to 30% for a unit step input and the ss error for a unit ramp is less than 1.0.

P5.6   A computer-compensated system is shown in Figure P5.3.

For $T = 1$ and $K = 1$

$$G(s) = \frac{K}{s(s + 1)}$$

Select the parameters A and r of $D(z)$ when

$$D(z) = \frac{A(z - 0.3678)}{(z + r)}$$

Select within the range: $1 < A < 2$ and $0 < r < 1$, then find the response of the compensated system and compare with the uncompensated system.

P5.7   A new suspended, mobile, remote-controlled camera system at a professional football stadium is shown in Figure P5.4.

The camera can be moved up and down the field as well as over the entire field. The motor control on each pulley is represented by the block diagram in Figure P5.5.

$$GH(s) = \frac{10}{s(s + 1)((s/10) + 1)}$$

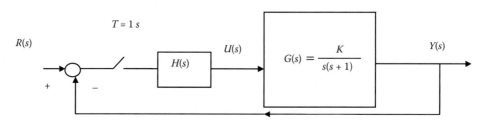

FIGURE P5.3
CL block diagram for a computer compensated system.

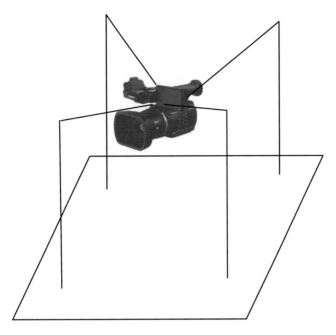

FIGURE P5.4
New suspended, mobile, remote-controlled camera system at a professional football stadium.

1. Design the system for a PM of 45°.
2. Select a suitable crossover frequency and a sampling period of $h = 0.05$ s to obtain $D(z)$.

P5.8 Adapted from FEEDBACK≪® manual 33–936 S Digital Pendulum Control Experiments courtesy FEEDBACK≪®.

A PID controller consists of three blocks: proportional, integral, and derivative. The equation governing the PID controller is as follows:

$$u(t) = P \cdot e(t) + I \cdot \int e(t)dt + D \cdot \frac{de(t)}{dt}$$

$$e(t) = y_{\text{desired}}(t) - y(t)$$

then

$$C(s) = \frac{U(s)}{E(s)} = \frac{Ds^2 + Ps + I}{s}$$

$$G(s) = \frac{10}{s(s+1)\left(\frac{s}{10}+1\right)}$$

FIGURE P5.5
Process $G(s)$ for a Digital Pendulum.

1. Using functions Gd and Gc from FEEDBACK≪® and $P = 27.84$, $I = 50$, and $D = 3.9$, find the RL.

2. Conduct the real-time PID control of the cart position.

3. Change values of P, I, and D so that I is increased to 75, and decrease D to 1.9 and $P = 35$; find the response and the cart position for these PID values.

P5.9 The discrete equivalent of the antenna positioning controller in continuous time is given by

$$G(s) = \frac{0.1}{s(s + 0.1)}$$

$$\tilde{G}(z) = \frac{0.048(z + 0.97)}{(z - 1)(z - 0.905)}$$

The compensated design is given by

$$H(w) = 1.0\frac{1 + (w/0.14)}{1 + (w/4.2)} = K\frac{1 + (w/\omega_2)}{1 + (w/\beta_2)} = KH_0(w)$$

1. Use lead compensation to increase bandwidth.

2. Design for ω_c as large as possible with a $\Phi_m \sim 55°$.

P5.10 Using backward transformation from w to z for

$$H(z) = 1.0\frac{1 + (w/0.14)}{1 + (w/0.42)}$$

by using

$$w = \frac{2}{h}\frac{(z - 1)}{(z + 1)}$$

with $h = 0.5$ and

1. Design a compensator with PO $\sim 15\%$ and $K_v = 1.0$.

2. Conduct a frequency domain evaluation of $LG(z)\big|_{z=e^{j\omega h}}$.

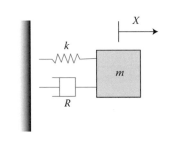

FIGURE 7.15
PC and HW controller. (Courtesy of FEEDBACK≪®.)

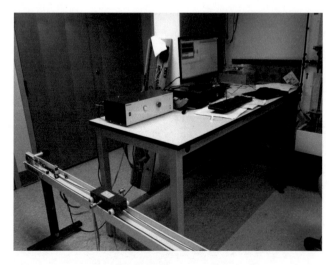

FIGURE 7.16
Inverted pendulum system. (Courtesy of FEEDBACK≪®.)

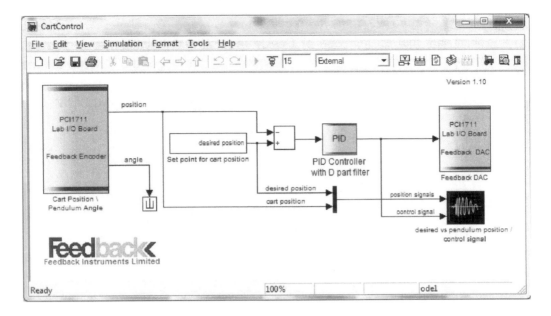

FIGURE AII.1
Implementation of real-time model using FEEDBACK≪® instrumentation.

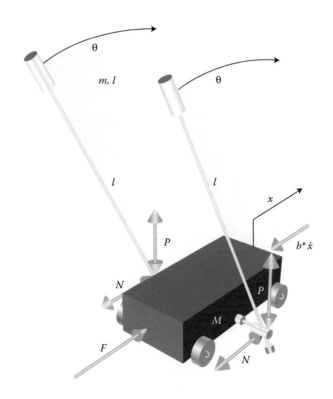

FIGURE AII.2
Model description example of an inverted pendulum.

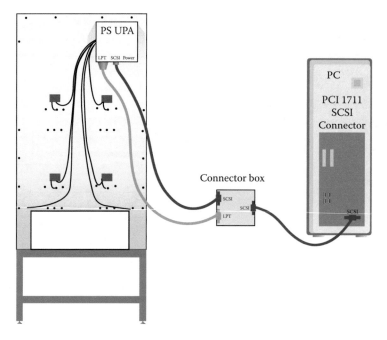

FIGURE AII.3
Coupled tanks connection diagram.

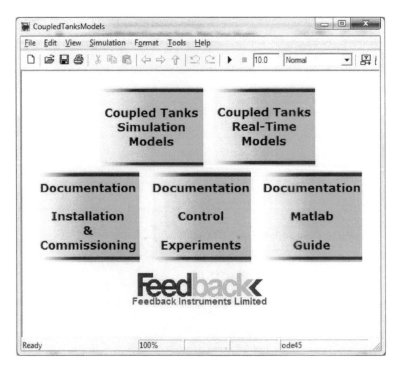

FIGURE AII.4
Simulink model menu for coupled tanks.

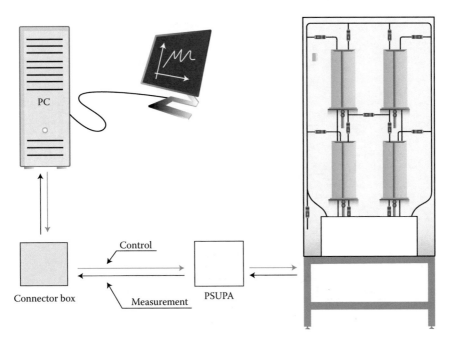

FIGURE AII.5
Control system schematic for coupled tanks.

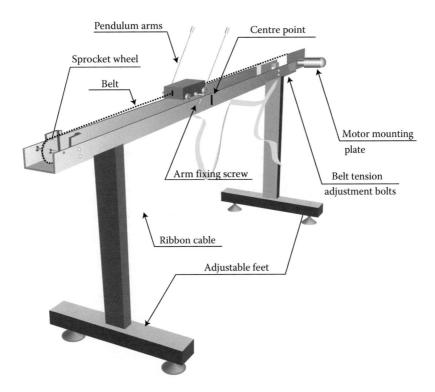

FIGURE AII.6
Inverted pendulum unit.

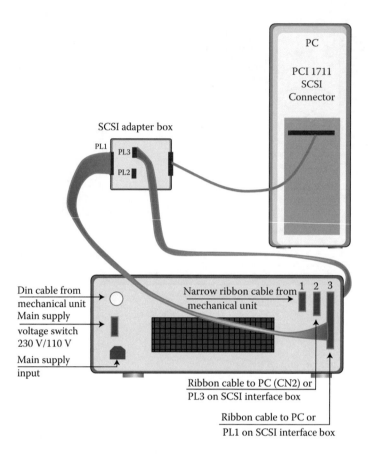

FIGURE AII.7
Connection diagram for inverted pendulum.

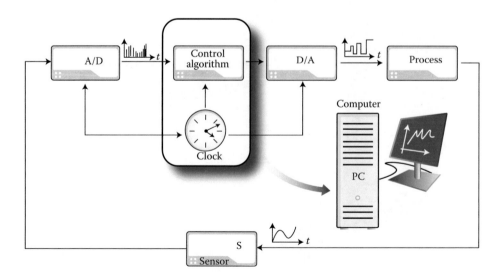

FIGURE AII.8
Computer control system diagram.

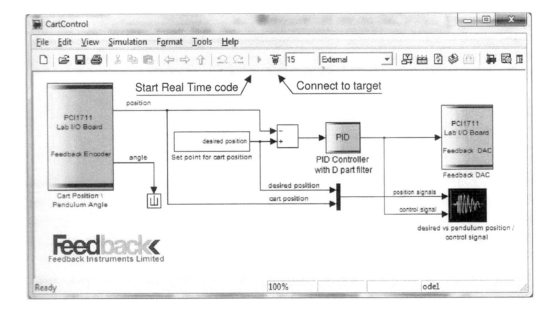

FIGURE AII.9
PID cart control.

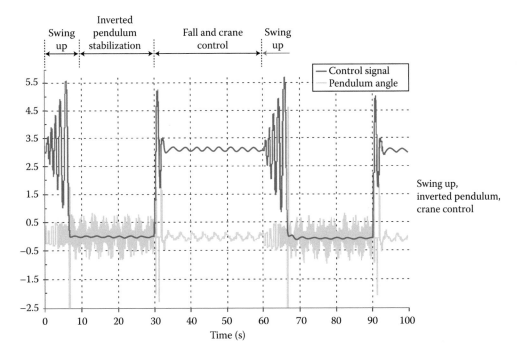

FIGURE AII.10
Pendulum PID control response.

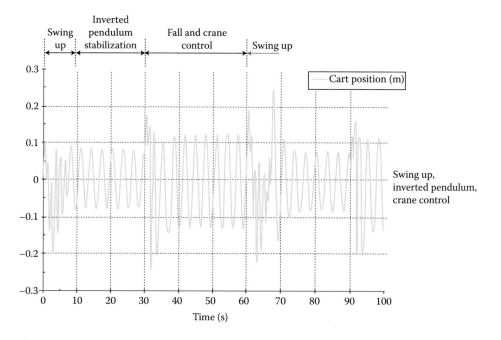

FIGURE AII.11
Pendulum cart position; PID control response.

6

State-Variable Feedback Design Methods

(See color insert.)

6.1 Linear State-Variable Feedback

1. Classical root locus (RL) methods:
   a. Here, we attempt to place n closed-loop (CL) poles using relatively few design parameters given $u(z)$ as

$$u(z) = K \frac{z^m + \beta_1 z^{m-1} + \cdots + \beta_m}{z^m + \alpha_1 z^{m-1} + \cdots + \alpha_m} e(z) \tag{6.1}$$

   The classical RL is shown in Figure 6.1.
   - CL poles are constrained to lie on the RL inside a unit circle to maintain stability.
   - Limited flexibility to place CL poles.
2. State-variable feedback:
   One should attempt to adjust "n" parameters as

$$u(k) = K_r r(k) - K \, X(k), \quad \text{where } K = [K_1, K_2, \ldots, K_n] \tag{6.2}$$

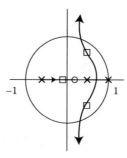

FIGURE 6.1
RL for $2m + 1$ design parameters, and $n + m$ CL poles.

to place n in CL. This is an unconstrained design, but should satisfy the following:

a. Must specify all CL pole locations.

b. $n > 2$ usually $\rightarrow$ need Computer Aided Design (CAD) techniques—a new software by "Feedback" is introduced in the sequel.

c. Can it be done?

d. Relation to compensator design?

6.2 Control in State-Space

What can be done with respect to controlling system states?

$$\text{System model: } X(k + 1) = \Phi\, X(k) + \Gamma u(k) \tag{6.3a}$$

$$\text{Measurement model: } y(k) = CX(k) \tag{6.3b}$$

$$X(0) = \text{known initial condition}$$

The equivalent discrete system matrices can be shown as

$$\Phi = e^{Ah}; \quad \Gamma = \int_{0}^{0h} e^{A\sigma} d\sigma B$$

With a corresponding state response given by

$$X(k) = \Phi^k X(0) + \sum_{i=0}^{k-1} \Phi^{k-1-i} \Gamma u(i) \tag{6.4}$$

Now consider $k = n$:

Can we find $u(0), u(1), \ldots, u(n - 1)$, so that $x(n) = \xi = $ arbitrary vector, starting at an initial condition $X(0)$? Let us find:

$$\xi - \Phi^k X(0) = \sum_{i=0}^{k-1} \Phi^{k-1-i} \Gamma u(i)$$ (6.5)

$$= \Gamma u(n-1) + \Phi \Gamma u(n-2) + \cdots + \Phi^{n-1} \Gamma u(0)$$

$$\xi - \Phi^k X(0) = \begin{bmatrix} | & | & & | \\ \Gamma & \Phi\Gamma & \cdots & \Phi^{n-1}\Gamma \\ | & | & & | \end{bmatrix} \begin{bmatrix} u(n-1) \\ u(n-2) \\ \vdots \\ u(0) \end{bmatrix}$$ (6.6)

where

$$H_c = \begin{bmatrix} | & | & & | \\ \Gamma & \Phi\Gamma & \cdots & \Phi^{n-1}\Gamma \\ | & | & & | \end{bmatrix}$$

If H_c is invertible, it is possible to find the requisite $\{u(i)\}$.
Note: state may not necessarily stay at ξ for $k > n$.

6.2.1 Controllability

A discrete system is completely controllable (CC) if and only if (iff):

$$\det \begin{bmatrix} | & | & & | \\ \Gamma & \Phi\Gamma & \cdots & \Phi^{n-1}\Gamma \\ | & | & & | \end{bmatrix} = \det(H_c) \neq 0$$ (6.7)

What is the physical interpretation of Equation 6.7? This can be answered as follows:

- Control $u(k)$ reaches all modes, either directly or indirectly.
- Controllability is a property of only $\{\Phi, \Gamma\}$.

For this interpretation to be valid, we need the continuous-discrete relationship.
Thus, the discrete equivalent system $\{\Phi, \Gamma\}$ is CC if the original continuous system is, and

$$h \neq M \frac{2\pi}{\omega_{c0}}; \quad M = \text{integer}$$

The imaginary ω_{c0} part of any purely imaginary eigenvalue of A is plotted in Figure 6.2.
Note: If sample time $h < \left((0.5 \rightarrow 1.0) / |\lambda_{\max}(A)| \right)$, then it is okay for suitable mathematical evaluations.

ω_{c0} = Imaginary part of
any purely imaginary
eigenvalue of A

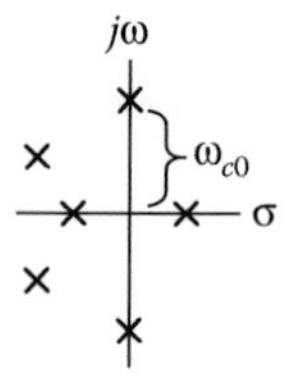

FIGURE 6.2
Imaginary part of any purely imaginary eigenvalue of matrix A of the original continuous system.

Recall the continuous system is CC iff.

$$\det \begin{bmatrix} | & | & & | \\ A & AB & \dots & A^{n-1}B \\ | & | & & | \end{bmatrix} \neq 0$$

Det (H_c) is sometimes used as a "measure of controllability." Also Det $(H_c, H_c') = [\det (H_c)]^2$ if the system has only one input.

6.2.2 Open-Loop versus CL Control

We know if the system is CC if the sequence $\{u(0), u(1),\dots, u(n-1)\}$ will drive

$$X(0) \rightarrow X(n) = \xi$$

Where the control sequence is given by

$$\begin{bmatrix} u(n-1) \\ u(n-2) \\ \vdots \\ u(0) \end{bmatrix} = H_c^{-1}\left[\xi - \Phi^n X(0)\right] \tag{6.8}$$

For an open-loop control one has

$$u(0) = [0 \quad 0 \quad \dots \quad 0 \quad 1] = H_c^{-1}\left[\xi - \Phi^n X(0)\right]$$
$$u(1) = [0 \quad 0 \quad \dots \quad 1 \quad 0] = H_c^{-1}\left[\xi - \Phi^n X(0)\right]$$

$X(0)$, not $X(k)$, is used here

$$u(n-1) = [1 \quad 0 \quad \dots \quad 0 \quad 0] = H_c^{-1}\left[\xi - \Phi^n X(0)\right]$$

Thus, the CL control via time invariance can be achieved by "turn system on" at time "k":

$$X(0) \leftrightarrow X(k), \quad u(0) \leftrightarrow u(k) \tag{6.9}$$

This accomplishes the same control sequence, but via state-variable feedback (SVFB).

As a special case, when $\xi = 0$, it yields:

$$u(k) = -[0 \quad 0 \quad \dots \quad 0 \quad 1] = H_c^{-1}\Phi^n X(k) \tag{6.10}$$

The above is an SVFB control that reduces any (initial) state to 0 in n steps and is described by a "deadbeat controller" (unique to discrete systems).

CL dynamics $X(k + 1) = (\Phi - \Gamma K)X(k)$;

$$X(n) = (\Phi - \Gamma K)^n X(0) = 0$$

$\rightarrow \Phi - \Gamma K$ has all eigenvalues at $z = 0$.

EXAMPLE 6.1   PURE INERTIA CONTROL (e.g., SATELLITE)

An example of a pure inertia control system is a satellite as shown in Figure 6.3, which has a state model and a measurement model given by

$$\text{State model: } \begin{bmatrix} \dot{x}_1(t) \\ \dot{x}_2(t) \end{bmatrix} = \begin{bmatrix} 0 & 1 \\ 0 & 0 \end{bmatrix}\begin{bmatrix} x_1(t) \\ x_2(t) \end{bmatrix} + \begin{bmatrix} 0 \\ 1 \end{bmatrix}u(t)$$

$$\text{Measurement model: } y(t) = \begin{bmatrix} 1 & 0 \end{bmatrix}X(t)$$

From the above two equations, an equivalent discrete system can be obtained as before:

$$X(k + 1) = \begin{bmatrix} 1 & h \\ 0 & 1 \end{bmatrix}X(k) + \begin{bmatrix} \dfrac{h^2}{2} \\ h \end{bmatrix}u(k) \tag{6.11}$$

$$= \Phi X(k) + \Gamma u(k)$$

where

$$\Phi = \begin{bmatrix} 1 & h \\ 0 & 1 \end{bmatrix} \quad \text{and} \quad \Gamma = \begin{bmatrix} \dfrac{h^2}{2} \\ h \end{bmatrix}$$

Now, pick $h = 1$.

Then the deadbeat controller is

$$u(k) = KX(k) = -K_1 x_1(k) - K_2 x_2(k)$$

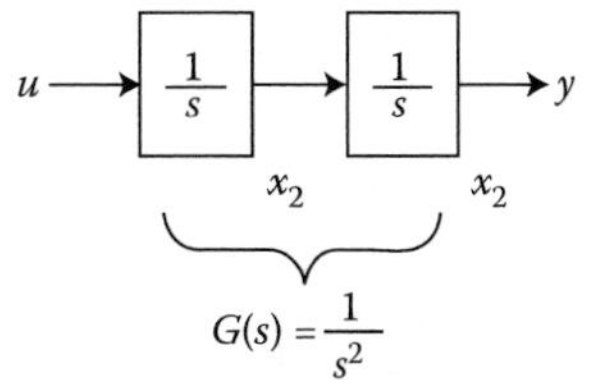

$$G(s) = \frac{1}{s^2}$$

FIGURE 6.3
Pure inertia control of a satellite.

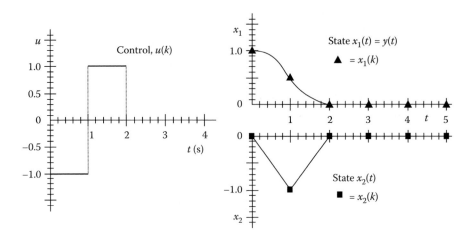

FIGURE 6.4
Input and output of a discrete system—satellite.

where

$$K = \begin{bmatrix} 0 & 1 \end{bmatrix} \begin{bmatrix} \dfrac{1}{2} & \dfrac{3}{2} \\ 1 & 1 \end{bmatrix}^{-1} \begin{bmatrix} 1 & 2 \\ 0 & 1 \end{bmatrix} = \begin{bmatrix} 1 & \dfrac{3}{2} \end{bmatrix}$$

Time response with $x_1(0) = 1$ and $x_2(0) = 0$ brings $X(0) \to 0$ in 2 time steps. This time response of the discrete system is shown in Figure 6.4 for the related input or controller $u(k)$.

6.3 Discrete SVFB Design Methods

1. Continuous → discrete equivalence methods:
   In this method, given a continuous FB control law,

   $$u(t) = K_r r(t) - KX(t)$$

   Develop from K_r and K an "equivalent" discrete FB control law:

   $$u(t) = \tilde{K}_r r(t) - \tilde{K}X(k)$$

   Examples of this method are also described using the "Feedback" software in the sequel.

   a. One idea can be to capitalize on the earlier design for

   $$\dot{X} = AX + Bu$$

   b. Compare continuous versus discrete response, phase margin, CL poles, and so on.

2. Direct digital controller design:

   In this method find $u(k) = K_r r(k) - K X(k)$ directly, to place poles at $z = z_1, z_2, \ldots, z_n$. Then,

   a. Select:

   $$z_i = e^{s_i h} \quad i = 1, 2, \ldots, n$$

   where s_i = desired pole location in s-plane or,

   b. Select:

   $$z_1, z_2, \ldots, z_n \text{ directly}$$

3. Evaluation:

   a. In this method, time response is evaluated via simulation.

   b. The phase margin via the Bode/Nyquist plot of the system loop gain (LG) is also evaluated:

   $$LG(j\omega) = K(zI - \Phi)^{-1}\Gamma \big|_{z=e^{j\omega h}} \tag{6.12}$$

   c. Sensitivity to parameters and return difference (RD) are analyzed.

6.3.1 Continuous → Discrete Gain Transformation Methods

In continuous time, the system is given by

$$\text{State model: } \dot{X}(t) = AX(t) + Bu(t)$$

$$\text{Controller model: } u(t) = K_r r(t) - KX(t)$$

which in discrete time is given by ⇩

$$\text{Discrete time: State model: } X(k+1) = \Phi X(k) + \Gamma u(k)$$

$$\text{Controller model: } u(k) = \tilde{K}_r r(k) - \tilde{K}X(k)$$

and the CL system is given by

$$\begin{array}{ccc}
\text{continuous time} & \rightarrow & \text{discrete time} \\
\dot{X}(t) = (A - BK)X(t) + K_r Br(t) & \rightarrow & X(k+1) = (\Phi - \Gamma\tilde{K})X(k) + \tilde{K}_r\Gamma r(k)
\end{array}$$

- For simplicity and accuracy:
  - Time response of discrete (D) $\overset{?}{\approx}$ Time response of continuous (C)
  - Eigenvalues of CL D $\overset{?}{\approx}$ exponential{h. eigenvalues of CL C}

- Start C at $X(0)$ with $r(t) = r_0$ and response at $t = h$. Then,

$$X(h) = e^{(A-BK)h}X(0) + \int_0^h e^{(A-BK)\sigma}K_rB\,d\sigma \cdot r_0 \tag{6.13}$$

- For a discrete response, this is given by

$$X(h) = (\Phi - \Gamma\tilde{K})X(0) + \tilde{K}_r\Gamma r_0 \tag{6.14}$$

1. To obtain time response equivalence, equate Equations 6.13 and 6.14 and find:
   a. $\Phi - \Gamma\tilde{K} = e^{(A-BK)h} \rightarrow$ "solve" for K?
      It is only possible to obtain equivalence to $O(h^2)$. Then,

$$\tilde{K} \approx K + \frac{K(A - BK)h}{2} \tag{6.15a}$$

   b. $\tilde{K}_r\Gamma = K_r\int_0^h e^{(A-BK)\sigma}Bd\sigma$
      One can only obtain equivalence to $O(h^2)$, to give:

$$\tilde{K}_r \approx \left\{1 - KB\left(\frac{h}{2}\right)\right\}K_r \tag{6.15b}$$

6.3.2 Average Gain Method [5]

1. Consider C over $(0, h]$ with $X(0)$ and $r(t) = r_0$.
   Then in continuous time:

$$X(t) = e^{(A-BK)t}X(0) + \int_0^t e^{(A-BK)\sigma}K_rBd\sigma \cdot r_0 \tag{6.16}$$

2. Control, $u_c(t)$ over $(0, h]$ in a continuous system with the control law given by

$$u_c(t) = K_r r_0 - KX(t)$$

$$u_c(t) = \left[1 - K\int_0^t e^{(A-BK)\sigma}Bd\sigma\right]K_r r_0 - Ke^{(A-BK)t}X(0) \tag{6.17}$$

with discrete control over $(0, h] = u(0)$ and the control law given by

$$u(0) = \tilde{K}_r r_0 - \tilde{K}X(0) \tag{6.18}$$

$$\bar{u}_c = \left[1 - \frac{K}{h}\int_0^h\int_0^t e^{(A-BK)\sigma}d\sigma dt B\right]K_r r_0 - \frac{K}{h}\int_0^h e^{(A-BK)t}dt X(0) \tag{6.19}$$

Thus, the discrete equivalent gains are

$$1.\ \tilde{K} = \frac{K}{h}\int_0^h e^{(A-BK)t}dt \tag{6.20a}$$

$$2.\ \tilde{K}_r = \left[1 - \frac{K}{h}\int_0^h\int_0^t e^{(A-BK)\sigma}d\sigma dt B\right]K_r$$

$$= \left[1 + (K - \tilde{K})(A - BK)^{-1}B\right]K_r \tag{6.20b}$$

6.3.2.1 Computing Average Gain

- Obtain $\tilde{K}$ using subroutine Dscrt, then compute $\tilde{K}_r$.
- Approximation for small h:

$$\tilde{K} = K\left[1 + \frac{(A - BK)h}{2} + (A - BK)^2\frac{h^2}{3!} + \cdots\right] \tag{6.21a}$$

$$\tilde{K}_r = \left\{1 - K\left[\frac{h}{2} + (A - BK)\frac{h^2}{3!} + \cdots\right]B\right\}K_r \tag{6.21b}$$

- The average gain scheme is "good" provided:

$$h \le \frac{1.0}{|\lambda_{max}(A - BK)|} \sim \frac{1.0}{||A - BK||}$$

CL system matrix

- Generally requires a smaller h than does the criterion:

$$h \le \frac{0.5 \to 1.0}{|\lambda_{max}(A)|}$$

- Using average gain $\tilde{K}$ is always better than just using K.
- Inverse procedure: given a discrete K_d, find continuous K.

Solve:

$$K = K_d h\left\{\left[\int_0^h e^{(A-BK)t}dt\right]\right\}^{-1}$$

Iteratively

$$K_{i+1} = K_d h \left\{ \left[\int_0^h e^{(A - BK_i)t} dt \right] \right\}^{-1}$$

with $K_0 = K_d$.

- Generally converges in 2 to 3 iterations.
- Useful when h is subject to, for example, $K_d(h_1) \to K \to K_d(h_2)$.

6.3.3 Example: Satellite Motor Control

The transfer function is given by

$$G(s) = 1/s^2$$

$$\text{State model:} \begin{bmatrix} \dot{x}_1(t) \\ \dot{x}_2(t) \end{bmatrix} = \begin{bmatrix} 0 & 1 \\ 0 & 0 \end{bmatrix} \begin{bmatrix} x_1(t) \\ x_2(t) \end{bmatrix} + \begin{bmatrix} 0 \\ 1 \end{bmatrix} u(t)$$

$$\text{Measurement model: } y(t) = \begin{bmatrix} 1 & 0 \end{bmatrix} X(t)$$

Thus, continuous SVFB control is shown to be:

$$u(t) = 1.0 r(t) - \begin{bmatrix} 1 & 1 \end{bmatrix} X(t), \quad \text{where } K_r = 1.0 \text{ and } K = \begin{bmatrix} 1 & 1 \end{bmatrix}$$

Gives the CL system for the satellite motor control as

$$\dot{X}(t) = \begin{bmatrix} 0 & 1 \\ -1 & -1 \end{bmatrix} X(t) + \begin{bmatrix} 0 \\ 1 \end{bmatrix} r(t); \quad y(t) = \begin{bmatrix} 1 & 0 \end{bmatrix} X(t)$$

The CL transfer function for this satellite system is

$$\frac{y(s)}{r(s)} = \frac{1}{s^2 + s + 1}$$

which yields CL poles at

$$s = -\frac{1}{2} \pm j\frac{\sqrt{3}}{2}; \quad (\zeta = 0.5)$$

Equivalent discrete gains, $\tilde{K}$, with $h = 0.5$.
Okay on "safety" requirements

$$h \leq \frac{1.0}{| \lambda_{\max}(A - BK) |} = \frac{1.0}{\sqrt{1}} = 1.0$$

Examine CL discrete eigenvalues of

$$\Phi - \Gamma \tilde{K}$$

"Expect" poles at $z_i = e_{sh}^i = e^{-0.25 \pm j0.433} = 0.707 \pm j0.327$.
Thus,

$$\Phi - \Gamma\tilde{K} = \begin{bmatrix} 1 & 0.5 \\ 0 & 1 \end{bmatrix} - \begin{bmatrix} 0.125 \\ 0.5 \end{bmatrix} \begin{bmatrix} 0.755 & 0.964 \end{bmatrix} = \begin{bmatrix} 0.906 & 0.379 \\ -0.378 & 0.518 \end{bmatrix}$$

Eigenvalues at $\lambda_i = 0.712 \pm j0.325$; $(\zeta \approx 0.5)$

Without gain equivalence, $\lambda_i(\Phi - \Gamma K) = 0.688 \pm j0.39$; $(\zeta = 0.41)$.

Figure 6.5a shows the CL simulation for the satellite control with $h = 0.5$; and average gain equivalence with equivalent discrete gains.
Figure 6.5b shows the CL simulation for the satellite control with $h = 0.5$; and average gain equivalence with unconverted gains.
Figure 6.6a shows the CL simulation for the satellite control with $h = 1.0$; and average gain equivalence with equivalent discrete gains.
Figure 6.6b shows the CL simulation for the satellite control with $h = 1.0$; and average gain equivalence with unconverted gains.

6.3.3.1 Satellite Motor Control CL simulation

The satellite motor control is illustrated with a CL simulation in Figure 6.7.

- Even when $h > \dfrac{1.0}{|\lambda_{max}(A - BK)|}$, the equivalent (average) gains will often give a "reasonable" CL system.
- With $h = 1.8$, $\tilde{K} = \begin{bmatrix} 0.261 & 0.683 \end{bmatrix}$, $\tilde{K}_r = 0.261$.
- Unconverted discrete system ($K = [1\ 1]$) becomes unstable as h increases.
- The CL system with average gains is still hanging in, with noticeable slow-down in step response.

6.3.3.2 Summary of Equivalent Gain Method

- Average gain is the best method to convert:

$$K \to \tilde{K}; \quad K_r \to \tilde{K}_r$$

- If h is small, use $\tilde{K}; \tilde{K}_r$
- Do not simply use $\tilde{K} = K; K_r = \tilde{K}_r$ (instability as h increases)
- Useful if we need to change h online frequently
  - Store K. K_r from continuous design.
  - Use series approximation to obtain $\tilde{K}; \tilde{K}_r$ for current value of h.

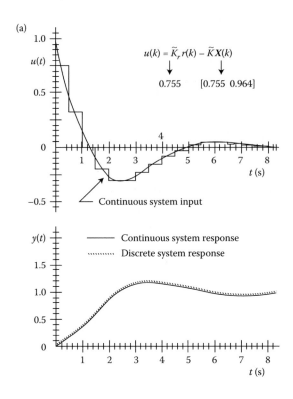

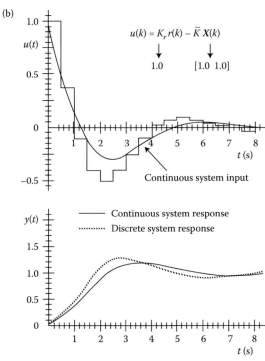

FIGURE 6.5
Average gain equivalence; $h = 0.5$; (a) with equivalent discrete gains; (b) with unconverted gains.

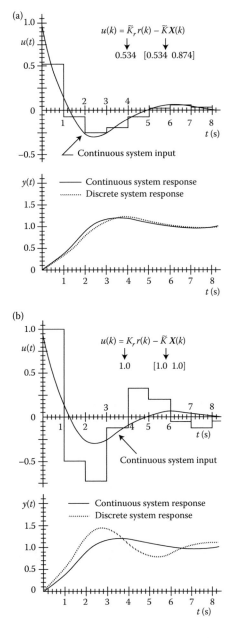

FIGURE 6.6
Average gain equivalence; $h = 1.0$; (a) with equivalent discrete gains; (b) with unconverted gains.

- Generally,
  - $\tilde{K}_i$ will be smaller in magnitude than K_i.
  - Gains $\tilde{K}$ will yield discrete CL poles with a slightly smaller ω_n than the original continuous system (i.e., slower response).
  - Eigenvalues of $\Phi - \Gamma\tilde{K} \sim \exp(h$ eigenvalues of $A - BK)$.

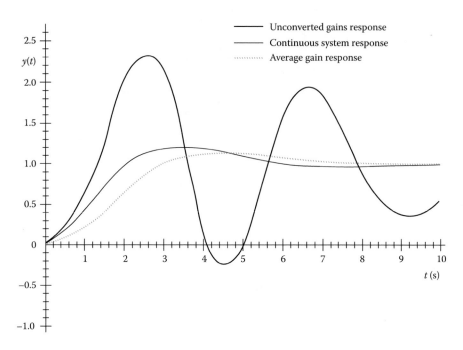

FIGURE 6.7
Satellite control by CL simulation—output with different gains.

- Phase margin of discrete system with average gain ~ phase margin of a discrete system with poles at $\exp[h.\ \lambda_i(A - BK)]$.
- DC gain $(r \rightarrow y)$ of equivalent system ~ same as C.
- If $h \neq$ small, design K, K_r directly for a discrete system.

6.4 State Variable Feedback Control: Direct Pole Placement

6.4.1 Discrete System Design

Given the state and measurement models in Equations 6.22a and 6.22b, respectively,

$$\text{State model: } X(k + 1) = \Phi X(k) + \Gamma u(k) \tag{6.22a}$$

$$\text{Measurement model: } y(k) = K_r r(k) - KX(k) \tag{6.22b}$$

and the linear feedback control law structure is

$$u(k) = K_r r(k) - KX(k) \tag{6.23}$$

The CL dynamics are given by

$$X(k + 1) = (\Phi - \Gamma K)X(k) + K_r \Gamma r(k); \quad y(k) = CX(k) \tag{6.24}$$

where $\bar{\Phi} = (\Phi - \Gamma K)$ as the discrete closed-loop matrix.
  Using the pole placement:

1. Find FB gains $K = [K_1, K_2, \ldots, K_n]$ so that the CL system matrix has eigenvalues (poles) at preselected locations, that is, $\bar{\Phi}$ has poles at $z_1, z_2, \ldots, z_n^*$ as required to satisfy user specifications on performance parameters such as over shoot (OS), settling time (TS), ω_c, and so on.

$$\Rightarrow |zI - \bar{\Phi}| = |zI - (\Phi - \Gamma K)| = (z - z_1) \cdots (z - z_n) = p_d(z) \tag{6.25}$$
$$p_d(z) = z^n + d_1 z^{n-1} + \cdots + d_n$$
$$\nearrow$$

desired closed loop (CL) characteristic polynomial (CP)

2. Adjust K_r to have a unity direct current (DC) gain from r to y:

$$\frac{y(z)}{r(z)} \bigg|_{z=1} = K_r \left[C(zI - \bar{\Phi})^{-1} \Gamma \right] \big|_{z=1} = 1 \tag{6.26}$$

Note: Usually $z_i = e_i^{sh}$ where the $\{s_i\}$ are desired pole locations in the s-plane.

6.5 Pole Placement Methods

- The same scheme should work for either continuous or discrete problems, with matrices given by

$$\Phi \Leftrightarrow A, \quad \Gamma \Leftrightarrow B$$

and

$$\Phi - \Gamma K \Leftrightarrow A - BK$$

- For the direct approach:
  - In the direct approach, we expand the determinant of the CP as

$$|zI - \Phi + \Gamma K| = z^n + f_1(k)z^{n-1} + \cdots + f_n(k) \tag{6.27}$$

with each f_i, linear in $K_1, K_2, \ldots, K_n$.

- Expand the CP further as

$$p_d(z) = (z - z_1)(z - z_2)\cdots(z - z_n) = z^n + d_1 z^{n-1} + \cdots + d_n \tag{6.28}$$

- Equate coefficients and solve n linear equations with n unknowns. Then,

$$f_i(k) = d_i; \quad i = 1, 2, \ldots, n \tag{6.29}$$

- This is useful in simple problems, where some structured $f_i(k)$ are avilable.

EXAMPLE 6.2

A discrete system has the following discrete poles:

$$z_1 = 0.5 + j0.3, \quad z_2 = 0.5 - j0.3$$

We can use the above analysis and find the discrete matrices as defined in Equation 6.27 as

$$\Phi = \begin{bmatrix} 1.0 & 0.5 \\ 0.2 & 1.0 \end{bmatrix} \Gamma = \begin{bmatrix} 1.0 \\ 0.5 \end{bmatrix} K = \begin{bmatrix} K_1 & K_2 \end{bmatrix}$$

$$\bar{\Phi} = \Phi - \Gamma K = \begin{bmatrix} 1.0 - K_1 & 0.2 - K_2 \\ 0.2 - 0.5K_1 & 1.0 - 0.5K_2 \end{bmatrix};$$

Then, CP is obtained as

$$p_d = (z - z_1)(z - z_2) = z^2 - 1.0z + 0.34$$

$$|zI - \bar{\Phi}| = z^2 + (-2 + K_1 + 0.5K_2)z + (0.96 - 0.3K_2 - 0.9K_1)$$

Thus, from Equation 6.27 the coefficients of the powers of z are

$$f_1 = -2 + K_1 + 0.5K_2 \text{ and } f_2 = 0.96 - 0.3K_2 - 0.9K_1$$

Thus, yielding after equating coefficients:

$$-2 + K_1 + 0.5K_2 = -1; \quad 0.96 - 0.9K_1 - 0.3K_2$$

Solving these two equations in two unknowns yields K matrix coefficients as

$$\Rightarrow K_1 = 0.067; \text{ and } K_2 = 1.867$$

We can now select K_r so that DC gain = 1, as required originally.

6.5.1 Transformation Approach for Pole Placement

1. Let $V(k) = T^{-1}X(k)$, where T transforms Φ, Γ to SCF.

   Then, with

   $$X(k + 1) = \Phi X(k) - \Gamma X(k)$$

   gives

   $$\Rightarrow V(k + 1) = T^{-1}\Phi T V(k) - T^{-1}\Gamma K T V(k) \tag{6.30}$$

   to give

   $$T^{-1}\Phi T = \begin{bmatrix} -a_1 & -a_2 & \cdots & -a_n \\ 1 & 0 & \cdots & 0 \\ 0 & 1 & & \\ \cdot & & \cdot & \cdot \\ \cdot & & \cdot & \cdot \\ \cdot & & \cdot & \\ 0 & \cdots & & 1 & 0 \end{bmatrix}; \quad T^{-1}\Gamma K = \begin{bmatrix} 1 \\ 0 \\ \cdot \\ \cdot \\ \cdot \\ 0 \end{bmatrix}$$

   a_i = coefficients of open-loop CP.

2. If

   $$KT = \begin{bmatrix} -a_1 + d_1, -a_2 + d_2, \ldots, -a_n + d_n \end{bmatrix}$$

   then

   $$T^{-1}\Phi T - T^{-1}\Gamma K T = T^{-1}(\Phi - \Gamma K)T = \begin{bmatrix} -d_1 & -d_2 & \cdots & -d_n \\ 1 & 0 & \cdots & 0 \\ 0 & 1 & & \cdot & \cdot \\ \cdot & & & \cdot & \cdot \\ \cdot & & & \cdot & \cdot \\ \cdot & & & & \\ 0 & \cdots & & 1 & 0 \end{bmatrix}$$

$$\Rightarrow \Phi - \Gamma K \text{ has desired } CP \text{ as } p_d(z)$$

with

$$K = \begin{bmatrix} -a_1 + d_1, -a_2 + d_2, \ldots, -a_n + d_n \end{bmatrix} T^{-1} \tag{6.31}$$

Here it is best to solve

$$T' \begin{bmatrix} K_1 \\ K_2 \\ . \\ . \\ . \\ K_n \end{bmatrix} = \begin{bmatrix} -a_1 + d_1 \\ -a_2 + d_2 \\ . \\ . \\ . \\ -a_n + d_n \end{bmatrix}$$

and obtain K.

6.5.1.1 Algorithm for Obtaining Transformation Matrix T

This method is useful not just for pole-placement problems:

$$T = \begin{bmatrix} | & | & & & & | \\ t_1 & t_2 & . & . & . & t_n \\ | & | & & & & | \end{bmatrix}$$

- Generate T columnwise; it is evaluated as follows:

$$t_1 = \Gamma$$
$$t_2 = \Phi t_1 + a_1 \Gamma$$
$$t_3 = \Phi t_2 + a_2 \Gamma$$
$$.$$
$$.$$
$$.$$
$$t_n = \Phi t_{n-1} + a_{n-1} \Gamma$$

- Requires computation of $\{a_i\}$—possible numerical problems, encountered in Chapra and Canale (1998) [6].
- If $T-1$ exists then the system is CC and column t_k is a linear combination of $\Gamma, \Phi \Gamma, \ldots, \Phi^{k-1} \Gamma$.
- If $y(k) = C\, X(k)$

$$y(k)\, CTV(k) = [b_1,\ b_2, \ldots,\ b_n]V(k) \tag{6.32}$$

with $b_i = C\, t_i$.

- CL transfer function with $u(k) = K_r r(k) - KX(k)$:

$$\frac{y(z)}{r(z)} = K_r \frac{b_1 z^{n-1} + \cdots + b_n}{z^n + d_1 z^{n-1} + \cdots + d_n}; \quad b_i = Ct_i \tag{6.33}$$

For unity DC gain, $K_r = \left(1 + \sum_{i=1}^{n} d_i\right) \bigg/ \sum_{i=1}^{n} b_i$

- LG:

$$K(zI - \Phi)^{-1}\Gamma = \frac{\gamma_1 z^{n-1} + \cdots + \gamma_n}{z^n + a_1 z^{n-1} + \cdots + a_n}; \quad \gamma_i = Kt_i \tag{6.34}$$

6.5.2 Ackermann Formula [7]

Circumvents requirement to compute a_j

$$p_d(z) = (z - z_1)(z - z_2)\cdots(z - z_n) = z^n + d_1 z^{n-1} + \cdots + d_n$$
$$= \text{desired CL CP}$$

$$K = \begin{bmatrix} 0 & 0 & \cdots & 1 \end{bmatrix} H_c^{-1} p_d(\Phi) \tag{6.35}$$

where

$$H_c = \begin{bmatrix} | & | & & | \\ \Gamma & \Phi\Gamma & \cdots & \Phi^{n-1}\Gamma \\ | & | & & | \end{bmatrix} = \text{Controllability matrix}$$

Then,

$$p_d(\Phi) = (\Phi - z_1 I)(\Phi - z_2 I)\cdots(\Phi - z_n I)$$

(z_i = desired poles; must be complex conjugate pairs).
Algorithm given below illustrates this method:

1. Set up H_c matrix one column at a time.

$$\text{Transpose } H_c \to H_{c'}$$

2. Solve:

$$H_{c'}q = \begin{bmatrix} 0 \\ 0 \\ \cdot \\ \cdot \\ \cdot \\ 1 \end{bmatrix} \quad \text{for } n\text{-vector } q$$

a. Use any available routine for solving $AX = \mathbf{b}$ [for eigenvalues; eigenvectors].

b. If solution fails stop. The system is not CC.

3. Evaluate $p_d(\Phi) = X$.

4. Obtain gains:

$$K = q'X = \begin{bmatrix} K_1 & K_2 & \cdots & K_n \end{bmatrix}$$

5. Compute K_r if needed.

6.5.3 Algorithm to Obtain $p_d(\Phi)$

Compute using complex conjugate pairs to avoid complex matrices.
For example, if $z_3 = a + jb$; $z_4 = a - jb$

$$(z - z_3)(z - z_4) = z^2 - 2az + (a^2 + b^2)$$

$$(\Phi - z_3 I)(\Phi - z_4 I) = \Phi^2 - 2a\Phi + (a^2 + b^2)I$$

Incorporate into iterative scheme:

1. Initialize $X = I, k = 1$.

2. Read and imaginary parts of roots $RA(i)$, $RB(i)$, $I = 1,2,\ldots,n$

3. If $RB(k) = 0$: $X = X * [\Phi - RA(k)I]$:

$$k = k + 1$$

4. If $RB(k) \neq 0$: $X = X * [\Phi^2 - 2RA(k)\Phi + (RA^2(k) + RB^2(k))I]$:

$$k = k + 2$$

5. If $k \leq n$, go to 3:

$$\text{if } k = n + 1 \text{ done}$$

- Develop as a subroutine GAINS (introduce code if available here).
  - Can be used for continuous or discrete models.
- Generally pick z_i via $e_i^{sh.}$
- No restriction on h other than usual $(0.5 \to 1.0)/|\lambda_{\max}(A)|$.
- Deadbeat response: all $z_i = 0 \to p_d(z) = z^n$; also $p_d(\Phi) = \Phi^n$.

$$K = [0 \quad 0 \quad \cdots \quad 0 \quad 1] H_c^{-1}\Phi^n \qquad (6.36)$$

- Deadbeat gains $K_i \to \infty$ as $h \to 0$.

6.5.4 CL System Zeros

- CL transfer function in general is given by

$$\frac{y(z)}{r(z)} = \frac{K_r C(zI - \Phi)^{-1}\Gamma}{1 + K(zI - \Phi)^{-1}} = \frac{K_r \cdot \text{Open-loop numerator}}{p_d(z)}$$

⇒ SVFB has no effect on system zeros, as we already know from the continuous time system evaluation.

- For a general state-space model:

$$\text{State model: } X(k + 1) = \Phi^* X(k) + \Gamma^* u(k)$$

$$\text{Measurement model: } y(k) = C^* X(k) + D^* u(k)$$

- Denominator $= |zI - \Phi^*|$ (values of transfer function in z, where system can have a response with no input).
- Numerator zeros = values of z, where output is always zero:

$$(zI - \Phi^*)X(z) - \Gamma u(z) = 0$$

$$\left.\begin{vmatrix} zI - \Phi^* & -\Gamma^* \\ C^* & D^* \end{vmatrix} = 0\right\} \Longrightarrow$$

$$C^* X(z) + Du(z) = 0$$

- Apply to SVFB system; then,

$$X(k + 1) = (\Phi - \Gamma K)X(k) + K_r \Gamma r(k);$$

$$y(k) = CX(k)$$

- CL system zeros via

$$\begin{vmatrix} zI - \Phi + \Gamma K & -K_r \Gamma \\ C & 0 \end{vmatrix} = \text{numerator polynomial of } \frac{y(z)}{r(z)} \text{ here;}$$

$$\left(\text{add } \frac{K}{K_r} * \text{ last column to first } n \text{ columns}\right)$$

$$\begin{vmatrix} zI - \Phi & -K_r \Gamma \\ C & 0 \end{vmatrix} = K_r \cdot \begin{vmatrix} zI - \Phi & C \\ C & 0 \end{vmatrix} = K_r \text{ numerator polynomial of open-loop } (K = 0) \text{ system}$$

⇨ CL zeros = OL zeros, as usual;

- Then for series compensator design:

CL zeros = OL zeros of $\tilde{G}(z)$ and those zeros added by $H(z)$.

EXAMPLE 6.3   SATELLITE MOTOR CONTROL/POINTING

For this example of satellite motor control, pointing state model is given by

$$X(k+1) = \begin{bmatrix} 1 & h \\ 0 & 1 \end{bmatrix} X(k) + \begin{bmatrix} \dfrac{h^2}{2} \\ h \end{bmatrix} u(k)$$

Desired CL CP

$$p_d(z) = z^2 + d_1 z + d_2$$

This yields the two roots of CP in general for

$$si = -\zeta\omega_n \pm j\omega_n(1 - \zeta^2)^{1/2} \quad \text{for } i = 1, 2$$

which yields

$$d_1 = -2e^{-\zeta\omega n}(1 - \zeta^2)^{1/2}; \quad d_2 = e^{-2\zeta\omega n}$$

Using Ackermann's algorithm:

$$K = \begin{bmatrix} 0 & 1 \end{bmatrix} H_c^{-1} p_d(\Phi)$$

$$H_c = \begin{bmatrix} | & | \\ \Gamma & \Phi\Gamma \\ | & | \end{bmatrix} = \begin{bmatrix} \dfrac{h^2}{2} & \dfrac{3h^2}{2} \\ h & h \end{bmatrix}; \quad \text{then solve for } \begin{bmatrix} \dfrac{h^2}{2} & h \\ \dfrac{3h^2}{2} & h \end{bmatrix} \begin{bmatrix} q_1 \\ q_2 \end{bmatrix} = \begin{bmatrix} 0 \\ 1 \end{bmatrix} \Rightarrow q = \begin{bmatrix} \dfrac{1}{h^2} \\ -\dfrac{0.5}{h} \end{bmatrix}$$

$$p_d(\Phi) = \begin{bmatrix} 1 + d_1 + d_2 & 2h + d_1 h \\ 0 & 1 + d_1 + d_2 \end{bmatrix} = \Phi^2 + d_1\Phi + d_2 I$$

$$K = q' p_d(\Phi) = \begin{bmatrix} \dfrac{1 + d_1 + d_2}{h^2} & \dfrac{3 + d_1 - d_2}{2h} \end{bmatrix} \tag{6.37}$$

- Here in this case for a deadbeat controller, $d_1 = d_2 = 0 \rightarrow K = [1/h^2 \; 3/2h]$.
  - As $h \rightarrow 0$, it requires excessive control energy. This is not practical.
  - Good control scheme when h is large.
- For $\zeta = 0.707$, $\omega_n = 0.707$ and $h = 0.5$:

$$s_i|_{i=1,2} = -0.5 \pm j0.5 \rightarrow z_i|_{i=1,2} = e_i^{sh} = 0.53 \pm j0.29$$

$$d_i = -1.06, d_2 = 0.365 \rightarrow K = [1.22 \; 1.575]$$

6.6 Inverted Pendulum on a Cart

An inverted pendulum on a cart is shown in Figure 6.8. The configuration consists of a small ball of mass "m," connected to a pivot on a moving cart with wheels of mass "M," by a

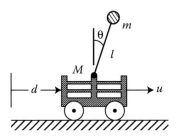

FIGURE 6.8
Inverted pendulum on a cart.

short arm of length "l." The angle of displacement from the vertical of the arm "l" is "θ." The horizontal distance of the cart from the vertical is "d." The force applied to the cart is "u."
   The small-angle equations are

$$\ddot{\theta}(t) = \frac{(m + M)g}{Ml} \theta(t) - \frac{ut}{Ml}$$

$$\ddot{d}(t) = -g \frac{m}{M} \theta(t) + \frac{u(t)}{M}$$

Thus, the state equation is: $X = [\theta, q, d, v]'$; $q = d\theta/dt$; $v = dd/dt$, which yields the state model as given below:

$$\dot{X} = \begin{bmatrix} 0 & 1 & 0 & 0 \\ \dfrac{(m + M)g}{Ml} & 0 & 0 & 0 \\ 0 & 0 & 0 & 1 \\ -\dfrac{g}{m} & 0 & 0 & 0 \end{bmatrix} \begin{bmatrix} \theta \\ q \\ d \\ v \end{bmatrix} + \begin{bmatrix} 0 \\ \dfrac{-1}{Ml} \\ 0 \\ \dfrac{1}{M} \end{bmatrix} u \qquad (6.38)$$

Let $m = 0.1$, $M = 1.0$, $l = 1$ m, $g \approx 10$ m/s^2, then

$$\dot{X} = \begin{bmatrix} 0 & 1 & 0 & 0 \\ 11 & 0 & 0 & 0 \\ 0 & 0 & 0 & 1 \\ -1 & 0 & 0 & 0 \end{bmatrix} \begin{bmatrix} \theta \\ q \\ d \\ v \end{bmatrix} + \begin{bmatrix} 0 \\ -1 \\ 0 \\ 1 \end{bmatrix} u$$

Open-loop eigenvalues are at: $s = 0, 0, +\sqrt{11}, -\sqrt{11}$.
*   The objective is to find SVFB, where $u = -KX$ such that any $X(0) \to 0$.

   $\to$ (regulator design, with $r = 0$).

*   Introduce "Feedback" inverted pendulum here.

FIGURE 6.9
CL system in continuous time.

- Continuous time design yields:

$$u(t) = -[77.9 \ -23.0 \ -16.9 \ -13.0] \ X(t)$$

gives a stable CL system $\dot{X} = (A - BK)X$ with CL eigenvalues:

$$\lambda_i(A - BK) = -2 \pm 3j, \ -3 \pm 2j$$

as shown in Figure 6.9.

- Examine discrete time design(s):
  - Equivalent (average) gains
  - Direct design with $z_i = e_i^{sh}$

6.6.1 Equivalent Discrete Design $u(k) = -\tilde{K}X(k)$

- Expect good performance for $h \leq \dfrac{1.0}{|\lambda_{max}(A - BK)|} = \dfrac{1.0}{\sqrt{13}} = 0.28.$
- System parameters versus h (s) are shown in Table 6.1.
- Large decreases in gain magnitudes as h increases. Variation of PM and ω_c plots are also shown in Figures 6.10 and 6.11, respectively.
- Stability analysis:
  - Discrete system has a delay of $\sim h/2$.
    → Reduces Φ_m by $\omega_c h/2$.
  - To avoid instability, the average gain lowers ω_c.
    → Lessens the destabilizing effect of discretization delay.

TABLE 6.1

System Parameters versus h (s)

h	$\tilde{K}_1$	$\tilde{K}_2$	$\tilde{K}_3$	$\tilde{K}_4$	ω_c	Φ_m
0.00	−77.9	−23.0	−16.9	−13.0	9.4	53.6°
0.02	−72.6	−21.5	−15.3	−11.9	9.1	46.7°
0.05	−65.1	−19.4	−13.0	−10.3	8.6	39.8°
0.10	−53.5	−16.0	−9.53	−7.92	7.4	30.0
0.15	−43.2	−13.0	−6.61	−5.81	6.5	23.1
0.20	−34.1	−10.4	−4.17	−3.98	5.4	17.7
0.25	−26.3	−8.05	−2.16	−2.42	4.6	14.8

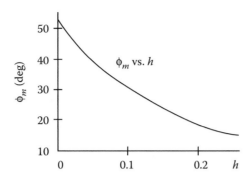

FIGURE 6.10
PM versus h.

$\Rightarrow$ Moves ω_c into a region, where $\angle LG(j\omega)$ is larger.

- Examine the Bode plot of LG to get ω_c and Φ_m:

$$LG(j\omega) = \tilde{K}(zI - \Phi)^{-1}\Gamma \big|_{z=e^{j\omega h}} \tag{6.39}$$

6.6.2 Direct Digital Design: Inverted Pendulum

- The selection of $h \le (1.0/|\lambda_{max}(A)|) = (1.0/\sqrt{11}) = 0.3\,\text{s}$ (alternate formula $h \le (1/||A||)$ gives $h \le (1/\sqrt{31}) = 0.18$).

  Pick $h = 0.18$ s.

1. Pole placement at $z_i = e_i^{sh} = \{0.60 \pm j0.36,\ 0.55 \pm j0.21\}$.
   $\rightarrow$ discrete gains $K = [-43.8\ -13.2\ -6.67\ -5.91]$.

2. Equivalent discrete design: $\tilde{K} = [-37.6\ -11.4\ -5.09\ -4.68]$ yields CL poles of $\Phi = \{0.67 \pm j0.41,\ 0.57 \pm j0.14\}$:
   - ω_c pole placement = 6.7 versus 5.9 for an equivalent design. Both have ~ same $\Phi_m \approx 19.3°$.

The Bode plot of the LG is plotted in Figure 6.12. The continuous system had 56.3°.

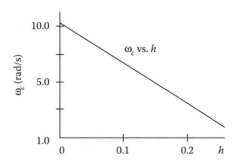

FIGURE 6.11
ω_c versus h.

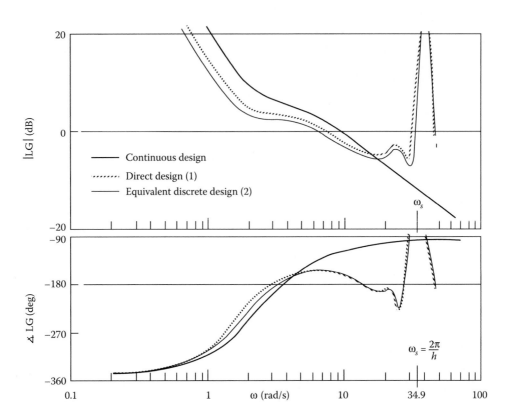

FIGURE 6.12
The Bode plot of loop-gain of digital design of inverted pendulum.

6.6.3 CL Simulation Inverted Pendulum $X(0) = [0.2, 0, 1, 0]'$

The CL simulation of the inverted pendulum for the initial state as above and $h = 0.18$ is shown in Figures 6.13 through 6.15, respectively, for control input, pendulum angle from vertical, and cart position for digital and continuous control.

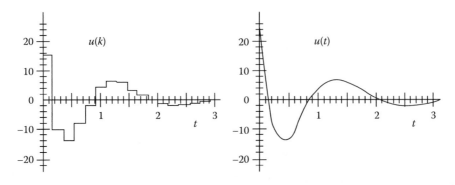

FIGURE 6.13
Control input for digital control and continuous control.

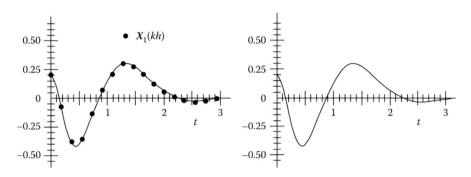

FIGURE 6.14
Pendulum angle from vertical $x_1(t)$ for digital control and continuous control.

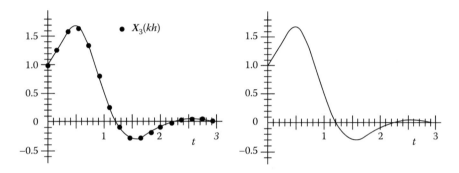

FIGURE 6.15
Cart position from "home" $x_3(t)$ for digital control and continuous control.

6.6.4 Deadbeat Controller Inverted Pendulum $X(0) = [0.2, 0, 1, 0]''$

- Impossible physically in this example, but interesting.

Input and the different state are shown in Figure 6.16.

- Excessive control and state overshoots, yet $X(4) = \mathbf{0}$.
- There is virtually no PM ($\omega_c \approx 13.3$, $\Phi_m \approx 40$).
- The Bode plot of LG for the deadbeat controller is plotted in Figure 6.17.
- CL system has virtually no robustness properties.
- Best to reserve deadbeat for "slow" systems with $h \sim$ large.

6.6.5 Summary of Pole Placement Design by SVFB

- Valid for continuous or discrete design.
  - Ackermann formula to find K.
  - Transform $K \rightarrow \tilde{K}$ if design developed on C and $h \sim$ small.
  - Need to select all n pole locations.
- SVFB does not modify system zeros.
  - Can combine compensator $H(z)$ and SVFB to adjust/move zeros.

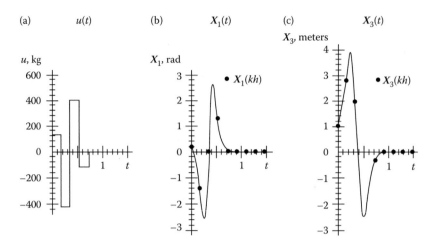

FIGURE 6.16
Input, and states for deadbeat controller.

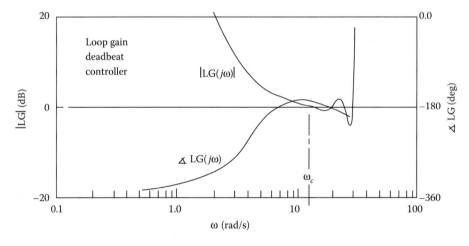

FIGURE 6.17
The Bode plot of LG for deadbeat controller.

Advantages:

- Straightforward design methodology
- Direct control over CL pole locations
- Uses all available information in the feedback
- Ability to design deadbeat control
- Possible to extend to MIMO systems but cumbersome

Disadvantages:

- Need to measure or estimate all states

- More complex design than series compensation
- No direct control over CL time response (still requires trial and error with CL simulation)
- Not always clear where to place all n poles
- No direct control over Φ_m, ω_c

6.7 SVFB with Time Delay in Control: $\tau = Mh + \varepsilon$

First design SVFB $u(k) = -KX(k)$ assuming $\tau = 0$.

6.7.1 State Prediction

Let us consider
   Case 1: $M > 0$, $\varepsilon = 0$
   Then, the state model is given by

$$X(k + 1) = \Phi X(k) + \Gamma u(k - M) \tag{6.40}$$

- For the predictor controller, use control law as

$$u(k) = -K\hat{X}(k + M) \tag{6.41}$$

$$\nwarrow \begin{cases} \text{the prediction of state at time } (k + M)h \\ \text{from } X(k) \text{ and } u(k - 1), \ldots, u(k - M) \end{cases}$$

Thus, at different "k," the predicted value of $X(k)$ is given by

$$\hat{X}(k + 1) = \Phi X(k) + \Gamma u(k - M)$$
$$\hat{X}(k + 2) = \Phi X(k + 1) + \Gamma u(k - M + 1)$$
$$= \Phi^2 X(k) + \Phi \Gamma u(k - M) + \Gamma u(k - M + 1)$$

     .

     .

$$\hat{X}(k + M) = \Phi^M X(k) + \sum_{i=1}^{M} \Phi^{i-1} \Gamma u(k - i) \tag{6.42}$$

- Present control $u(k)$ will have its first effect on $x(k + 1 + M)$.
- Need to store past controls in a pushdown stack.
- Requires a good knowledge of Φ, Γ to perform accurate propagation of $X(k)$.

Case 2: $M = 0$, $\varepsilon > 0$ $(0 \le \varepsilon < h)$

- Use

$$u(k) = -K\hat{X}(kh + \varepsilon)$$

$$\hat{X}(kh + \varepsilon) = e^{A\varepsilon}X(k) + \int_0^\varepsilon e^{A\sigma}d\sigma Bu(k-1)$$

$$\Rightarrow u(k) = -K_x X(k) - K_u u(k-1) \tag{6.43}$$

- Modification to the structure only, and propagation is "hidden."
- Identical to Equations 6.41 and 6.42 when $\varepsilon = h^-$ (corresponding to $M = 1$).

6.7.2 Implementation of the Delay Compensator: General Case

- Basic idea: construct $u(k)$ so that τ seconds later,

$$u(kh + \tau) \sim -KX(h + \tau)$$

$$\Rightarrow u(k) = -K\hat{X}(kh + \tau) \text{ now}$$

- Algorithm: enter with

$$X(k) = \text{current state measurement}$$

$$u(k-1) = \text{last control generated}$$

- Need to set up a delay stack (initialized to zero) as

$$V = \begin{bmatrix} v_0 & v_1 & \cdot & \cdot & \cdot & v_M \end{bmatrix} = \begin{bmatrix} u(k-1-M) & u(k-M) & \cdot & \cdot & \cdot & u(k-1) \end{bmatrix}$$

- Propagate current state ahead ε seconds: $X_e = \hat{X}(kh + \varepsilon)$

$$\hat{X}(kh + \varepsilon) = e^{A\varepsilon}X(k) + \int_0^\varepsilon e^{A\sigma}d\sigma Bu(k-1-M) \tag{6.44}$$

- Propagate X_e ahead M time steps and apply control $u = -K X_e$. This algorithm is shown in Figure 6.18.
- The algorithm can be rearranged for greater efficiency (need to store $\Phi^j\Gamma$; $j = 1$, $2, \ldots, n$).

6.7.3 Example—Inverted Pendulum

Let $h = 0.18$ s with gains obtained by pole placement as

$$K = [-43.8 \ -13.2 \ -6.67 \ -5.91]$$

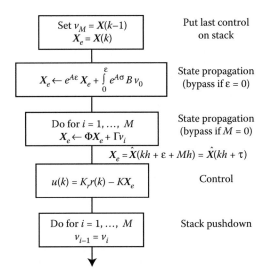

FIGURE 6.18
Flow chart of algorithm for implementation of delay compensator.

$$\omega_c = 6.7, \Phi_m = 19.3° \Rightarrow \tau_{max} = \frac{\Phi_m}{\omega_c} \approx 0.05\,\text{s}$$

Select $\tau = 0.18$ (corresponds to $M = 1$, $\varepsilon = 0$).

The system is highly unstable unless delay is compensated. The simulation results with these end-user specifications are shown in Figure 6.19.

- Simulation $X(0) = [0.1\ 0\ 1.0\ 0] = [\theta, d\theta/dt, d, dd/dt]$. Compare with the response of the system without delay.
- System "drifts" for the first τ seconds, then is controlled to zero.

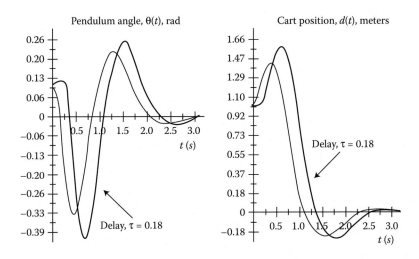

FIGURE 6.19
Inverted pendulum angle θ from vertical and cart position $d(t)$ from home.

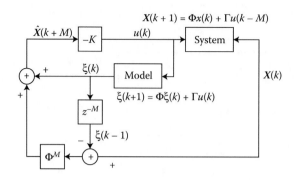

FIGURE 6.20
Loop structure of SVFB with delay.

- In ideal cases, the state response for $k > M$ is identical to an undelayed response with an initial condition $X(M) = \Phi M\, X(0)$ and shifted by Mh s.

  $\rightarrow$ from $k > M$, the predictor control is "perfect" (assuming one knows Φ and Γ)

6.7.4 Comparison with Smith Predictor Structure ($\varepsilon = 0$)

- Define system "model":

$$\xi(k+1) = \Phi\xi(k) + \Gamma u(k); \quad \xi(k) \sim \text{estimate of } X(k) \tag{6.45}$$

$$\Rightarrow \xi(k) = \Phi^M\xi(k-M) + \sum_{i=1}^{M}\Phi^{i-1}\Gamma u(k-i) \tag{6.46}$$

$$\Rightarrow \sum_{i=1}^{M}\Phi^{i-1}\Gamma u(k-i) = \xi(k) - \Phi^M\xi(k) \tag{6.47}$$

- $\hat{X}(k+M)$ prediction estimate in Equation 6.42:

$$\hat{X}(k+M) = \Phi^M[X(k) - \xi(k-M)] + \xi(k) \tag{6.48}$$

Control $u(k) = -KX(k+M)$.

- SVFB with delay loop structure is shown in Figure 6.20.
- Nearly identical to Smith predictor ($X \sim y$).
  - Preferable to use the state propagation formula, especially if the system is open-loop unstable, and Φ, Γ are not perfectly known.

6.8 Command Inputs to SVFB Systems

- Consider continuous case for simplicity:

$$\text{State model: } \dot{X}(t) = AX(t) + Bu(t) \tag{6.49a}$$

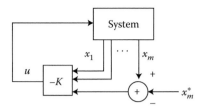

FIGURE 6.21
Control system configuration.

$$\text{Control law: } u(t) = -KX(t) \tag{6.49b}$$

- CL $A - BK$ has poles at desired locations.
- Desire $x_m = m$-th component of $X(t) \to x_m^*$ in ss.
- Idea: use control.

As shown in Figure 6.21,

$$u(t) = -\left[K_1 x_1(t) - \cdots - K_m \left[x_m(t) - x_m^* \right] - \cdots - K_n x_n(t) \right] \tag{6.50}$$

$$u(t) = K_m x_m^* - KX(t) \tag{6.51}$$

Like $K_r r(t)$ with $r(t) = $ constant

- Steady-state X: $X_{ss} = -(A - BK)^{-1} BK_m x_m^*$

$$X_{m,ss} = -e_m'(A - BK)^{-1} BK_m x_m^*; \; e_m' = \begin{bmatrix} 0 & 0 & \cdots & 1 & \cdots & 0 \end{bmatrix} \tag{6.52}$$

m-th element.
Generally, $x_{m,ss} \neq x_m^* \to \beta x_m^*$ and pick β so that $x_{m,ss} = x_m^*$ or consider the use of integral control

$$u(t) = -[K_1 x_1(t) - \cdots - K_m[x_m(t) - x_m^*] - \cdots - K_n x_n(t)] - K_{n+1} \int_0^{t+1} \left[x_m(\sigma) - x_m^* \right] d\sigma \tag{6.53}$$

6.8.1 Integral Control in SVFB

- Define $x_m(t)|_{\text{new}} = x_m(t)|_{\text{old}} - x_m^* \triangleq$ error in state x_m

$$\dot{X}(t) = AX(t) + a_m x_m^* + Be(t) \tag{6.54}$$

a_m = m-th column of A (often $a_m = 0$, especially if x_m is a position variable)

- Define $x_{n+1}(t) = \int_0^t x_m(\sigma)d\sigma$ = {integral of error in x_m from desired stady-state (ss) value:

$$\dot{X}_{n+1}(t) = x_m(t); \quad x_{n+1}(0) = 0$$

- Augmented $(n + 1)$-st order system, $X_a = [X, x_{n+1}]'$:

$$\begin{bmatrix} \dot{X} \\ \dot{X}_{n+1} \end{bmatrix} = \begin{bmatrix} A & 0 \\ e'_m & 0 \end{bmatrix} \begin{bmatrix} X \\ x_{n+1} \end{bmatrix} + \begin{bmatrix} a_m \\ 0 \end{bmatrix} x_m^* + \begin{bmatrix} B \\ 0 \end{bmatrix} u \qquad (6.55)$$

- Augmented system may not be controllable. Examine:

$$\begin{bmatrix} B_a & A_a B_a & \cdots & A_a^n B_a \end{bmatrix} = \begin{bmatrix} B & AB & \cdots & A^n B \\ 0 & e'_m B & \cdots & e'_m A^{n-1} B \end{bmatrix}$$

$$= (n + 1) \times (n + 1) \text{ controllability matrix}$$

- Selection of control law is given by

$$u(t) = -K_a X_a(t) = -K X(t) - K_{n+1} x_{n+1}(t)$$

  - Design K as before, to place poles of $A - BK = \bar{A}$.
  - K_{n+1} ~ small gain on integral error.
- CL CP, $|sI - A_a + B_a K_a|$:

$$\begin{vmatrix} sI - \bar{A} & BK_{n+1} \\ -e'_m & s \end{vmatrix} = |sI - \bar{A}| \cdot |s + e'_m(sI - \bar{A})^{-1} BK_{n+1}| = |sI - \bar{A}_a|$$

where the second determinant is ~ $\left(s - e'_m(\bar{A})^{-1} BK_{n+1}\right)$ for small K_{n+1}.

- Select K_{n+1} so that the pole is in left half plane (LHP).

EXAMPLE 6.4   INTEGRAL CONTROL

$$\text{State model: } \dot{X}(t) = \begin{bmatrix} \dot{x}_1 \\ \dot{x}_2 \end{bmatrix} = \begin{bmatrix} -1 & 0 \\ 1 & 0 \end{bmatrix} X(t) + \begin{bmatrix} 1 \\ 0 \end{bmatrix} u; \quad X = \begin{bmatrix} x_1 \\ x_2 \end{bmatrix}$$

$$\text{Control law: } u = -[1 2] X(t) \text{ places closed loop poles at } s = -1 \pm j$$

$$x_1 = \text{Motor shaft velocity}; \quad x_2 = \text{Shaft angular position}$$

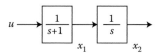

FIGURE 6.22
Motor system.

The block diagram of the motor system is shown in Figure 6.22.

1. Desire $x_1(t) \to x_1^*$ in ss (obvious problem since $x_2 \to \infty$).
   Introduce integral control:

$$x_1 \triangleq x_{1,e}$$

Obtain augmented state equation

Augmented state model: $\dot{X}_a = \begin{bmatrix} -1 & 0 & 0 \\ 1 & 0 & 0 \\ 1 & 0 & 0 \end{bmatrix} X_a + \begin{bmatrix} 1 \\ 0 \\ 0 \end{bmatrix} u; \quad |H_c| = \begin{vmatrix} 1 & -1 & 1 \\ 0 & 1 & -1 \\ 0 & 1 & 1 \end{vmatrix} = 0!$

Uncontrollable, cannot have a stable system if $x_1 =$ constant.

2. Desire $x_2(t) \to x_2^*$ in ss. Introduce integral control, $x_2 \to x_{2,e}$.

New state model: $\dot{X}_a = \begin{bmatrix} -1 & 0 & 0 \\ 1 & 0 & 0 \\ 1 & 0 & 0 \end{bmatrix} X_a + \begin{bmatrix} 0 \\ 0 \\ 0 \end{bmatrix} x_2^* + \begin{bmatrix} 1 \\ 0 \\ 0 \end{bmatrix} u; \quad |H_c| \neq 0$

Let $\begin{bmatrix} K_1 & K_2 \end{bmatrix} = \begin{bmatrix} 1 & 2 \end{bmatrix} =$ same as before for primary poles; $K_3 = \varepsilon \sim$ small

$$|sI - \bar{A}_a| = s(s^2 + 2s + 2) + \varepsilon \sim (s^2 + 2s + 2)\left(s + \frac{\varepsilon}{2}\right)$$

Too much gain on integral term is no good (NG) when simply added in as shown in Figure 6.23.

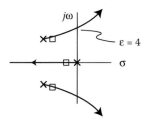

FIGURE 6.23
Alternate approach—to place all three CL poles in LHP.

PROBLEMS

P6.1   A servo system is described in Figure P6.1, and its state equations are given by

$$X(k + 1) = \begin{bmatrix} 1 & 0.0828 \\ 0 & 0.91 \end{bmatrix} X(k) + \begin{bmatrix} 0.00336 \\ 0.0828 \end{bmatrix} u(k)$$

$$y(k) = \begin{bmatrix} 1 & 0 \end{bmatrix} X(k)$$

$x_1(k)$ is the position of the motor shaft and $x_2(k)$ is the shaft velocity.

1. Design a deadbeat controller.
2. Find the time response with $x_1(0) = 5$ radians and $x_2(0) = 0$ for $X(0)$ is brought to 0 in two time steps.

P6.2 1. Repeat Problem 6.1 using the pole-placement via SVFB design method, and
2. Using Ackermann formula/algorithm.
3. Display results using MATLAB or your own routines.

P6.3 1. Repeat the above Problem 6.1 using average gain method of Kleinman (1978) [5] and
2. Using subroutine Dscrt for discrete gains, examine discrete eigenvalues and display response.
3. Using subroutine Dscrt for unconverted discrete gains, examine discrete eigenvalues and display response.
4. Display results using MATLAB or your own routines.

P6.4   A servo system as described in Figure P6.1 and its state equations are given by

$$X(k + 1) = \begin{bmatrix} 1 & 0.0828 \\ 0 & 0.91 \end{bmatrix} X(k) + \begin{bmatrix} 0.00336 \\ 0.0828 \end{bmatrix} u(k)$$

$$y(k) = \begin{bmatrix} 1 & 0 \end{bmatrix} X(k)$$

$x_1(k)$ is the position of the motor shaft and $x_2(k)$ is the shaft velocity.

1. Design a CL controller using direct pole-placement design method.
2. Find the time response with $x_1(0) = 5$ rad and $x_2(0) = 0$ for $X(0)$ is brought to 0 in two time steps.

P6.5   A servo system is described in Figure P6.1, and its state equations are given by

$$X(k + 1) = \begin{bmatrix} 1 & 0.0828 \\ 0 & 0.91 \end{bmatrix} X(k) + \begin{bmatrix} 0.00336 \\ 0.0828 \end{bmatrix} u(k)$$

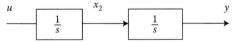

FIGURE P6.1
Block diagram representation of a servo system.

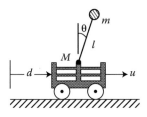

FIGURE P6.2
Inverted Pendulum.

$$y(k) = \begin{bmatrix} 1 & 0 \end{bmatrix} X(k)$$

$x_1(k)$ is the position of the motor shaft and $x_2(k)$ is the shaft velocity.

1. Design a series compensator design with $\zeta = 0.707$, $\omega_n = 0.707$, and $h = 0.5$.
2. Find the time response with $x_1(0) = 5$ rad and $x_2(0) = 0$ for $X(0)$ is brought to 0 in two time steps

P6.6 An inverted pendulum is described in Figure P6.2.

With state equations given by

$$\dot{X} = \begin{bmatrix} 0 & 1 & 0 & 0 \\ 11 & 0 & 0 & 0 \\ 0 & 0 & 0 & 1 \\ -1 & 0 & 0 & 0 \end{bmatrix} X + \begin{bmatrix} 0 \\ -1 \\ 0 \\ 1 \end{bmatrix} u$$

for $h = 0.3$.

1. Find system parameters like ω_c and Φ_m.
2. The discrete time system has a delay of $h/2$ s; find the reduction in Φ_m.
3. Use FEEDBACK$\ll^{\circledR}$ system 33-936S (Digital Control Experiment) to illustrate this inverted pendulum performance and stabilization.

P6.7 Compare the inverted pendulum system used in Problem 6.6 to the Smith Predictor structure, with $\varepsilon = 0$.

P6.8 As shown in Figure P6.3, consider a motor with current constant K_i that drives a load consisting of two masses coupled with a spring constant, K. The input signal is the motor current I. The angular velocities and the angles of the masses are ω_1 and ω_2, θ_1 and θ_2, respectively. The moments of inertia for these masses

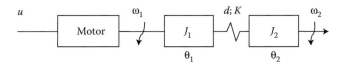

FIGURE P6.3
Two mass systems coupled by a spring and excited by a motor current.

are J_1 and J_2, respectively. It is assumed that there is a damping, d, in the spring and that the first mass may be disturbed by a torque τ. Finally, the output of the process is the angular velocity ω_2.

Introduce states as follows:

$$x_1 = \theta_1 - \theta_2$$

$$x_2 = \frac{\omega_1}{\omega_0}$$

$$x_2 = \frac{\omega_2}{\omega_0}$$

where

$$\omega_0 = \sqrt{K\frac{(J_1 + J_2)}{J_1 J_2}}$$

Choose:

$$\omega_0 = 1;$$

$$J_1 = \frac{10}{9}$$

$$J_2 = 10$$

$$K = 1$$

$$d = 0.1$$

$$K_i = 1$$

1. Find the poles and zeros of the system.
2. Find the damping coefficient and the natural frequency.
3. Plot the Bode plot of the process using MATLAB/Simulink.
4. It is desired that the CL system has a response from the reference signal such that the dominating modes have a natural frequency $\omega_n = 0.5$ (rads/s) and a damping $\zeta = 0.7$.
5. Assuming that all the states can be measured, design an SVF controller to satisfy these conditions.
6. Plot the input response of the process.
7. Plot the behavior of the CL system when the state feedback is used for $\alpha_1 = 2$.
8. The reference signal at time $t = 0$ and the disturbance τ is a pulse at $t = 25$ of height $= -10$ and a duration of 0.1 time units. Use MATLAB/Simulink for all simulations and plotting.

The controller for this system can be:

$$u(k) = -LX(k) + l_c u_c(k)$$

Desired poles: $\left(s^2 + 2\zeta\omega_n s + \omega_n^2\right)(s + \alpha_1\omega_n) = 0$

The parameter l_c is determined such that the ss given from u_c to y is unity. Please watch the number of integrators used in the controller.

7

Advanced Design Methods

(See color insert.)

7.1 Lyapunov Stability Theory Preliminaries

The Lyapunov theorem is a general theory for studying the stability of linear and nonlinear systems. It was developed in 1900 and advanced in the United States in 1960 [8].

In this chapter we consider only linear system cases. This proves to be a useful lead-in to optimal control.

- To study this further, we form a quadratic form given by

$$v(X) = X'PX = p_{11}x_1^2 + 2p_{12}x_1x_2 + \cdots + p_{nn}x_n^2$$

Thus, $v(X)$ is a quadratic form on X if P is a positive definite (PD).
Further an $n \times n$ matrix P is PD ($P > 0$), if

  i. $X'PX \geq 0$ for any $X \varepsilon R^n$

 ii. $X'PX = 0$ if and only if $X = 0$

iii. $P = P'$ (i.e., symmetric)

The following are some properties of a PD matrix:

1. All eigenvalues are real, $>0 \rightarrow P^{-1}$ exists.
2. Eigenvectors are orthogonal, $\xi_i'\xi_j = 0$, $i \neq j$.

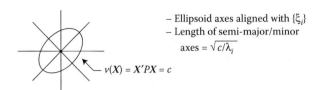

- Ellipsoid axes aligned with $\{\xi_i\}$
- Length of semi-major/minor
  axes = $\sqrt{c/\lambda_i}$

$v(X) = X'PX = c$

FIGURE 7.1
Ellipsoid in R^n for a quadratic form $X'PX$.

3. Can find $S'S = P$ (e.g., the Cholesky decomposition) with S invertible.
4. For any X:

$$0 < \lambda_{\min}(P)X'X \leq X'PX \leq \lambda_{\max}(P)X'X$$

- The equation $X'PX = c$ defines an ellipsoid in R^n as shown in Figure 7.1.

7.1.1 Application to Stability Analysis

- Study stability of an unforced system

$$X(k + 1) = \Phi X(k) + \Gamma u(k) \tag{7.1}$$

The second term in the RHS of Equation 7.1 is equal to zero and $X(0)$ is the initial state.

- Suppose we found a quadratic form $v(X) = X'PX$ such that when we monitor $v(X(k))$ at any sequence of increasing k:

$$v(X(0)) > v(X(1)) > v(X(2)) > \cdots$$

When the $v(X(k)) = c_k$ functions are plotted in Figure 7.2:

Implication: $X(k) \rightarrow 0$ as $k \rightarrow \infty$.

- Result—if we can find a positive scalar (quadratic) function:

$v(X)$ such that $v(X)$ is always decreasing, that is,

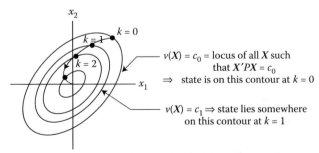

$v(X) = c_0$ = locus of all X such that $X'PX = c_0$
$\Rightarrow$ state is on this contour at $k = 0$

$v(X) = c_1 \Rightarrow$ state lies somewhere on this contour at $k = 1$

Implication: $X(k) = 0$ as $k \rightarrow \infty$

FIGURE 7.2
c_k contours of $v(X(k) = c_k) \rightarrow$ state lies on this contour.

if $k_2 > k_1$,

$$v(X(k_2)) < v(X(k_1)) \text{ then } X(k) \to 0$$

- Such a $v(X(k))$ is called a Lyapunov function.
- Analogous to a generalized "stored energy."

7.1.2 Main Theorem for Linear Systems

- Existence of a Lyapunov function $\Rightarrow$ stability and vice versa
- Consider $v(X) = X'PX, P > 0$, determine

$$\Delta v(X) = v(X(k + 1)) - v(X(k)) \tag{7.2}$$

Along the system response trajectory $X(k + 1) = \Phi X(k)$

$$\Delta v(X) = X'(k)\Phi'P\Phi X(k) - X'(k)PX(k)$$

$$= -X'(k)[P - \Phi'P\Phi]X(k); \quad Q = [P - \Phi'P\Phi] \tag{7.3}$$

if $Q > 0$, $v(X) \downarrow$ and $x(k) \to 0$, but if Q is not > 0 no conclusions can be drawn.
$\to$ Use reverse procedure:

Pick $Q > 0$ and solve

$$P = \Phi'P\Phi + Q \xrightarrow{\text{as}} \tag{LEqn}$$

Then,
Theorem: $X(k + 1) = \Phi X(k)$ is stable if and only if given any positive definite (PD), Q, the solution P of the equation:

$$P = \Phi'P\Phi + Q \text{ is PD, with } \theta$$

as defined in Equation 7.3 above.
LEqn represents a set of $n(n + 1)/2$ linear equations:

- Expand right hand side (RHS) term $\Phi'P\Phi$.
- Solve for $p_{ij} = p_{ji}$ for $i = 1, \ldots, n; j = 1, \ldots, n$.
- A solution exists provided $\lambda_i(\Phi)\lambda_j(\Phi) \neq 1$.
- Test if P is PD.
- A slightly weaker condition is Q positive semidefinite ($Q \geq 0$), as long as $X'QX \neq 0$ along a system response trajectory. The trajectories are shown in Figure 7.2.

7.1.3 Practical Use of the Lyapunov Theorem

1. To test the stability of Φ pick a $Q > 0$ and solve for P. If P is not PD, the system is unstable. If $P > 0$ system is stable:
   - Need only do this for one Q
   - Not very practical (there are easier ways to test stability)
   - But useful in developing/proving further results

2. If the system is stable, the theorem gives an easy way to find a Lyapunov function. Pick any $Q > 0$ (e.g., $Q = \beta I$) and solve LEqn for P. Then $v(X) = X'PX$ is a Lyapunov function and $\Delta v(X) = -X'QX$. Different Q yields different P.
   - Our major efforts will involve finding a $v(X)$ for a stable system, and using it to develop a state variable feed back (SVFB) control.

EXAMPLE 7.1

$$X(k + 1) = \begin{bmatrix} 0.2 & -0.2 \\ 0.1 & 0.5 \end{bmatrix} X(k); \quad \text{pick } Q = \begin{bmatrix} 2 & 0 \\ 0 & 2 \end{bmatrix}$$

$$\Phi = \begin{bmatrix} 0.2 & -0.2 \\ 0.1 & 0.5 \end{bmatrix} \text{ has stable poles at } 0.3 \text{ and } 0.4$$

Solve $P = \Phi'P\Phi + Q$, which yields

$$\begin{bmatrix} p_{11} & p_{12} \\ p_{12} & p_{22} \end{bmatrix} = \begin{bmatrix} 0.2 & 0.1 \\ -0.2 & 0.5 \end{bmatrix} \begin{bmatrix} p_{11} & p_{12} \\ p_{12} & p_{22} \end{bmatrix} \begin{bmatrix} 0.2 & -0.2 \\ 0.1 & 0.5 \end{bmatrix} + \begin{bmatrix} 2 & 0 \\ 0 & 2 \end{bmatrix}$$

$$= \begin{bmatrix} 0.04p_{11} + 0.04p_{12} + 0.01p_{22} + 2 & -0.04p_{11} + 0.08p_{12} + 0.05p_{22} \\ -0.04p_{11} + 0.08p_{12} + 0.05p_{22} & 0.04p_{11} - 0.2p_{12} + 0.25p_{22} + 2 \end{bmatrix}$$

Thus,

$$0.96p_{11} - 0.04p_{12} - 0.01p_{22} = 2$$
$$0.04p_{11} + 0.92p_{12} - 0.05p_{22} = 0$$
$$-0.04p_{11} + 0.2p_{12} + 0.75p_{22} = 2$$

$$\Rightarrow p_{22} = 3.0, \; p_{12} = -0.833 = p_{21}, \; p_{11} = 2.083$$

$$P = \begin{bmatrix} 2.083 & -0.833 \\ -0.833 & 3.0 \end{bmatrix} > 0$$

$$X'PX = p_{11}x_1{}^2 + 2p_{12}x_1x_2 + p_{22}x_2{}^2 \text{ is a Lyapunov function}$$

7.2 Numerical Solution of the Lyapunov Equation

Setting up and solving the $n(n + 1)/2$ system is not practical, since it requires $O(n^6)$ operations for large n. One can desire an algorithm requiring $O(n^3)$ operations.

Thus, if $|\lambda_i(\Phi)| < 1$, that is, the system is stable, and

$$P = \sum_{i=0}^{\infty} (\Phi')^i Q \Phi^i \qquad (7.4)$$

satisfies $P = \Phi'P\Phi + Q$. This can be easily checked by direct substitution.

- If the system is unstable, the sum in Equation 7.4 diverges.
- An efficient way to sum the series is to use the doubling algorithm as illustrated in the flow diagram of Figure 7.3. Equation 7.4 is further expanded as shown below to become the basis of the doubling algorithm described below:

$$P = \sum_{i=0}^{\infty} (\Phi')^i Q \Phi^i = Q + \Phi'Q\Phi + (\Phi')^2 Q\Phi^2 + (\Phi')^3 Q\Phi^3 + (\Phi')^4 Q\Phi^4 + \cdots$$

The doubling algorithm consists of several passes; at the end of the first pass the first two terms of the RHS are initiated. Then, in the second pass the first four terms of the series in Equation 7.4 are generated, in the third pass the first eight terms are generated, and so forth until the ultimate k is reached, thus yielding 2^k terms of the series.

- Stop when $P_k = P_{k+1}$ or when diagonals $(p_{ii})k = (p_{ii})k - 1$ $i = 1, \ldots, n$.

7.2.1 Algorithm to Solve the Lyapunov Equation (DLINEQ)

- Embed a stability test if $P \rightarrow \infty$.

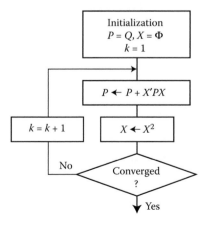

FIGURE 7.3
Doubling algorithm.

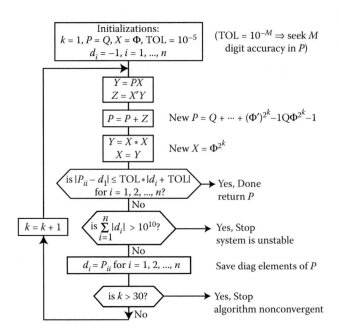

FIGURE 7.4
Flow diagram to solve the Lyapunov equation, DLINEQ.

The flow diagram for algorithm to solve the Lyapunov equation (DLINEQ) is shown in Figure 7.4.

- The algorithm generally converges in $K \sim 10$ iterations.
  - Requires $\sim K \times (2.5) n^3$ multiply and add (MADD) operations
  - When $Q = 10$, P has $1024 = 2^{10}$ terms, and $||\Phi^{1024}|| < 10^{-5}$ provided all $|\lambda_i(\Phi)| < 0.99$
- By all counts, this is an extremely versatile algorithm since it satisfies the $O(n^3)$ requirement for algorithm efficiency and also converges in 10 iterations. Thus, the computer is not engaged for a longer period of time. An open-loop system response is shown in Figure 7.5.

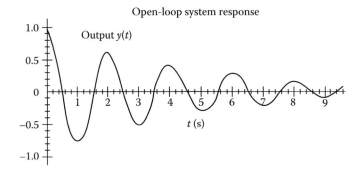

FIGURE 7.5
Open-loop system response of a lightly damped system.

7.3 Constructive Application of the Lyapunov Theorem to SVFB

If

$$X(k + 1) = \Phi\, X(k) + \Gamma\, u(k)$$

is completely controllable, and a control law, $u(k) = -K_0 X(k)$ results in a stable closed-loop (CL) system, where, [9]

$$K_0 = \Gamma' W_M^{-1}\Phi, \quad \text{where } M \text{ is arbitrary} \geq n \tag{7.5}$$

Then,

$$W_M = \sum_{i=0}^{M} \Phi^{-i}\Gamma\Gamma'(\Phi')^{-1} > 0 \text{ via controllability condition} \tag{7.6}$$

- An outline of proof

$$\text{Let } \tilde{\Phi} \triangleq \Phi - \Phi^{-i}\Gamma\Gamma'W_M^{-1}$$

1. Since

$$\Phi W_M \Phi' = W_M + \Phi\Gamma\Gamma'\Phi' - \Phi^{-M}\Gamma\Gamma'(\Phi')^{-M}$$

Here we can show:

$$W_M = \tilde{\Phi}W_M\tilde{\Phi}' + \Phi\Gamma[I - \Gamma'W_M^{-1}\Gamma]\Gamma'\Phi' + \Phi^{-M}\Gamma\Gamma'(\Phi')^{-M}$$

2. $I - \Gamma'W_M^{-1}\Gamma = I - \Gamma'[\Gamma\Gamma' + \Phi^{-1}W_{M-1}(\Phi')^{-1}]^{-1}\Gamma \triangleq Q_1$
   with

$$W_M = \Gamma\Gamma' + \Phi^{-1}W_{M-1}(\Phi')^{-1}$$

3. Via a matrix inversion lemma, thus

$$Q_1 = [I + \Gamma'\Phi'W_{M-1}^{-1}\Phi\Gamma]^{-1} \Rightarrow Q_1 > 0$$

4. Establish that $X'WX$ is a Lyapunov function for $\tilde{\Phi}'$

$$W_M = \tilde{\Phi}W_M\tilde{\Phi}' + \Phi\Gamma Q_1\Gamma'\Phi' + \Phi^{-M}\Gamma\Gamma'(\Phi')^{-M}$$

by showing that $X'\Phi\Gamma Q_1\Gamma'\Phi'X > 0$ along the system response trajectory; thus, $X(k + 1) = \tilde{\Phi}'\, X(k)$ (okay if system is controllable).

5. By Lyapunov $\tilde{\Phi}'$, $\tilde{\Phi}$ has all eigenvalues with $|\lambda_i(\tilde{\Phi})| < 1$, then

6. $\tilde{\Phi} = \Phi(\Phi - \Gamma\Gamma'W_M^{-1}\Phi)\Phi^{-1} \Rightarrow \tilde{\Phi}$ and $\Phi - \Gamma K_0$ have the same eigenvalues, where

$$K_0 = \Gamma'W_M^{-1}\Phi$$

- *Corollary:* If the system is not completely controllable, then the control law is given by

$$u = -\Gamma'W_M^{-\#}\Phi$$

(# in the exponent denotes pseudo inverse that will stabilize the controllable modes.)

7.3.1 Discussion of Stabilization Result

$$K_0 = \Gamma'W_M^{-1}\Phi$$

- Applicable to multi-input systems, where $\Gamma = \Psi B \rightarrow n \times n$ matrix and $K_0 = m \times n$ gain matrix (m = number of inputs).
- If R = arbitrary $m \times m$ PD matrix, then

$$K_0 = -R^{-1}\Gamma'W_{R,M}^{-1}\Phi \text{ is stabilizable} \tag{7.7}$$

where

$$W_{R,M} = \sum_{i=0}^{M} \Phi^{-i}\Gamma R^{-1}\Gamma'(\Phi')^{-i} \tag{7.8}$$

- Gives additional degrees of freedom
- Alternate representation can be shown as

$$K_0 = (R + \Gamma'V_{R,M-1}\Gamma)^{-1}V_{R,M-1}\Phi \tag{7.9}$$

with

$$V_{R,M-1} = \Phi'W_{R,M}^{-1}\Phi \tag{7.10}$$

- Computing

$$V_{R,M-1} = \Phi'W_{R,M}^{-1}\Phi = (\Phi')^{M}\left[\sum_{i=0}^{M}\Phi^{i}\Gamma R^{-1}\Gamma'(\Phi')^{i}\right]^{-1}\Phi^{M} \tag{7.11}$$

1. Pick $M = 2^p \geq n$ (best to pick min p such that $2^p > n$).
2. Go through the doubling algorithm p times: $\Phi \to \Phi'$, and; $Q = \Gamma R^{-1}\Gamma'$.

$$P = \sum_{i=0}^{2^P-1} \Phi^i Q(\Phi')^i; \quad X = (\Phi')^{2^p}$$

3. Use the Cholesky decomposition $P = S'S$, then, we have:

$$V_{R,M-1} = (XS^{-1}) \cdot (XS^{-1})'$$

- CL eignevalues are inside the unit circle, but otherwise unspecified. This is not a design method for feedback control, but rather a starting point for SVFB design and application.

7.3.1.1 Examples of System Stabilization, with SCALAR a, and R = 1

$$K_0 = \frac{\Gamma'V_{M-1}\Phi}{1 + \Gamma'V_{M-1}\Gamma} \quad \text{where } V_{M-1} = (\Phi')^M \left[\sum_{i=0}^{M} \Phi^i \Gamma R^{-1}\Gamma'(\Phi')^i \right]^{-1} \Phi^M$$

- Satellite system (with double integrator):

$$A = \begin{bmatrix} 0 & 1 \\ 0 & 0 \end{bmatrix}; \quad B = \begin{bmatrix} 0 \\ 1 \end{bmatrix} \overset{h=1}{\to} \Phi = \begin{bmatrix} 1 & 1 \\ 0 & 1 \end{bmatrix}; \quad \Gamma = \begin{bmatrix} 0.5 \\ 1 \end{bmatrix}$$

Choose $M = 2^2 = 4$, which yields as under:

$$V_3 = \begin{bmatrix} 0.20 & 0.40 \\ 0.40 & 1.05 \end{bmatrix}; \quad K_0 = \begin{bmatrix} 0.20 & 0.70 \end{bmatrix}$$

CL poles of $\Phi - \Gamma K_0 = 0.6 \pm j\,0.2 \Rightarrow \zeta = 0.82, \omega_n = 0.56$
As M is increased, K_0 decreases and CL poles $\to 1, 1$.
For

$$M = 8: K_0 = [0.067 \quad 0.411] \Rightarrow z_i = 0.78 \pm j0.13; \quad \zeta = 0.82, \omega_n = 0.29$$

$$M = 16: K_0 = [0.02 \quad 0.225] \Rightarrow z_i = 0.88 \pm j0.076; \quad \zeta = 0.82, \omega_n = 0.15$$

EXAMPLE 7.2

Inverted pendulum on cart, $h = 0.18$

$$A = \begin{bmatrix} 0 & 1 & 0 & 0 \\ 11 & 0 & 0 & 0 \\ 0 & 0 & 0 & 1 \\ -1 & 0 & 0 & 0 \end{bmatrix}, B = \begin{bmatrix} 0 \\ -1 \\ 0 \\ 1 \end{bmatrix} \rightarrow \Phi = \begin{bmatrix} 1.18 & 0.19 & 0 & 0 \\ 2.10 & 1.18 & 0 & 0 \\ -0.017 & -.001 & 1 & 0.18 \\ -.19 & -.017 & 0 & 0 \end{bmatrix}, \Gamma = \begin{bmatrix} -.017 \\ -.19 \\ .016 \\ .181 \end{bmatrix}$$

Continuous system's open-loop poles at $0, 0, \pm\sqrt{11}$

$$\text{Pick } M = 8 = 2^3: K_0 = \begin{bmatrix} -47.4 & -14.7 & -6.89 & -6.75 \end{bmatrix}$$

$$\text{CL poles } \Phi - \Gamma K_0 = \begin{cases} 0.57 \pm j0.30 & (\zeta = 0.67, \omega_n = 3.62) \\ 0.48 \pm j0.065 & (\zeta = 0.98, \omega_n = 4.08) \end{cases}$$

If M is increased to 12:

$$K_0 = \begin{bmatrix} -31.5 & -9.67 & -1.99 & -2.90 \end{bmatrix} \rightarrow \begin{cases} 0.74 \pm j0.16 & (\zeta = 0.79, \omega_n = 2) \\ 0.54 \pm j0.12 & (\zeta = 0.94, \omega_n = 3.5) \end{cases}$$

If M is decreased to 5:

$$K_0 = \begin{bmatrix} -87.9 & -26.7 & -25.8 & -17.0 \end{bmatrix} \rightarrow \begin{cases} 0.31 \pm j0.39 & (\zeta = 0.61, \omega_n = 6.3) \\ 0.54 \pm j0.12 & (\zeta = 0.97, \omega_n = 6.0) \end{cases}$$

$\Rightarrow$ as M increases, K_0 decreases and the result is slower CL response!

7.3.2 Lyapunov ("Bang-Bang") Controllers

We consider a stable system with bounded control:

$$X(k + 1) = \Phi X(k) + \Gamma u(k); \quad |u(k)| \le c_1 \tag{7.12}$$

We then obtain a Lyapunov function $v(X) = X'PX$ for a free part, such that

$$P = \Phi'P\Phi + Q \tag{7.13}$$

where, Q = arbitrary PD matrix.

- Along the trajectory of the controlled system, we have an incremental $v(X)$ function as

$$\Delta v(X) = X'(k + 1)PX(k + 1) - X'(k)PX(k)$$

$$= X'(k)[\Phi'P\Phi - P]X'(k) + 2u(k)\Gamma' P\Phi X(k) + u^2(k)\Gamma' P\Gamma, \text{ where } - Q = \Phi'P\Phi - P$$

- *Idea:* Pick $u(k)$ to drive $x(k) \to 0$ even faster than the open loop.
- Make $\Delta v(X)$ as negative as possible. Set $(\partial[\Delta v(X)]/\partial u(k)) = 0$;

$$\text{so that; } u(k) = -(\Gamma'P\Gamma)^{-1}P\Phi X(k) \quad \text{if } |u(k)| < c_1 \tag{7.14a}$$

$$\text{Yielding } u(k) = -c_1 \cdot \text{sgn}[(\Gamma'P\Gamma)^{-1}\Gamma' \, P\Phi X(k)] \quad \text{if } |u(k)| \geq c_1 \tag{7.14b}$$

- Algorithm:
  With $K = [(\Gamma'P\Gamma)^{-1}\Gamma'P\Phi]$
  1. Compute $w = -KX(k)$
  2. If $|w| < c_1$ set $u(k) = w$, else $u(k) = c_1\text{sgn}(w)$
- As $h \to 0$, $u(k) \to$ "bang-bang" controller
- For $u(k) = \pm c_1$
- Different $Q \to$ different P and $K \to Q =$ "design" parameters:

$$\Delta v(X) = -X'[Q + \Phi'P\Gamma(\Gamma' \, P\Gamma)^{-1}\Gamma'P\Phi]X \text{ in linear region}$$

- Does increasing q_{ii} speed up the response? (i.e., drive $x_i \to 0$ faster)

EXAMPLE 7.3
(Lightly damped system w/unconstrained "u")
  State model and measurement model of a lightly damped system is given by

$$\text{State Model: } \dot{X} = \begin{bmatrix} 0 & 1 \\ -10 & -0.5 \end{bmatrix} X + \begin{bmatrix} 0 \\ 10 \end{bmatrix} u$$

$$\text{Measurement Model: } y = [1 \quad 0]X;$$

and u is unconstrained.

- The open-loop system response is shown in Figure 7.6 with $\zeta = 0.08$, $\omega_n = \sqrt{10}$; $X(0) = [1 \ 0]'$.
- The Lyapunov digital design is designed with 'h' as

$$h = \frac{0.5}{|\lambda_{\max}(A)|} \approx 0.15$$

Here, we pick $Q = \begin{bmatrix} 1 & 0 \\ 0 & 1 \end{bmatrix} \Rightarrow P = \begin{bmatrix} 74.0 & 0.448 \\ 0.488 & 7.85 \end{bmatrix}$

$$K = (\Gamma'P\Gamma)^{-1}\Gamma'P\Phi = [-0.469 \quad 0.631] \Rightarrow \{\text{poles of } \Phi - \Gamma K @ 0, 0.887\}$$

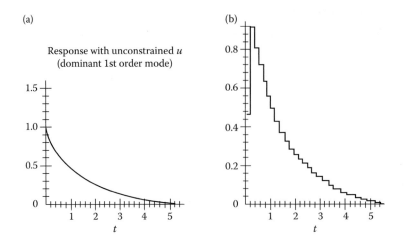

FIGURE 7.6
Response of a lightly damped system with unconstrained u. (a) Input $u(kh)$, (b) output $y(t)$.

- Response with unconstrained u (dominant first-order mode) is shown in Figure 7.6.

We will now illustrate the previous example with a Constrained Response, $X(0) = [1\ 0]'$.

$$|u(k)| \leq 0.30$$

For the same lightly damped system from the previous example, the output is plotted in Figure 7.7 showing both the input and output of the system for

$$|u(k)| \leq 0.15$$

For the same lightly damped system from the previous example, the output is plotted in Figure 7.8 showing both the input and output of the system.

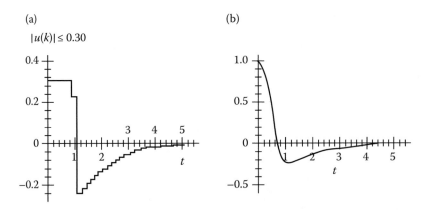

FIGURE 7.7
Input ($u(k) \leq 0.3$) and output of the system under constrained conditions. (a) Input $u(kh)$, (b) output $y(t)$.

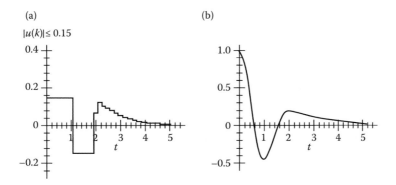

FIGURE 7.8
Input ($u(k) \leq 0.15$) and output of the system under constrained conditions.

- "Bang-bang" controller has a peculiar behavior until $|KX(k)| \leq c_1$, whereupon CL linear response takes over.
- Modification of q_{11}, q_{22} has a small effect on response.

General useful comments:

- The system must be open-loop stable to compute P.
- Very little control over time response or poles.
- The Lyapunov controller is useful in cases where $h \sim$ large.
- Applicable to multi-input case, $u_i = -c_i \, \text{sat}[K_i X(k)/c_i]$.
- Assures system stability even when control is limited.

Not necessarily true in other SVFB design approaches.

7.4 Introduction to Least-Squares Optimization

Say the state model is given by

$$X(k + 1) = \Phi\, X(k) + \Gamma u(k); X(0) = \text{initial state} \tag{7.15}$$

and control law is given by

$$u(k) = \text{unconstrained}$$

- *Objective:* Determine an SVFB control $u(k) = -KX(k)$ so that $X(k) \to 0$ "nicely" $\to$ stability and more can be achieved.
- Pole placement approach
  - Don't necessarily know where good pole locations are!
  - Resulting system may have low $|RD|$ or Φ_m.
  - Needed gains are often too big $\Rightarrow$ need to manage $|u(k)|$.
- Optimal control approach

We express mathematically the methodology to achieve an end result:

1. Each $x_i(k) \to 0$ nicely: consider minimizing a function given by

$$ISE = \sum_{k=0}^{\infty} q_{11}x_1^2(k) + \cdots + q_{nn}x_n^2(k) = \sum_{k=0}^{\infty} X'(k)QX(k) \qquad (7.16)$$

q_{ii} ~ scale factors to weigh relative importance of different errors, $q_{ii} > 0$

Q = positive (semi) definite which is usually diagonal

2. Do not want $u(k)$ to be too large: conserve energy, say energy is given by

$$E = \sum_{k=0}^{\infty} u^2(k) \qquad (7.17)$$

3. Combine into a composite criterion, $J = ISE + R*E$

$$J = \sum_{k=0}^{\infty} [X'(k)QX(k) + Ru^2(k)] \qquad (7.18)$$

$R > 0$ adjusts the tradeoff between speed of response and magnitude of control input.

7.4.1 Problem Definition General Comments

Let us study the linear quadratic (LQ) optimal control problem, where we need to generate quadratic cost function. Then, find $u(k) = -K\,X(k)$ to minimize

$$J = \sum_{k=0}^{\infty} \left[X'(k)QX(k) + Ru^2(k) \right] \quad Q \geq 0, R > 0$$

- General quadratic cost functional:
  - Historical use (from Gauss, Wiener, Kalman, etc.)
  - Physical appeal: larger deviations from nominal are weighted more heavily
  - Physical interpretation: energy is generally ~ x_i^2, u^2
  - Mathematically tractable ("easy" to take $\partial/\partial K$)
  - Most overworked problem in modern control theory
- Properties of J:
  - $J \geq 0$; zero only if $X(0)$ is such that free response satisfies $X'(0)(\Phi')^k Q(\Phi)^k X(0) \equiv 0$
  - Any feedback control that gives a finite value to J must necessarily be stabilizing
  - If $R \to 0$, "optimal" control would try to place CL poles at $z_j \to 0$

(drive $X(0) \to 0$ as fast as possible)

Special case: If there is only concern about output deviations, consider minimizing

$$J = \sum_{k=0}^{\infty} \alpha y^2(k) + Ru^2(k) \tag{7.19}$$

Since $y(k) = CX(k)$, $y^2(k) = X'(k)C'CX(k)$, and

$$J = \sum_{k=0}^{\infty} X'(k)[\alpha C'C]X(k) + Ru^2(k), \quad \text{where } Q = \alpha C'C \tag{7.20}$$

=> a "special" case of general state weightings.

7.4.2 Optimization Approach and Algorithm

An expression for J let $u(k) = -KX(k)$ be any FB control such that the CL system $X(k + 1) = (\Phi - \Gamma K)X(0)$ is stable, then

$$X(k + 1) = (\Phi - \Gamma K)^k X(0), \quad u(k) = -KX(k)$$

$$J = X'(0)\left[\sum_{k=0}^{\infty} (\Phi - \Gamma K)'^k (Q + K'R(k))(\Phi - \Gamma K)^k\right]X(0) \tag{7.21}$$

The term in the brackets [.] is equivalent to P_k. Since the CL system is stable,

1. P satisfies the linear (the Lyapunov equation)

$$P_k = (\Phi - \Gamma K)'P_k(\Phi - \Gamma K) + Q + K'RK \tag{7.22}$$

2. P_k is positive (semi-) definite symmetric. P_k is called the cost matrix associated with gain K

$$J = X'(0)P_kX(0) \quad \text{for any } X(0)$$

P_k does not depend on $X(0)$, but only on feedback gains K, $P_k \leftrightarrow K$.

Design approach. Find the gain K^* that gives the "smallest" cost matrix in a PD sense, that is, if $K^* \leftrightarrow P_k^* \equiv P^*$ then for any K with $K \leftrightarrow P_k$

$$X'P^*X \leq X'P_kX \quad \text{for all } X$$

Develop an iterative approach to find K^*. Start with gain $K_0 \leftrightarrow P_0$, try to find $K_1 \leftrightarrow P_1$ so that $P_1 < P_0$, that is, K_1 is optimal over K_0.

7.4.3 Continued Method for Obtaining K_1 from P_0

The following steps are followed:

- Start with a stabilizing gain $K_0 \leftrightarrow P_0$

$$P_0 = (\Phi - \Gamma K_0)' P_0 (\Phi - \Gamma K_0) + Q + K_0' R K_0 \tag{7.23}$$

If $K_1 \leftrightarrow P_1$ (assuming K_1 is stabilizing), then

$$P_1 = (\Phi - \Gamma K_1)' P_1 (\Phi - \Gamma K_1) + Q + K_1' R K_1 \tag{7.24}$$

Find if the difference $\partial P = P_0 - P_1$ satisfies the following:

$$\partial P = (\Phi - \Gamma K_1)' \partial P (\Phi - \Gamma K_1) + (K_0 - K_1)'(R + \Gamma' P_0 \Gamma)(K_0 - K_1)$$
$$+ (K_0 - K_1)[(R + \Gamma' P_0 \Gamma)K_1 - \Gamma' P_0 \Phi] + [(R + \Gamma' P_0 \Gamma)K_1 - \Gamma' P_0 \Phi]'(K_0 - K_1)$$
$$\Rightarrow \text{if true then select: } K_1 = (R + \Gamma' P_0 \Gamma)^{-1} \Gamma' P_0 \Phi \tag{7.25}$$

Then by the Lyapunov function (if the CL matrix $\Phi - \Gamma K_1$ is stable): $\partial P > 0$; that is, $P_1 < P_0$ ($X' P_1 X \leq X' P_0 X$), so K_1 is "better," than K_0.

- If K_1 is selected as shown then

$$\Phi - \Gamma K_1 \text{ is stable}$$

- Rewrite equation for P_0 as

$$P_0 = (\Phi - \Gamma K_1)' \partial P (\Phi - \Gamma K_1) + Q + K_1' R K_1 + (H K_0 - \Gamma' P_0 \Phi)' H^{-1} (H K_0 - \Gamma' P_0 \Phi);$$

for

$$H \triangleq R + \Gamma' P_0 \Gamma$$

- Since $Q_{eff} > 0$ and $P_0 > 0 \rightarrow \Phi - \Gamma K_1$ is stable by Lyapunov.
- Continue the process $K_1 \rightarrow P_1 \rightarrow K_2 \rightarrow P_2 \rightarrow \cdots$.
- Each $P_1 < P_{i-1} \Rightarrow \{P_i\}$ converges to P^*:

$$\{K_i\} \text{ converges to } K^*$$

- Each P_i is positive (semi-) definite; no P can be $< P^* \Rightarrow K^*$ is unique.

7.4.4 The Discrete Riccati Equation

The main algorithm to finding optimal gains is illustrated below. Select any K_0 such that

$$\Phi - \Gamma K_0 \text{ is stable, then}$$

$$K^* = \lim_{i \to \infty} K_i = \text{optimal gain} \tag{7.26}$$

where

$$K_{i+1} = (R + \Gamma'P_i\Gamma)^{-1}\Gamma'P_i\Phi; \quad i = 0, 1, \dots \tag{7.27}$$

and P_i is the cost matrix associated with gain K_i:

$$P_i = (\Phi - \Gamma K_i)'P_i(\Phi - \Gamma K_i) + Q + K_i'RK_i \tag{7.28}$$

At convergence, we can show:

$$K^* = (R + \Gamma'P^*\Gamma)^{-1}\Gamma'P^*\Phi \tag{7.29}$$

Thus,

$$J_{\min} = X'(0)P^*X(0) \tag{7.30}$$

and

$$P^* = (\Phi - \Gamma K^*)'P^*(\Phi - \Gamma K^*) + Q + K^{*'} RK^* \tag{7.31}$$

$$P^* = \Phi'[P^* - P^* \Gamma(R + \Gamma'P^* \Gamma)^{-1}\Gamma'P^*]\Phi + Q$$

Equation 7.31 is referred to as the "Discrete Riccati Equation" (DRE).

- Alternate schemes, besides the iterative one, exist for solving the DRE directly.
- P^* = the unique PD solution of the DRE.
- Computing P^* via the iterative algorithm.
- Requires only a subroutine to solve the Lyapunov equation.
  - $K_i \to K^*$ quadratically, $||K_{i+1} - K^*|| < c \, ||K_i - K^*||^2$.
  - Convergence occurs in typically ~ 10 iterations (depends upon how close $|\lambda_j(\Phi - \Gamma K_i)|$ are to 1).
  - If desire N digit accuracy in P^*, need to solve the Lyapunov equation to $N + 1$ digit accuracy.
  - Use stabilization algorithm to obtain K_0.

7.4.5 Comments and Extensions

All previous results are valid for $m > 1$ inputs ($\Gamma = n \times m$)

$$R u^2(k) \to U'(k)RU(k); \quad R = m \times m \text{ matrix} > 0$$

$$K^* = m \times n \text{ optimal FB gain matrix}$$

Cross-weights in cost functional ($M = n \times m$) are

$$J = \sum_{k=0}^{\infty} [X'(k)QX(k) + 2X'(k)MU(k) + U'(k)RU(k)] \tag{7.32}$$

Usually arises when weighting a "generalized" output

$$Y(k) = F\,X(k) + D\,U(k)$$

Optimal control is

$$U(k) = -(R + \Gamma'\tilde{P}^*\Gamma)^{-1}\Gamma'\tilde{P}^*\Phi X(k) - R^{-1}M'X(k) \tag{7.33}$$

where $\tilde{P}^*$ satisfies DRE.

$$\tilde{P}^* = \tilde{\Phi}'[\tilde{P}^* - \tilde{P}^*\Gamma(R + \Gamma'\tilde{P}^*\Gamma)^{-1}\Gamma'\tilde{P}^*]\tilde{\Phi} + \tilde{Q} \tag{7.34}$$

$$\tilde{\Phi} = \Phi - \Gamma R^{-1}M'; \quad \tilde{Q} = Q - MR^{-1}M' \geq 0$$

Translation of the continuous cost function is given by

$$J_c = \int_0^{\infty} [X'(t)Q_1 X(t)] + U'(t)R_1 U(t)dt \tag{7.35}$$

$$= \sum_{k=0}^{\infty} [X'(k)QX(k) + 2X(k)MU(k) + U'(k)RU(k)]$$

$$Q = \int_0^h e^{A'\sigma}Q_1 e^{A\sigma}d\sigma \sim \frac{h}{2}[\Phi'Q_1\Phi + Q_1] \tag{7.36a}$$

$$M = \int_0^h e^{A'\sigma}Q_1 \int_0^h e^{A\xi}Bd\xi d\sigma \sim \frac{h}{2}[\Phi'Q_1\Phi] \tag{7.36b}$$

$$R = hR_1 + \int_0^h \left[\int_0^\sigma e^{A\xi}Bd\xi d\sigma\right]' Q_1 \left[\int_0^\sigma e^{A\xi}Bd\xi\right]d\sigma \sim hR_1 + \frac{h}{2}\Gamma'Q_1\Gamma$$

$$\tag{7.36c}$$

(It is easier to use gain equivalence $K^*|_{\text{continuous}} \to \tilde{K}^*|_{\text{discrete}}$).

7.5 Application of the Optimal Control

- We can show $u(k) = K^* X(k)$ is the optimal control, not just the linear optimal one.
- The CL $X(k + 1) = \Phi X(k) - \Gamma K^* X(k)$ must be stable, selection of weightings.
  - Major design step in method's application
  - Initial design:

$$q_{ii} = \text{relative weighting on state } x_i$$

$$= \frac{1}{\mid x_{i,max} \mid^2}$$

  where $x_{i,max}$ = maximum desired (or anticipated) value of $x_i(k)$. If unconcerned about x_i, deviations from zero, set $q_{ii} = 0$.

- Adjust control weighting R to achieve desired balance between control usage and response speed. Initially, "tune" q_{ii}, R to obtain desired CL time response starting with representative $X(0)$s
- Increase q_{jj} to decrease RMS x_j
  - Decrease R to increase the CL response speed
  - Trade-off errors in $x_j \leftrightarrow x_i$ via q_{jj} versus q_{ii}
- Basically, the approach is time-domain oriented, but
  - Examine CL pole locations, Φ_m, ω_c, etc.
- Other techniques and rules exist for picking weights

EXAMPLE 7.4: SATELLITE (DOUBLE INTEGRAL) SYSTEM, $h = 1$

State model:

$$X(k + 1) = \begin{bmatrix} 1 & 1 \\ 0 & 1 \end{bmatrix} X(k) + \begin{bmatrix} 0.5 \\ 1.0 \end{bmatrix} u(k);$$

Measurement model:

$$y(k) = x_1(k);$$

Initial condition:

$$X(0) = \begin{bmatrix} 1 \\ 0 \end{bmatrix}$$

$q_{11} = 1$, $q_{22} = 0$ (interested in output $\rightarrow 0$), R = design parameter. The results for $R = 1, 0.1$, and 0.01 are illustrated in Figure 7.9.
  $\rightarrow$ Examine CL pole locations as a function of Q, R.

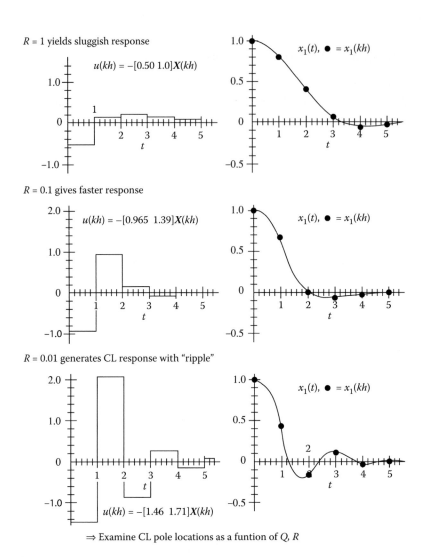

FIGURE 7.9
Satellite system at $h = 1$ results, showing effect due to different values of R.

7.5.1 Properties of the Optimal CL System-1

1. CL pole locations:
   - CL poles are the n roots inside the unit circle of

$$\det\left[R\Gamma'(z^{-1}I - \Phi')^{-1}Q(zI - \Phi)^{-1}\Gamma \right] = 0 \qquad (7.37)$$

   - In a single input case, if $Q = C'C$ (output weighting only), CL poles satisfy

$$R + \tilde{G}(z^{-1})\tilde{G}(z) = 0 \qquad (7.38)$$

   $\rightarrow$ CL poles of $\Phi - \Gamma K^*$ are not arbitrary.

EXAMPLE 7.5

Satellite system, $\tilde{G}(z) = (1/2)((z + 1)/(z - 1)^2)$, has output weighting only.

The RL of CL poles is shown in Figure 7.10.

As $R \to 0$, CL poles follow a locus of constant damping $\zeta = 0.707$, until $R = R_0 = 0.025$. Then, for $R < R_0$ have two real roots on $(-1, 0)$!

$\Rightarrow$ Too small a value of R will give oscillatory CL responses.

- General property as $R \to 0$: (single input with $Q = C'C$)
  - Assume $\tilde{G}(z)$ has r zeros $\delta_1, \delta_2, \ldots, \delta_r$.
  - As $R \to 0$, r CL poles $\to r$ zeros of $\tilde{G}(z)\tilde{G}(z - 1)$ inside or on the unit circle. The remaining $n-r$ poles $\to z = 0$ (in ex., $r = 1$, $\delta_1 = -1$); that is, if δ_i is a zero of $\tilde{G}(z)$, a CL pole $\to \delta_i$ or $1/\delta_i$ (whichever has magnitude < 1) as $R \to 0$.

7.5.2 Properties of the Optimal CL System-2

Return difference and phase margin:

- Loop gain properties, LG $(z) = K^*(zI - \Phi)^{-1}\Gamma$
- Via algebraic manipulations on the Riccati equation:

$$[I + LG(z^{-1})]'(R + \Gamma'P^*\Gamma)[I + LG(z)] = R + \Gamma'(z^{-1}I - \Phi')^{-1}Q(zI - \Phi)^{-1}\Gamma \quad (7.39)$$

- In a single input case, factor $Q = S'S$

$$\tilde{G}_{eff} \triangleq S(zI - \Phi)^{-1}\Gamma$$

$$[\tilde{G}_{eff} = \tilde{G}(z) \text{ if } Q = C'C]$$

$$|1 + LG(z)|^2 = \frac{R + \tilde{G}'_{eff}(z^{-1})\tilde{G}_{eff}(z)}{R + \Gamma'P^*\Gamma}$$

$$\geq \frac{R}{R + \Gamma'P^*\Gamma} = \rho^2 \ (\rho < 1)$$

- Phase margin, Φ_m, properties are shown in Figure 7.11

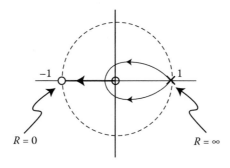

FIGURE 7.10
RL of CL poles for a satellite system.

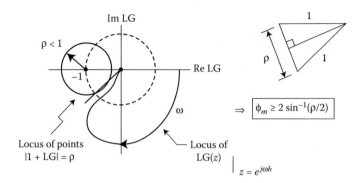

FIGURE 7.11
Phase-margin properties of an optimal CL system.

7.6 Examples and Applications

EXAMPLE 7.6: INVERTED PENDULUM ON A CART, $h = 0.18\ s$

$$X = \begin{bmatrix} \theta, & \dot{\theta}, & d, & \dot{d} \end{bmatrix}$$

Initial design:

$$\left.\begin{array}{r} \theta_{max} \\ d_{max} \end{array}\right\} \begin{array}{l} \approx 0.5\ \text{rad} \Rightarrow Q = \text{diag}\ [4\ 0\ 1\ 0] \\ \approx 1\ \text{m} \end{array}$$

Select $R = 1 \Rightarrow K^* = [-22.9\ -6.98\ -0.487\ -1.08]$:

$$\text{CL poles} = \begin{cases} 0.55 \pm j0.03\ (\zeta \approx 1,\ \omega_n = 3.3) \\ 0.88 \pm j0.11\ (\zeta \approx 0.7,\ \omega_n = 0.95) \end{cases}$$
$$\rho = 0.487 \Rightarrow \Phi_m \geq 28.2°$$

The cart position and the pendulum angle are shown in Figure 7.12.
Reduce weighting on u to speed response $\theta(t) \rightarrow 0$ (will require more control input).

Inverted Pendulum II
Second design iteration, $R = 0.1$:

$$K^* = [-28.4\ -8.67\ -1.39\ -2.20]$$

$$\text{CL poles} = \begin{cases} 0.53 \pm j0.09\ (\zeta \approx 0.96,\ \omega_n = 3.5) \\ 0.80 \pm j0.16\ (\zeta \approx 0.73,\ \omega_n = 1.59) \end{cases}$$
$$\rho = 0.439 \Rightarrow \Phi_m \geq 25.4°$$

The cart position and the pendulum angle are shown in Figure 7.13 for the second iteration.

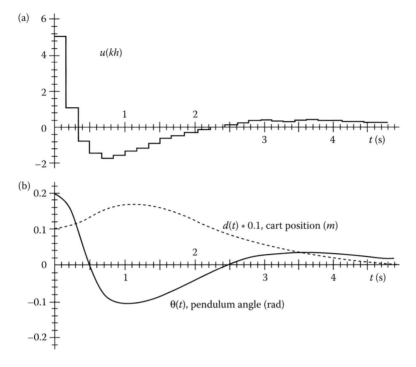

FIGURE 7.12
(a) Input to the inverted pendulum and the cart position and pendulum angle for $h = 0.18$ s for the digital optimal controller. (b) Cart position and pendulum angle for $h = 0.18$ s.

- Further possibilities:
  - Further decrease R (e.g., $R = .01$ yields ~ 3 s setting time with 1 1/2 to 2 times the amount of control)
  - Modify $\theta{:}d = 0.5{:}1$ ratio (minor effect)
  - As $R \to 0$: $\Phi_m \downarrow$, $|u(kh)| \uparrow$, $t_s \downarrow$ and $\theta(t)$ overshoot $\uparrow$

Inverted Pendulum, Phase Margin Analysis

$R = 0.1$

$$LG(z) = K^*(zI - \Phi)^{-1}\Gamma = \frac{1.708z^3 - 4.084z^2 + 3.197z - 0.8066}{z^4 - 4.367z^3 + 6.734z^2 - 4.367z + 1.0}$$

The loop gain Bode plot is shown in Figure 7.14.

- Formula $\Phi_m \geq 2 \sin^{-1}(\rho/2)$ is reasonably tight (25.4 vs. 25.6!)
- As $h \to 0$, $\rho \to 1$ and $\Phi_m \geq 60°$ for optimal continuous design
- Generally as gains increase Φ_m decreases
  - For single input systems, optimal control design $\equiv$ pole placement design with the same poles
- Note that pole placement can achieve CL pole locations where an optimal design will not/cannot achieve successfully

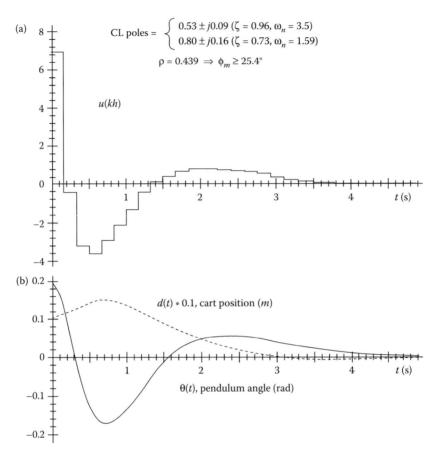

FIGURE 7.13
Input to the inverted pendulum and the cart position and pendulum angle for $h = 0.18$ s for the digital optimal controller for the second iteration.

7.6.1 Examples with FEEDBACK≪® Hardware and Software Package

An introduction to the FEEDBACK≪® system as used in my course offerings in the Electrical and Computer Engineering Department, College of Engineering, Technology and Architecture (CETA) at the University of Hartford, Connecticut is illustrated in Appendix II in the sequel. Here, we will illustrate two specific examples with some aspects of laboratory experiments for Digital Pendulum (33-936S) and Coupled Tanks System (33-041-IC). A MATLAB Guide which is used in the Feedback Control Instrumentation is also illustrated. This material and other information are courtesy of FEEDBACK≪® Instruments Ltd. Park Road, Crowborough, East Sussex YN6 2QR, the United Kingdom. For further information and use of these control systems, please visit FEEDBACK≪® at www.feedabck-instruments.com. We have been offering courses in Automatic Control and Digital Control for the past two decades both at the undergraduate and graduate level under my leadership as the Director of Graduate Studies for the College of Engineering since 1992. Over the past few years, we have introduced the hands-on aspect of controls with a dedicated system purchased from FEEDBACK≪®. For more information on course offerings, please visit the website of the University of Hartford at www.hartford.edu/ceta/: http://catalog.hartford.edu/preview_program.php?catoid=11&poid=2734&returnto=830.

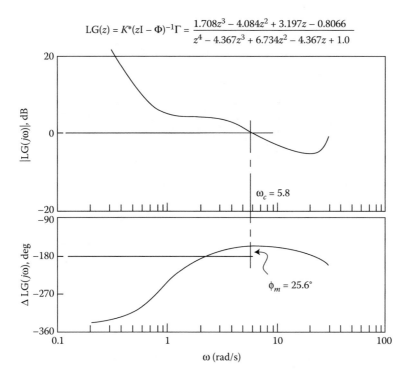

$$LG(z) = K^*(zI - \Phi)^{-1}\Gamma = \frac{1.708z^3 - 4.084z^2 + 3.197z - 0.8066}{z^4 - 4.367z^3 + 6.734z^2 - 4.367z + 1.0}$$

FIGURE 7.14
LG Bode plot for an inverted pendulum with PM analysis.

The PC with the HW controller is shown in Figure 7.15. For the inverted pendulum, the system is illustrated in Figure 7.16.

The inverted pendulum open-loop system dynamics is described by the following equations for small-angle deviations that are

$$\sin \theta = \theta$$

$$\cos = 1$$

$$\dot{\theta}^2 = 0$$

The motion dynamics are given by (page 10 of 33-936S FEEDBACK≪® document). Using the earlier modifications, we have the following linearized equations of motion:

$$(m + M)\ddot{x} + b\dot{x} + ml\ddot{\theta} = F$$

$$(I + ml^2)\ddot{\theta} - mgl\theta + ml\ddot{x} + d\dot{\theta} = 0$$

The related OL transfer function is given by

$$G(s) = \frac{\theta(s)}{F(s)} = \frac{s}{d_3s^3 + d_2s^2 + d_1s + d_0}$$

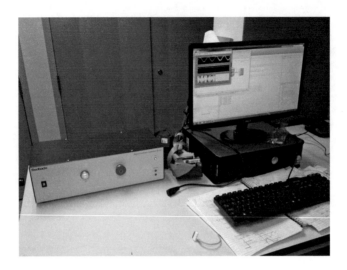

FIGURE 7.15
(**See color insert.**) PC and HW controller. (Courtesy of FEEDBACK≪®.)

The results of the inverted pendulum real-time response in OL is shown in Figure 7.17. The results of the inverted pendulum in real time in CL is shown in Figure 7.18.

7.6.2 Summary of Optimal Control Design Method

- Basically, a "smart" pole placement SVFB design
  - SVFB does not modify system zeros
- Based on minimizing a quadratic criterion
  - Function of state and control deviations

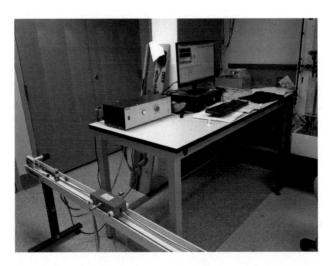

FIGURE 7.16
(**See color insert.**) Inverted pendulum system. (Courtesy of FEEDBACK≪®.)

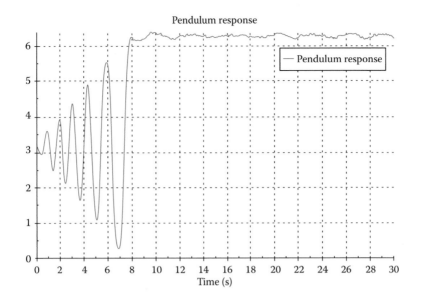

FIGURE 7.17
OL real-time response for an inverted pendulum output. (Courtesy of FEEDBACK≪®.)

Advantages:

- Straightforward design methodology
- Design parameters (Q, R) relate to CL response
- Directly applicable to MIMO systems
- Small number of design parameters

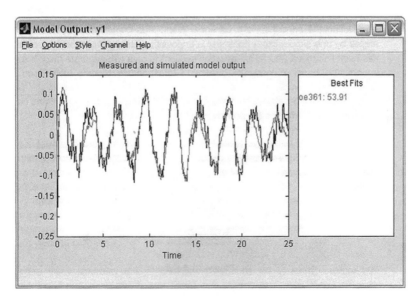

FIGURE 7.18
CL real-time response for an inverted pendulum.

- Has a guaranteed lower bound on Φ_m
- CL system is always stable
- Numerous extensions can/have been done. For example, integral FB via a small weight on

$$|\text{Return difference}| = |1 + K^*(zI - \Phi')^{-1}\Gamma| \; |_{z=e^{i\omega h}} > \rho \tag{7.40}$$

$$x_I^2(k) = \left[\sum_{i=1}^{k} x_m(i)\right]^2 \tag{7.41}$$

as compared to, command following via weighting $[CX(k) - r(k)]^2$ as shown in Equation 7.41.

Disadvantages:

- Requires fairly extensive software to do design
- Do not have direct control over CL pole locations (some choices of Q, R can give poles on $z < 0$)
- Weighting the selection process is largely trial and error
- The quadratic criterion is not always best
- Need to measure or estimate all states

7.7 Rate Weighting

7.7.1 Weighting of Control Rate

- Usual optimal FB control has high bandwidth
  - Can give problems if actuators are rate limited
  - Often unnecessary if system dynamics are "slow"
- Weight $\Delta(k) = [u(k) - u(k-1)]/h$ in cost functional

$$J = \sum_{k=0}^{\infty} \left[X'(k)QX(k) + Ru^2(k-1) + G\Delta^2(k)\right] \tag{7.42}$$

- Develop augmented system dynamics, $x_{n+1}(k) = u(k-1)$

$$u(k) = u(k-1) + h\,\Delta(k)$$

$$X(k+1) = \Phi X(k) + \Gamma u(k-1) + h\Gamma\Delta(k) \tag{7.43a}$$

$$x_{n+1}(k+1) = x_{n+1}(k) + h\Delta(k) \tag{7.43b}$$

Let

$$\chi(k) = [X(k), u(k-1)]'$$

$$\chi(k+1) = \begin{bmatrix} \Phi & \Gamma \\ 0 & 1 \end{bmatrix} \chi(k) + h \begin{bmatrix} \Gamma \\ 1 \end{bmatrix} \Delta(k); \quad \chi(0) = \begin{bmatrix} \chi(0) \\ 0 \end{bmatrix}$$

$$\Phi_a = \begin{bmatrix} \Phi & \Gamma \\ 0 & 1 \end{bmatrix}; \quad \Gamma_a = \begin{bmatrix} \Gamma \\ 1 \end{bmatrix}$$

$$J = \sum_{k=0}^{\infty} \left[\chi'(k) Q_a \chi(k) + G\Delta^2(k) \right] \tag{7.44}$$

$$Q_a = \text{diag} \begin{bmatrix} Q & R \end{bmatrix}$$

- Solve "augmented" optimal control problem:

$$\chi(k) <=> X(k), \Delta(k) <=> u(k)$$

- The augmented system is controllable with respect to Δ, if the original system was controllable with respect to u:

$$\Delta(k) = -K_a\chi(k) = -K_x X(k) - K_u u(k-1) \tag{7.45}$$

- Alternate structure:

$$u(k) = (1 - hK_u)u(k-1) - hK_x X(k) \tag{7.46}$$

7.7.2 Properties of a Rate-Weighted Controller

$$u(k) = (1 - hK_u)u(k-1) - hK_x X(k)$$

- Analagous to FB $v = -1/a\,(K_x X)$ put through a first-order filter $(a)/(s+a)$, with $a \sim K_u$
- As $G \to 0$, $K_u \to 1/h$, $K_x \to K^*/h$ and the original SVFB control is recovered
- Highly recommended for all physical systems
  - Adds robustness to design
  - Generally gives slightly smaller ω_c.
  - Provides ability to manage CL bandwidth
  - Effects trade-off between $\dot{u}$ and u, X

EXAMPLE 7.7

Example for inverted pendulum on a cart,

$$X = \begin{bmatrix} \theta & \dot{\theta} & d & \dot{d} \end{bmatrix}$$

add a rate weighting to previous design

$$Q = \text{diag}\begin{bmatrix} 4 & 0 & 1 & 0 \end{bmatrix}, R = 0.1$$

$$G = 0.0081 = \left| \frac{h}{\Delta u_{max}} \right|^2 \text{ with } \Delta u_{max} = 2$$

FB control with rate weighting:

$$u(k) = (1 - hK_u)u(k - 1) - hK_x X(k)$$

$K_u = 4.97; K_x = [-109.5\ -33.4\ -3.61\ -6.29\ ];$
$K_x/K_u = \{-22.0\ -6.72\ -0.726\ -1.27\}.$

By analogy recall K^* for $G = 0$; $K^* = [-28.4\ -8.67\ -1.39\ -2.20]$. Gains decrease to compensate for added filtering.

Simulation Results—Rate Weighting Controller

Simulation results are shown in Figure 7.19.

$$X(0) = [0.2\ 0\ 1\ 0]$$

- Decrease in magnitude of Δu_{max} (from 7.4 to ≈ 4.5).
- Accompanied by slower system response, but with ~ same overshoots.
- Similar to a time scaling effect.
- Increasing G to 0.09 shows a further response.

$$t_s \sim 6\ s, \Delta u_{max} \sim 3.2$$

- We soon reach a point of diminished return.

Inverted Pendulum Stabilization: There are two prominent techniques for stabilization of an inverted pendulum.

1. LQ controller
2. Design two separate PID controllers and sum their outputs to produce the final signal

The details of these schemes in real-time are illustrated in 33-936S of FEEDBACK≪® document on pages 37 through 45, with some graphical illustrations in Figures 7.20 and 7.21.

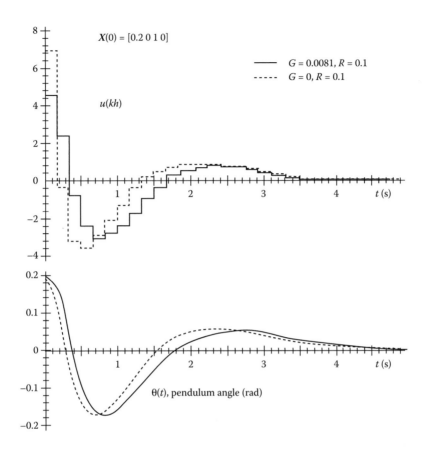

FIGURE 7.19
Rate weighting controller simulation results with an initial state vector showing pendulum angle.

7.7.3 Compensation for Fractional Time Delay

$$T = Mh + \varepsilon; \quad M = 0$$

- Recall the model for <1 step (computational delay):

$$\chi(k) \triangleq \begin{bmatrix} X(k) & u(k-1) \end{bmatrix}' = \text{augmeneted state}$$

$$\chi(k+1) = \begin{bmatrix} \Phi & \Gamma_1 \\ 0 & 1 \end{bmatrix} \chi(k) + h \begin{bmatrix} \Gamma_0 \\ 1 \end{bmatrix} u(k) \tag{7.47}$$

- Can apply optimal control design directly to the augmented model when $G = 0$;
  $Q_a = \text{diag } [Q, 0]$ [Gives same results as $u(kh) = -K^* \hat{X}(kh + \varepsilon)$.]
- Alternate time delay model

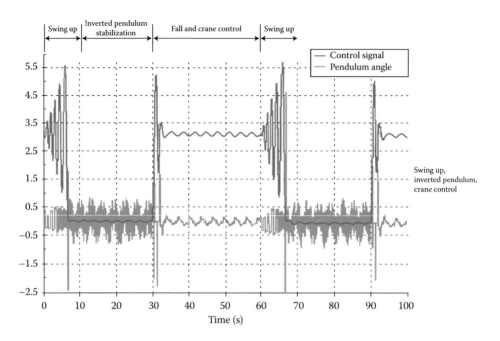

FIGURE 7.20
Control signal and pendulum angle.

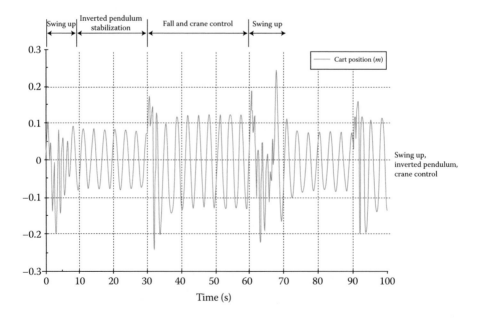

FIGURE 7.21
Cart position.

- Replace $u(k) => u(k-1) + h\,\Delta(k)$; note: $\Gamma_0 + \Gamma_1 = \Gamma$:

$$\chi(k+1) = \begin{bmatrix} \Phi & \Gamma \\ 0 & 1 \end{bmatrix}\chi(k) + h\begin{bmatrix} \Gamma_0 \\ 1 \end{bmatrix}\Delta(k) \qquad (7.48)$$

- In a desired form for weighting $\Delta(k)$
- Identical to augmented model Equation 7.43 but with a modified Γ_a (when $\varepsilon = h^-$, $\Gamma_0 = 0$)

$$\Delta(k) = -K_u u(k-1) - K_x X(k)$$

=> Natural fit between fractional delay model and weighting of control rate. Excellent for $\varepsilon < h$ that is, compensation of up to one time-step delay.
- For $M \geq 1$ apply state prediction ideas:

$$\Delta(k) = -K_u u(k-1) - K_x \hat{X}(k+M)$$

- $X(k+M)$: prediction of X at step $k+M$, obtained by propagating:

$$X(k+1) = \Phi X(k) + \Gamma_1 u(k-1-M) + \Gamma_0\, u(k-M)$$

PROBLEMS

P7.1 An inverted pendulum is described in Figure P6.2.

With state equations given by

$$\dot{X} = \begin{bmatrix} 0 & 1 & 0 & 0 \\ 11 & 0 & 0 & 0 \\ 0 & 0 & 0 & 1 \\ -1 & 0 & 0 & 0 \end{bmatrix} X + \begin{bmatrix} 0 \\ -1 \\ 0 \\ 1 \end{bmatrix} u$$

The state vector is defined as

X_1 = cart position
X_2 = cart velocity
X_3 = pendulum angle
X_4 = pendulum angular velocity

Let an LQ controller control signal be determined by

$$u = -(K_1 e_1 + K_2 e_2 + K_3 e_3 + K_4 e_4)$$

where
e_1 = cart position error
e_2 = cart velocity error

e_3 = pendulum angle error
e_4 = pendulum angular velocity error

for $h = 0.18$.

1. Find system parameters like ω_c and Φ_m.
2. Discrete time system has a delay of $h/2$ s. Find the reduction in Φ_m.
3. Use FEEDBACK≪® system 33-936S (Digital Control Experiment) to illustrate this inverted pendulum performance, stabilization and DRE.

P7.2 A damped system is shown below:

$$\dot{X} = \begin{bmatrix} 0 & 1 \\ -10 & -0.5 \end{bmatrix} X + \begin{bmatrix} 0 \\ 10 \end{bmatrix} u$$

$$y = \begin{bmatrix} 1 & 0 \end{bmatrix} X$$

1. Use the Lyapunov digital design method with $h = 0.2$ to obtain a controller.
2. Find the response with unconstrained u.
3. Repeat the response with a constrained $|u(k)| < 0.28$.
4. Repeat the response with a constrained $|u(k)| < 0.13$.

P7.3 A satellite system is defined as

$$X(k+1) = \begin{bmatrix} 1 & 1 \\ 0 & 1 \end{bmatrix} X(k) + \begin{bmatrix} 0.5 \\ 1.0 \end{bmatrix} u(k)$$

$$y(k) = x_1(k)$$

$$X(0) = \begin{bmatrix} 1 \\ 0 \end{bmatrix}; \text{ with } q_{11} = 1, q_{22} = 0$$

Find the output with different values of $R = 1, 0.5,$ and 0.02.

P7.4 In Problem 7.1, choose:

X_{1max} = cart position maximum = $1\ m$
X_{3max} = pendulum angle max = 0.5 rad

Resulting in $Q = [1\ 0\ 4\ 0]$; and for $R = 1$ $K^* = [-0.487\ -1.08\ -22.9\ -6.98]$
With $h = 0.3$, conduct the optimal control design, then:

1. Find ρ and Φ_m.
2. Find the cart position response and pendulum angle response.
3. Conduct the phase margin analysis for the $|LG(j\omega)|$.
4. Change $R = 0.1$ and repeat parts 1 through 3.

P7.5 For Problem 7.1, use suitable weighting of control rate and augmented optimal control with

$$X(0) = [1\ 0\ 0.2\ 0]$$

1. With G and R in Equation 7.42 as 0.00081 and 0.1 respectively, plot the response of the pendulum angle.
2. With G and R in Equation 7.42 as 0 and 0.1, respectively, plot the response of the pendulum angle.
3. If G is reduced to 0.09, find the effect on the response of the pendulum angle. Do you reach a point of diminished return?

8

Estimation of System State

8.1 State Estimation

- State $X(k)$ is often not measurable directly:
  - Measure $y(k) = CX(k)$ as a linear combination of states.
  - We will consider single-output case here.
- Assume measurements made with no noise/error for now. In general, this is not true in practical situations.
- Objective:
  - Develop an estimate $\hat{X}(k)$ of the state $X(k)$ for use in State Variable Feedback (SVFB) or for other purposes like for an observer.
  - Use available information about system input and output:

$$\{u(j), j < k\}, \{y(j), j \leq k\}$$

  - We need to generate state estimate online. The block diagram of this process is shown in Figure 8.1.
  - Will using $\hat{X}(k)$ as a substitute for $X(k)$ work? And to what extent of accuracy?
- Generally, design issues involve the following:
  - Desired properties of state estimator.
  - Expect/force a linear estimator. (The system is linear, so why not the estimator?)
  - How fast must $\hat{X}(k) \rightarrow X(k)$?; This is a practical computational issue.
- State estimate is useful even in noncontrol applications (e.g., decision making using state (X)).

8.1.1 "Observation" of System State

- What can be done to estimate $X(k)$?
  - Consider $u(k) = 0$: given

$$\text{State Model: } X(k + 1) = \Phi X(k)$$

$$\text{Measurement Model: } y(k) = CX(k)$$

$$X(0) = \text{unknown initial condition}$$

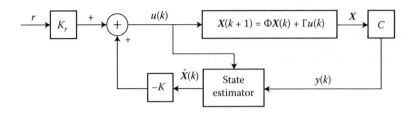

FIGURE 8.1
Generation of state estimate online.

- Estimate $X(0) = [x_1(0), \ldots, x_n(0)]'$ from the output measurements $\{y(0), y(1), \ldots, y(n-1)\}$, thus as seen below:

$$y(0) = CX(0)$$

$$y(1) = CX(1) = C\Phi X(0)$$

$$y(2) = CX(2) = C\Phi^2 X(0)$$

$$\vdots$$

$$\vdots$$

$$\begin{bmatrix} y(0) \\ y(1) \\ \vdots \\ y(n-1) \end{bmatrix} = \begin{bmatrix} --C-- \\ --C\Phi-- \\ \vdots \\ -C\Phi^{n-1}- \end{bmatrix} \begin{bmatrix} x_1(0) \\ x_2(0) \\ \vdots \\ x_n(0) \end{bmatrix} \qquad (8.1)$$

where we define:

$$H_0' = \begin{bmatrix} --C-- \\ --C\Phi-- \\ \vdots \\ -C\Phi^{n-1}- \end{bmatrix}$$

and

$$X(0) = \begin{bmatrix} x_1(0) \\ x_2(0) \\ \vdots \\ x_n(0) \end{bmatrix}$$

- If H_0' is invertible, it is possible to find $X(0)$.
- Obtain $X(0)$ after n measurements at step $k = n - 1$.

- Once $X(0)$ is obtained, $X(k) = \Phi^k X(0)$ for $k > 0$.
  - Eventually, we would like to obtain state estimates recursively.

8.1.2 System Observability Requirement

- A discrete system is completely observable (CO), if

$$\det \begin{bmatrix} | & | & & | \\ C' & \Phi'C' & \cdots & (\Phi')^{n-1}C' \\ | & | & & | \end{bmatrix} = \det(H_0) \neq 0 \tag{8.2}$$

- Physical interpretation:
  - All modes show up in the output, either directly or indirectly.
- Observability is a property of only $[\Phi, C]$.
- Controllability–observability duality is thus seen as

$$\{CC \to CO\} \to \Phi \to \Phi', \Gamma \to C'$$

- Continuous–discrete relationship:
  If the original continuous system is observable,

$$\det \begin{bmatrix} | & | & & | \\ C' & A'C' & \cdots & (A')^{n-1}C' \\ | & | & & | \end{bmatrix} \neq 0$$

  then the equivalent discrete system is also observable.

$$h \neq M(2\pi/\omega_{c0}), M = \text{integer}$$

  where
  ω_{c0} = imaginary part of any eigenvalue of A that is on $j\omega$-axis.

- Observability will be a necessary condition for state estimation.
  - $\text{Det}(H_0)$ and or $\text{Det}(H_0'H_0)$ are often used as a "measure of observability."

8.1.2.1 A Question of Notation

- We must consider state estimate in two parts:
  - The estimate will undergo a discontinuity at measurement point k.
  - We need to distinguish between the estimate $\hat{X}^-(k)$ prior to making the measurement of $y(k)$ and the estimate $\hat{X}^+(k)$ after making the measurement $y(k)$.

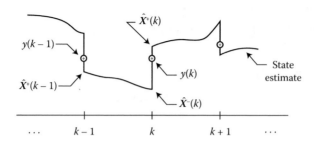

FIGURE 8.2
State estimation process.

This is illustrated in Figure 8.2, which ultimately evaluates the state estimate $X(k + 1)$ at time $(k + 1)$.

We now define:

- $\hat{X}(k \mid k - 1)$ = estimate of $X(k)$ prior to obtaining the measurement $y(k)$ at time $t = kh$
  = estimate of $X(k)$ from $\{y(k - 1), y(k - 2), \dots\} = \hat{X}^{-}(k)$.

- $\hat{X}(k \mid k)$ = estimate of $X(k)$ after obtaining and processing the measurement $y(k)$ at $t = kh$ = estimate of $x(k)$ from $\{y(k), y(k - 1), \dots\} = \hat{X}^{+}(k)$.

  - From the above, obviously $\hat{X}(k \mid k)$ is the better estimate of $X(k)$.
  - Thus, we desire a recursive estimation scheme as shown in Figure 8.3.

8.1.2.2 Structure of the Estimator

- "Prediction" is defined as an estimate of $\hat{X}(k \mid k - 1)$ from $\hat{X}(k - 1 \mid k - 1)$
  - Since no measurements are made over $(k - 1, k)$, the only way to estimate $\hat{X}(k \mid k - 1)$ is via the state equation:

$$\text{State Model: } X(k) = \Phi X(k - 1) + \Gamma u(k - 1); \tag{8.3}$$

  where $u(k - 1)$ is known input over $(k - 1, k)$

$$\Rightarrow \hat{X}(k \mid k - 1) = \Phi\hat{X}(k - 1 \mid k - 1) + \Gamma u(k - 1) \tag{8.4}$$

- Alternate notation $\hat{X}^{-}(k \mid k - 1) = \Phi\hat{X}^{+}(k - 1) + \Gamma u(k - 1)$

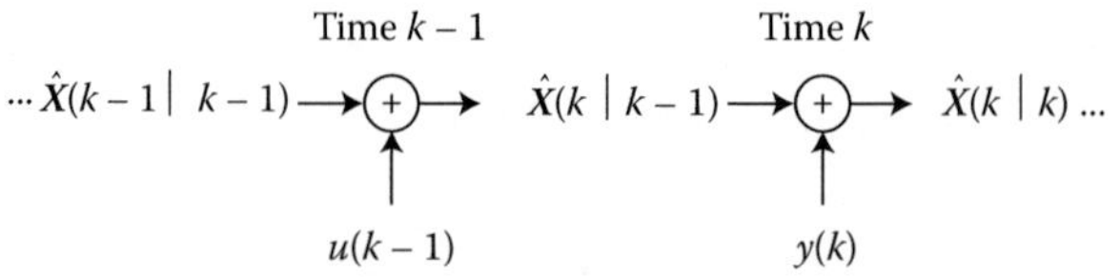

$$\cdots \hat{X}(k - 1 \mid k - 1) \longrightarrow \boxed{+} \longrightarrow \quad \hat{X}(k \mid k - 1) \longrightarrow \boxed{+} \longrightarrow \quad \hat{X}(k \mid k) \cdots$$

Time $k - 1$: $u(k - 1)$

Time k : $y(k)$

FIGURE 8.3
Recursive state estimation scheme.

- "Update" is defined as an estimate, $\hat{X}(k \mid k)$ from $\hat{X}(k \mid k - 1)$.
  - How to include the measurement $y(k)$. This is illustrated by the following:

$$\left.\begin{array}{l}\hat{X}(k \mid k - 1)\end{array}\right\} \longrightarrow \hat{X}(k \mid k)$$

$$y(k) \quad \text{ALGORITHM}$$

Thus,

$$\Rightarrow \hat{X}(k \mid k) = \hat{X}(k \mid k - 1) + L[Y(K) - C\hat{X}(k \mid k - 1)]$$

where

$$\hat{y}(k \mid k - 1) = C\hat{X}(k \mid k - 1) \tag{8.5}$$

where the estimate of $y(k \mid k - 1)$ is given by

$$\hat{y}(k \mid k - 1) \triangleq C\hat{X}(k \mid k - 1)$$

= the best prediction of what the measurement at step k should be given by

$$v(k) \triangleq y(k) - \hat{y}(k \mid k - 1)$$

= difference between what is actually measured at step k and what we expect to measure (innovation), given prediction of measurements up to $(k - 1)$.

Then, $L = n \times 1$ arbitrary gain vector, to be determined from the above information. Thus, we can have an alternate notation as

$$\hat{X}^{+}(k) = \hat{X}^{-}(k) + L[y(k) - C\hat{X}^{-}(k)]$$

Equations 8.4 and 8.5 are called a dynamic "observer."

- Requires a model of system $[\Phi, \Gamma, C]$.

8.1.2.3 The Estimation Error

- Observer "starts" at $k = 0$ with $\hat{X}(O \mid -1)$
  - $\hat{X}(O \mid -1) \triangleq$ is the estimate of initial state $X(0)$ based on all prior information
  - Usually, $\hat{X}(O \mid -1) = 0$
- Now obtain the evolution of the prediction error, which is given by

$$\tilde{e}(k \mid k - 1) \triangleq X(k) - \hat{X}(k \mid k - 1)$$

Thus,

$$\hat{X}(k+1\,|\,k) = \Phi\hat{X}(k\,|\,k)\Gamma u(k)$$

where

$$\hat{X}(k\,|\,k) = \hat{X}(k\,|\,k-1) + L[y(k) - C\hat{X}(k\,|\,k-1)]$$

$$\Rightarrow \hat{X}(k+1\,|\,k) = \Phi\hat{X}(k\,|\,k-1)\Gamma u(k) + \Phi L[y(k) - C\hat{X}(k\,|\,k-1)] \qquad (8.6)$$

Now subtract Equation 6.6 from the system equation $X(k+1) = \Phi X(k) + \Gamma u(k)$, yielding the estimate. Thus the prediction error is now at time "$k+1$" given at time "k" as

$$\tilde{e}(k+1\,|\,k) = X(k+1) - X(k+1\,|\,k)$$

$$\tilde{e}(k+1\,|\,k) = \Phi\tilde{e}(k\,|\,k-1)\Gamma u(k) - \Phi L[y(k) - C\hat{X}(k\,|\,k-1)]$$

where

$$C\tilde{e}(k\,|\,k-1) = [y(k) - C\hat{X}(k\,|\,k-1)]$$

$$\Rightarrow \tilde{e}(k+1\,|\,k) = (\Phi - \Phi LC)^k \tilde{e}(k\,|\,k-1) \qquad (8.7)$$

- Initial condition:

$$\tilde{e}(0\,|\,-1) = X(0) - X(0\,|\,-1) = X(0), \left[\text{If } \hat{X}(0\,|\,-1) = 0\right]$$

then

$$\tilde{e}(k\,|\,k-1) = (\Phi - \Phi LC)^k X(0) \qquad (8.8)$$

- Selection of observer gain L
  - Want $\tilde{e} \to 0$ rapidly
  - Rate at which $\tilde{e} \to 0$ depends on eigenvalues of $[\Phi - \Phi LC]$
  - Choose L so that eigenvalues of $\Phi - \Phi LC$ are within the unit circle
  - Since $\Phi - \Phi LC = \Phi(\Phi - \Phi LC)\Phi^{-1}$, eigenvalues of $\Phi - \Phi LC \equiv$ eigenvalues of $\Phi - LC\Phi$

Thus, the updated error is $\tilde{e}(k\,|\,k) \triangleq X(k) - \hat{X}(k\,|\,k) = (\Phi - LC\Phi)\tilde{e}(k-1\,|\,k-1)$.

8.1.3 Observer Pole Placement Problem

- Select L so that eigenvalues of $\Phi - LC\Phi$ are at preselected locations within the unit circle, such that:

$$\tilde{\lambda}_1, \tilde{\lambda}_2, \ldots, \tilde{\lambda}_n \to p_e(z) = z^n + \tilde{d}_1 z^{n-1} + \cdots + \tilde{d}_n$$

= estimator desired characteristic polynomial

$$= |zI - (\Phi - LC\Phi)| \tag{8.9}$$

- Reformulate as a "control" problem
    - Select L' so that the eigenvalues of $\Phi' - [\Phi'C']L'$ are at the desired locations.
    - Like pole placement for $\Phi\text{-}\Gamma K$ with associations.

$$\Phi \Leftrightarrow \Phi', \Gamma \Leftrightarrow \Phi'C', K \Leftrightarrow L'$$

- The Ackermann formula [8] can be used to show

$$L' = \begin{bmatrix} 0 & 0 & \cdots & 1 \end{bmatrix} \begin{bmatrix} | & & | & & | \\ \Phi'C' & (\Phi')^2C' & \cdots & (\Phi)^nC' \\ | & & | & & | \end{bmatrix}^{-1} p_e(\Phi') \tag{8.10}$$

- Inverse exists if system is CO.
- Take the transpose of Equation 8.10.

$$L = p_e(\Phi) \begin{bmatrix} - & C\Phi & - \\ - & C\Phi^2 & - \\ & \vdots & \\ - & C\Phi^n & - \end{bmatrix}^{-1} \begin{bmatrix} 0 \\ 0 \\ \vdots \\ 1 \end{bmatrix} = p_e(\Phi)(H_0'\Phi)^{-1} \begin{bmatrix} 0 \\ 0 \\ \vdots \\ 1 \end{bmatrix} \tag{8.11}$$

$$p_e(\Phi) = \Phi^n + \tilde{d}_1\Phi^{n-1} + \cdots + \tilde{d}_nI$$

- Computational approach:
    - Use previous subroutine 'GAINS' with $\Phi \to \Phi'$, $\Gamma \to \Phi'C'$; obtain $K \to L'$
    - Obtain q by solving $(H_0'\Phi)q = [0 \ 0 \ \ldots \ 1]'$

8.1.4 Selection of Observer CL Poles

- Observer *CL* pole placement depends on what you need to do with the estimate
    - Will it be used for SVFB or not? This needs to be determined.

1. Generally one is interested in a good state estimate, $\hat{X} \to X$
    - No tie-ins or constraints imposed by SVFB
    - Place poles within the unit circle depending on how fast we desire $\hat{X} \to X$

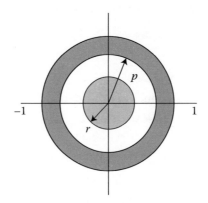

FIGURE 8.4
Location of observer poles within the magnitude of primary control poles "p."

- Example: if $|\tilde{\lambda}_i| \leq r < 1$ then error $\to 0$ as r^k (if $r = 0.5$, error decreases by 50% each step, with ~12% error after three steps)

2. Anticipating using $\hat{X}$ for X in SVFB control:
   - What matters is how fast $\tilde{e} \to 0$ compared to how fast $X(k) \to 0$
   - Desire $\tilde{e} \to 0$ faster by approximately two to three times
   - Example: if primary poles of $\Phi - \Gamma K$ satisfy $p \leq |\lambda_i|$ as shown in Figure 8.4, then place observer poles $\tilde{\lambda}_i$ inside the circle of radius $r = p^2$ to $r = p^3$ (p = magnitude of primary control poles)
   - Best to "uniformly" $\tilde{\lambda}_i$ on semi-circle of radius r, with $\text{Re}\{\tilde{\lambda}_i\} > 0$; as shown in Figure 8.4
   - Deadbeat observer is a special case when $r = 0 \Rightarrow$ all observer poles at $z = 0$
   - Any initial error $\tilde{e}(0 \mid -1) \to 0$ in n steps
     - obtain perfect estimate after n measurements $y(0), y(1), \ldots, y(n-1)$
     - $\hat{X}(n-1 \mid n-1) = X(n-1)$, and all subsequent estimates are exact

8.1.5 Example of State Estimation

- Satellite model $G(s) = 1/s^2$ is shown in Figure 8.5
  - Can only measure $y(kh) = x_1(kh)$; build estimator for $X(k)$

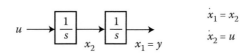

$$\dot{x}_1 = x_2$$
$$\dot{x}_2 = u$$

FIGURE 8.5
Satellite system.

- Equivalent discrete system:

$$X(k + 1) = \begin{bmatrix} 1 & h \\ 0 & 1 \end{bmatrix} X(k) + \begin{bmatrix} \dfrac{h^2}{2} \\ h \end{bmatrix} u(k); \quad y(k) = \begin{bmatrix} 1 & 0 \end{bmatrix} X(k)$$

where

$$X = \begin{bmatrix} x_1(k + 1) \\ x_2(k + 1) \end{bmatrix} \quad C = \begin{bmatrix} 1 & 0 \end{bmatrix}$$

- Check observability:

$$H_0 = \begin{bmatrix} C' & \Phi'C' \end{bmatrix} = \begin{bmatrix} 1 & 1 \\ 0 & h \end{bmatrix} \Rightarrow \text{observable (as long as } h \neq 0)$$

Design observer as follows:
- Desired characteristic polynomial,

$$p_e(z) = z^2 + \tilde{d}_1 z + \tilde{d}_2$$

- Observer gains are given by

$$L = p_e(\Phi) \begin{bmatrix} - & C\Phi & - \\ - & C\Phi^2 & - \end{bmatrix}^{-1} \begin{bmatrix} 0 \\ 1 \end{bmatrix} \tag{8.12}$$

- Deadbeat observer

$$\tilde{d}_1 = \tilde{d}_2 = 0 \, (\text{poles} @ z = 0)$$

$$L = \begin{bmatrix} 1 \\ 1/h \end{bmatrix} (\text{as } h \to 0, \text{need large } L)$$

8.1.6 Mechanics of Observer Dynamics

- Observer algorithm; initialize $\widehat{X}^-(0) = 0$ (usually)
  1. Measure $y(k)$ and form $v(k) = y(k) - \hat{y}(k)$
  2. Update $\widehat{X}^+(k) = \widehat{X}^-(k) + Lv(k)$
  3. Propagate $\widehat{X}^-(k + 1) = \Phi\widehat{X}^+(k) + \Gamma u(k)$
  4. $\hat{y}(k + 1) = C\widehat{X}^-(k + 1)$
  5. $k = k + 1$

- In the previous example, $h = 1$, then the "L" gains are given by

$$L = \begin{bmatrix} L_1 \\ L_2 \end{bmatrix} = \begin{bmatrix} 1 - \tilde{d}_2 \\ 1 + \tilde{d}_1 + \tilde{d}_2 \end{bmatrix}$$

$$v(k) = y(k) - \hat{x}_1^-(k)$$

$$\left. \begin{array}{l} \hat{x}_1^+(k) = \hat{x}_1^-(k) + L_1 v(k) \\ \hat{x}_2^+(k) = \hat{x}_2^-(k) + L_2 v(k) \end{array} \right\} \text{update}$$

$$\left. \begin{array}{l} \hat{x}_1^-(k+1) = \hat{x}_1^+(k) + x_2^+(k) + 1/2u(k) \\ \hat{x}_2^-(k+1) = \hat{x}_1^+(k) + u(k) \end{array} \right\} \text{Propagate}$$

- Deadbeat case: $L_1 = 1$, $L_2 = 1$. Open-loop $u = \{-1, 0.5, -03.\ 0.4, \ldots\}$.
- $U(k) =$ control input over time interval $(k, k + 1)$.
- Need only two measurements to obtain X exactly.

$$\hat{X}^+(1) = X(1)$$

- Subsequent $\hat{X}^+$, $\hat{X}^-$ are correct as long as we know state equations and system inputs.
- The actual system and the observer variables are shown in Table 8.1.

8.1.7 Implementation of the Observer-Controller Pair

- Use feedback control as

$$u(k) = K_r r(k) - K\hat{X}(k \mid k) \tag{8.13}$$

TABLE 8.1

Actual System and Observer Variables: k, x_1, x_2, u, y

	Actual System				Observer					
k	x_1	x_2	u	y	x_1^-	x_2^-	v	$\rightarrow$	x_1^+	x_2^+
0	1.0	−0.3	−1.0	1.0	0.0	0.0	1.0	$\rightarrow$	1.0	1.0
1	0.2	−1.3	0.5	0.2	1.5	0.0	−1.3	$\rightarrow$	0.2	−1.3
2	−0.85	−0.8	−0.3	−0.85	−0.85	−0.8	0.0	$\rightarrow$	−0.85	−0.8
3	−1.8	−1.1		−1.8	−1.8	−1.1	0.0	$\rightarrow$	−1.8	−1.1
.							.			
.							.			
and so on							.			

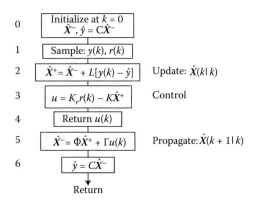

FIGURE 8.6
Flow diagram of algorithm for implementing an observer with a controller pair, at any particular "*k*."

- $\hat{X}(k \mid k)$ is the best estimate of $X(k)$ at step k.
- Includes the latest information $y(k)$:

$$\hat{X}(k \mid k-1) \text{ is not as good as } \hat{X}(k \mid k)$$

- K_r, K is obtained via the usual SVFB control design as seen earlier in Chapter 6.
- The algorithm at any particular k is as shown in the flow diagram in Figure 8.6:
  - Steps 2 and 3 require $2n$ Multiply and ADDS (MADDS) to obtain a new control.
  - Steps 5 and 6 set up $\hat{X}^-$ and $\hat{y}$ for the next cycle. This shifts $n^2 + 2n$ MADDS to the wait portion of cycle.

8.2 Implementation: Some Practical Considerations

- Propagation in step 5 assumes that the $u(k)$ computed will actually be applied to the system as shown in the algorithm:
  - Apply software limits to u, Δu, and so on, to match any system or hardware constraints/nonlinearities, or else conduct the following:
  - Modify algorithm to use actual control:
    1. Sample $y(k)$, $r(k)$, $u(k-1)$
       $5 \to 1a.\ \hat{X}^- = \Phi\hat{X}^+ + \Gamma u \leftarrow$ obtains $\hat{X}(k \mid k-1)$ at step
       $6 \to 1b.\ \hat{y} = C\hat{X}^+$
    2. $\hat{X}^+ = \hat{X}^- + L[y(k) - \hat{y}]$
    3. $u = K_r r(k) - K\,\hat{X}^+$
    4. Return $u(k)$

- Requires significantly more computation before $u(k)$ is obtained
    $\Rightarrow$ larger computational delay
- Any system time delay must be modeled in step 5

$$\hat{X}^- = \Phi\hat{X}^+ + \Gamma_1 u(k - M - 1) + \Gamma_0 u(k - M)$$

- Try to keep observer gains with $|L_i|$ small
    - Minimize amplification of the $y(k)$ measurement error
    - $\|L\|$ increases as observer poles $\to 0$
- Observer requires a "model" of system $\{\Phi, \Gamma, C\}$
    - Mismatch will yield estimation error
    - In a CL application, the feedback will reduce some of the effects of mismatch between the "model" and the system
- Large modeling errors can cause the estimate to diverge
- If $u(k)$ must be returned prior to sampling $y(k)$ use

$$u(k) = K_r r(k) - K\hat{X}(k \mid k - 1)$$

8.2.1 Composite CL Observer and Controller

A composite CL observer and controller block diagram is shown in Figure 8.7a.

- Composite system is of order $2n$, n-th-order system and n-th-order observer
- Alternative representation ($r = 0$, $\hat{X}^-(0) = 0$)
- Single input y to observer-controller and single output u
    - Can compute the transfer function as shown in Figure 8.7b

$$H_{eq}(z) = \frac{-u(z)}{y(z)}$$

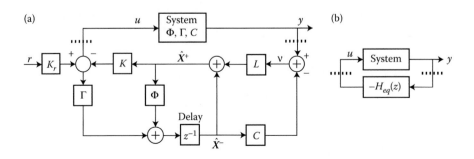

FIGURE 8.7
(a) Composite CL observer and controller. (b) Input y to observer-controller and single output u.

- Develop equation for $\hat{X}(k \mid k)$ in terms of $y(k)$

$$\Rightarrow \hat{X}(k+1 \mid k+1) = (I - LC)\hat{X}(k+1 \mid k) + Ly(k+1)$$

where

$$\hat{X}(k+1 \mid k) = \Phi \hat{X}(k \mid k) + \Gamma u(k) = (\Phi - \Gamma K)\hat{X}(k \mid k)$$

$$\Rightarrow \hat{X}(k+1 \mid k+1) = \tilde{\Phi} \hat{X}(k \mid k) + Ly(k+1); \quad \tilde{\Phi} = (I - LC)(\Phi - \Gamma K)$$

and

$$-u(k) = K\hat{X}(k \mid k)$$

- Take z-transform $[y(k+1) \to zy(z)]$

$$\Rightarrow H_{eq}(z) = zK(zI - \tilde{\Phi})^{-1}L = \frac{n\text{th order}}{n\text{th order}} \tag{8.14}$$

- Provides a "modern" control approach to the design of the "classical" series/Feedback (FB) compensators
- Use $LG_{ain}(z) = \tilde{G}(z) H_{eq}(z)$ for stability analysis (Φ_m, ω_c)

8.2.1.1 Command Inputs to Observer-Controller System

- Alternate representation in output feedback form (when $r \neq 0$) is shown in Figure 8.8:

$$H_{eq}(z) = zK(zI - \tilde{\Phi})^{-1}L$$

Includes a feedforward compensator $H_r(z)$ on $r(k)$.

- Observer-controller implementation generally preferable.
- Recall, for SVFB loop, $u(k) = K_r r(k) - KX(k)$
  - K_r is chosen so that DC gain $r \to y = 1$ in steady state

$$K_r = \left[C(zI - \Phi + \Gamma K)^{-1}\Gamma \right]^{-1} \Big|_{z=1} \tag{8.15}$$

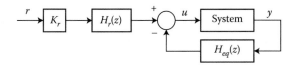

FIGURE 8.8
Alternate representation in output feedback form for $r \neq 0$.

In observer-controller implementation

$$u(k) = K_r r(k) - K\hat{X}(k \mid k)$$

- In steady state, $\hat{X}(k \mid k) \to X(k) \Rightarrow K_r$ is the same as in the SVFB case.
- A common situation:

$$y = x_j = \text{some variable position}$$

then generally $K_r = K_j = $ gain on x_j, that is, control law is given as

$$u = -K_1\hat{X}_1 - \cdots - K_j(\hat{X}_j - r) - \cdots - K_n\hat{X}_n$$

Redefine:

$$x_j \triangleq x_{je} = \text{error in } x_j$$

and we measure only $x_{je} \to$ error model; here
- All previous results (with $r = 0$) are applicable to the error model.

EXAMPLE 8.1: RADAR POSITIONING SYSTEM

The continuous time radar positioning system is shown in Figure 8.9.

- Design the digital control ($h = 1$) so that an initial offset $\theta_e(0) \to 0$ with $t_{s|1\%} \sim 10$ s and PO $\sim 10\%$, $x_1(0) = 0$, and $x_2(0) = -r_0$. One can only measure $\theta_e = x_2$.
  - Equivalent to an input command system when only the error:

$$e(t) \triangleq \theta_e(t) = \theta(t) - r$$

  is measurable and $\theta(0) = 0$.
- Digital control design in continuous time is given by

$$\text{State Model:} \begin{bmatrix} \dot{x}_1(t) \\ \dot{x}_2(t) \end{bmatrix} = \begin{bmatrix} -0.1 & 0 \\ 1 & 0 \end{bmatrix} \begin{bmatrix} x_1(t) \\ x_2(t) \end{bmatrix} + \begin{bmatrix} 0.1 \\ 1 \end{bmatrix} u(t)$$

- The equivalent discrete system with $h = 1$:

$$X(k + 1) = \Phi X(k) + \Gamma u(k)$$

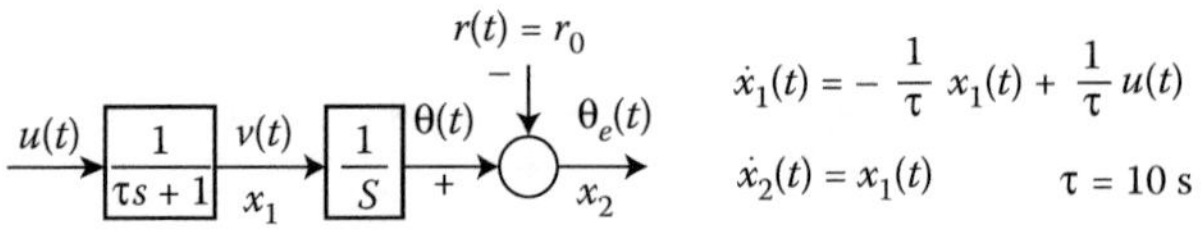

$$\dot{x}_1(t) = -\frac{1}{\tau} x_1(t) + \frac{1}{\tau} u(t)$$

$$\dot{x}_2(t) = x_1(t) \qquad \tau = 10 \text{ s}$$

FIGURE 8.9
Radar positioning system block diagram in continuous time. (a) Shaft offset angle, $\theta_e(t)$, $\theta_e(kh)$, (b) control input, $u(k)$.

$$\text{with } \Phi = \begin{bmatrix} 0.905 & 0 \\ 0.952 & 1 \end{bmatrix} \text{ and } \Gamma = \begin{bmatrix} 0.095 \\ 0.048 \end{bmatrix}$$

- Pick desired CL poles $\lambda_i = e^{s_i h}$, $s_i = -0.5 \pm j0.5$:

$$p_d(z) = z^2 - 1.06z + 0.367 \quad (\lambda_i = 0.53 \pm j0.29)$$

and

- SVFB gains $K = [7.21\ 3.19]$
- Only the shaft offset $\theta_e(k) = x_2(k)$ is available for measurement
  - Design observer to estimate $X(k)$
  - With $C = [0\ 1]$, the system is observable
  - Observer poles $\tilde{\lambda}_i \leq | \lambda_{prim}|^3 = 0.6^3 \approx 0.22$
  - Pick $\tilde{\lambda}_i = 0.2 \pm j\,0.1$ for good damping on $\tilde{e}$
- Observer gain $L = \begin{bmatrix} 0.588 \\ 0.945 \end{bmatrix}$
- Simulate CL response without initial condition as

$$X(0) = \begin{bmatrix} 0 \\ -1 \end{bmatrix}; \quad \hat{X}(0|-1) = \begin{bmatrix} 0 \\ 0 \end{bmatrix}; \quad r = \text{unit step}$$

8.2.1.2 Simulation Results

Simulation results are shown in Figure 8.10.

- Observer design uses significantly more control than does SVFB, especially at $k = 0$
- After first measurement $y(0) = \theta_e(0) = -1$; here examine the observer design uses significantly more control

$$\hat{x}_1(0|0) = -0.588, \quad \hat{x}_2(0|0) = -0.945 \Rightarrow -K\hat{X}(0|0) = 7.25$$

8.2.1.2.1 Alternate Representation in Output Feedback Form

An alternate form of output feedback control is shown in Figure 8.11.

- Feedback loop dynamics is given by

$$H_{eq}(z) = zK(zI - \tilde{\Phi})^{-1}L = \frac{\text{2nd order}}{\text{2nd order}}; \quad L = \begin{bmatrix} 0.588 \\ 0.945 \end{bmatrix}$$

$$K = \begin{bmatrix} 7.21 & 3.19 \end{bmatrix}$$

$$\tilde{\Phi} = (I - LC)(\Phi - \Gamma K) = \begin{bmatrix} -0.136 & -0.801 \\ 0.0333 & 0.0466 \end{bmatrix} \tag{8.16}$$

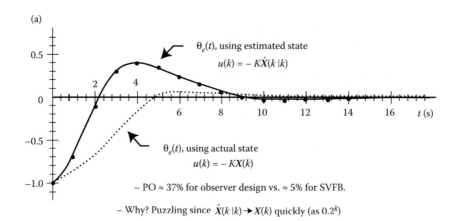

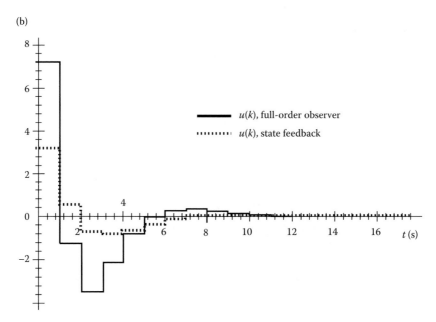

FIGURE 8.10
Shaft offset angle and control input for observer of radar positioning system.

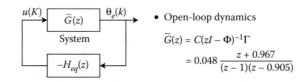

FIGURE 8.11
Alternate form of feedback control for radar positioning system.

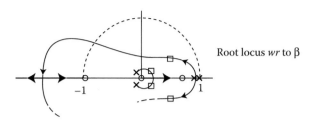

Root locus *wr* to β

FIGURE 8.12
RL of CL CP: $1 + \beta \tilde{G}(z)H_{eq}(z)$.

$$H_{eq}(z) = \frac{7.25z^2 - 5.1z}{z^2 + 0.0894z + 0.0204}$$

- We can now examine Φ_m, ω_c via $LG_{ain}(z) = \tilde{G}(z)H_{eq}(z)$.
  - This results in $\omega_c = 0.62$ rad/s and $\Phi_m = 40°$.
- Examine the root locus (RL) of the CL system with $H_{eq} \rightarrow \beta H_{eq}$.
- CL characteristic polynomial RL is shown in Figure 8.12 with respect to variation in β.
- When $\beta = 1$, CL poles are at $z_i = 0.2 \pm j\,0.1$ (observer poles) and $0.53 \pm j\,0.29$ (controller poles).

8.2.1.3 Possible Modifications to Improve Response

- There are problems when $X(0) \neq 0$, or if a sudden large ΔX, then

$$\hat{X}(0|-1) = 0, u(0) = -K\hat{X}(0|0) = -KLCX(0), \quad \text{where } y(0) = CX(0)$$

  - Initial $u(0)$ can be far from $-KX(0)$ in such cases
  - (Equation 7.25 vs. Equation 3.19) since $\hat{X}(0|0)$ is off
- These problems are typical of command input systems when we only measure the error $e(k) = y(t) - r(t)$. In other words, is a change in e due to a change in X or a change in r?

1. Initialize $\hat{x}_2(0|-1) = -r_0 = -1$:
   - Resulting $\theta_c(t) \equiv \theta_c(t)$ with SVFB (provided $x_1(0) = 0$)
   - Plausible to do in an input command system when $r(k)$ is known

$$(x_2 \sim \theta - r \quad \text{or} \quad \theta \sim x_2 + r, \theta = \text{shaft angle})$$

$$\hat{x}_2(k^+|k-1) = \hat{x}_2(k^-|k-1) - r(k) + r(k-1)$$

where

$$r(k-1) = 0 @ k = 0$$

- Not possible in general if $X(0)$ has an unknown structure
2. Slow down observer at $k = 0$:
   - Phase in observer gain $L(k)$: $0 \rightarrow L$
   - Use slower observer poles; for example, $\leq \rho^2 \Rightarrow \lambda_i = 0.3 \pm j0.2$ (results in smaller L, but slower CL response)
3. Slow down the control law by

$$u(k) = \alpha u(k-1) - (1-\alpha)K\hat{X}(k \mid k) \tag{8.17}$$

- Best obtained by using $\Delta(k) = u(k) - u(k-1)$ as the "control"
  - A very popular scheme in practice
4. Slow down the input command:
   - Use $r(k) = a$ sequence of smaller changes $\Rightarrow$ bound $|\Delta r(k)|$
5. Reformulate as a command input problem:

$$u(k) = r(k) - L \; \hat{X}(k \mid k)$$

where we measure $\theta(k)$

8.2.2 Example Satellite Control with Command Input

Formulation as a command input structure:

$$\dot{x}_1 = x_2, \; \dot{x}_2 = u, \text{ measure } y(k) = x_1(k)$$

and with

$$h = 0.5, \; r(k) = 1 \text{ (input)}$$

Select SVFB gains to place control poles at $z_i = 0.85 \pm j0.3$ (gives $\omega_c \sim 0.78$ rad/s, $\Phi_m \sim 32.50$). Select L to place observer poles at $z_j = 0.55 \pm j0.15$.

$$X(0) = [-0.50, 0.0]', \; \hat{X}(0 \mid -1) = \begin{bmatrix} 0.0. & 0.0 \end{bmatrix}'$$

$$u(k) = K_r r(k) - K \; \hat{X}(k \mid k)$$

The results are shown for the output with state estimation control and state determinate control in Figure 8.13.

If $X(0) = 0$, the response of the system using observer is identical to the system response using actual state.

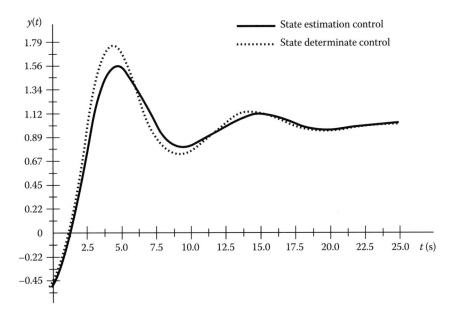

FIGURE 8.13
Output $y(k)$ for satellite control with command input.

8.3 Transfer Function of Composite CL Observer and Controller

- This is a $2n$-th-order system ($\Rightarrow 2n$ poles)
- Examine transfer function $T(z) = y(z)/r(z)$
  - Obtain via state equations ($2n$)
- State $x(k)$ evolution can be shown as

$$X(k + 1) = \Phi X(k) - \Gamma K \hat{X}(k \mid k) + K_r \Gamma r(k);$$

$$\hat{X}(k \mid k) = (I - LC)\hat{X}(k \mid k - 1) + LCX(k); y(k) = LCX(k)$$
$$\Rightarrow X(k + 1) = (\Phi - \Gamma KLC)X(k) - \Gamma K\hat{X}(k \mid k - 1) + K_r \Gamma r(k) \qquad (8.18)$$

- State estimate $\hat{X}(k \mid k - 1)$ evolution:

$$\hat{X}(k + 1 \mid k) = \Phi\hat{X}(k \mid k) + \Gamma u(k)$$

$$\hat{X}(k + 1 \mid k) = (\Phi - \Gamma K)\hat{X}(k \mid k) + K_r \Gamma r(k)$$

and

$$\hat{X}(k \mid k) = (I - LC)\hat{X}(k \mid k - 1) + LCX(k)$$

$$\Rightarrow \hat{X}(k + 1 \mid k) = (\Phi - \Gamma K)(I - LC)\hat{X}(k \mid k - 1) + (\Phi - \Gamma K)LC\hat{X}(k) + K_r \Gamma r(k) \qquad (8.19)$$

- Augmented system can be evaluated as follows:

$$X_a(k) \triangleq [X(k), \ \hat{X}(k|k-1)]'$$

$$\begin{bmatrix} X(k+1) \\ \hat{X}(k+1|k) \end{bmatrix} = \begin{bmatrix} \Phi - \Gamma KLC & -\Gamma K(I-LC) \\ (\Phi - \Gamma K)LC & (\Phi - \Gamma K)(I-LC) \end{bmatrix} \begin{bmatrix} X(k) \\ \hat{X}(k|k-1) \end{bmatrix} + \begin{bmatrix} K_r\Gamma \\ K_r\Gamma \end{bmatrix} r(k) \quad (8.20)$$

where

$$\Phi^* = \begin{bmatrix} \Phi - \Gamma KLC & -\Gamma K(I-LC) \\ (\Phi - \Gamma K)LC & (\Phi - \Gamma K)(I-LC) \end{bmatrix}; \quad X_a(k) = \begin{bmatrix} X(k) \\ \hat{X}(k|k-1) \end{bmatrix}; \quad \Gamma^* = \begin{bmatrix} K_r\Gamma \\ K_r\Gamma \end{bmatrix}$$

with

$$y(k) = \begin{bmatrix} C & 0 \end{bmatrix} X_a(k)$$

such that

$$C^* = \begin{bmatrix} C & 0 \end{bmatrix}$$

$$T(z) = C^{*'}(zI - \Phi^*)^{-1}\Gamma^* = \frac{N(z)}{D(z)} \quad (8.21)$$

- This is messy! It is easier to obtain $D(z)$ first, then get $N(z)$.

8.4 Poles and Zeros of Composite $T(z)$

- Let us consider:

$D(z) = |zI - \Phi^*|$ = characteristic polynomial of a *CL* system

where

$$\begin{bmatrix} zI - \Phi + \Gamma KLC & \Gamma K(I-LC) \\ -\Phi LC + \Gamma KLC & zI - (\Phi - \Gamma K)(I-LC) \end{bmatrix} \rightarrow \begin{bmatrix} zI - \Phi + \Gamma K & \Gamma K(I-LC) \\ 0 & zI - \Phi(I-LC) \end{bmatrix}$$

where

$$-(\Phi - \Gamma K)(I-LC) = -\Phi(I-LC) + \Gamma K(I-LC)$$

$$\Rightarrow D(z) = |zI - \Phi + \Gamma K| \cdot |zI - \Phi(I-LC)| = p_d(z) \cdot p_e(z) \quad (8.22)$$

- Poles of the composite system are those of the controller $(\lambda_1, \ldots, \lambda_n)$, and the observer poles are

$$(\tilde{\lambda}_1, \tilde{\lambda}_2, \ldots, \tilde{\lambda}_n)$$

$$N(z) = \begin{vmatrix} zI - \Phi^* & -\Gamma^* \\ C^* & 0 \end{vmatrix}$$

$$= \begin{vmatrix} zI - \Phi^* & \begin{matrix} -K_r\Gamma \\ -K_r\Gamma \end{matrix} \\ C \; 0 & 0 \end{vmatrix} \quad \xrightarrow[\text{via row and column manipulation}]{} \quad \begin{vmatrix} zI - \Phi & -K_r\Gamma & \Gamma K(I - LC) \\ C & 0 & 0 \\ 0 & 0 & zI - \Phi(I - LC) \end{vmatrix}$$

$$\Rightarrow N(z) = \begin{vmatrix} zI - \Phi & -K_r\Gamma \\ C & 0 \end{vmatrix} \cdot |zI - \Phi(I - LC)| \tag{8.23}$$

where

$$K_r^* = \begin{vmatrix} zI - \Phi & -K_r\Gamma \\ C & 0 \end{vmatrix} \rightarrow \text{numerator of open-loop system;}$$

and

$$p_e(z) = |zI - \Phi(I - LC)|$$

$$N(z) = K_r B(z) p_e(z), \quad B(z) = \text{open-loop system}$$

- Transfer function

$$\Rightarrow T(z) = \frac{K_r B(z) p_e(z)}{p_d(z) p_e(z)} = \frac{K_r B(z)}{p_d(z)} \tag{8.24}$$

- Same as SVFB case.
- Observer dynamics are "transparent" in a steady state (after initial transient in-state estimate $\tilde{e} \rightarrow 0$).
- SVFB with the observer does not modify system zeros.

8.5 Reduced-Order Observers

- Redefine states so that $y(k) = x_1(k)$
  - Use standard observable form, or
  - Use SVFB transformation $V = T^{-1}X$ with $T^{-1} = \begin{bmatrix} - & C & - \\ & T_{n-1} & \end{bmatrix}$

    $T_{n-1} = (n - 1)x\ n$, arbitrary (need only $CT = e_1'$)

- Idea: if $y(k) = x_1(k)$ is the measure, need only to estimate $X_b = \begin{bmatrix} x_2 \\ \cdot \\ \cdot \\ \cdot \\ x_n \end{bmatrix}$

  $\Rightarrow$ Need only build an $(n-1)$st-order estimator
  (in general, if there are only p measurements; then it implies this is a $(n-p)$-th-order observer)

- Decompose the state equation as given below:

$$X(k) = \Phi X(k-1) + \Gamma u(k-1); \; y(k) = x_1(k)$$

$$\begin{bmatrix} x_1(k) \\ X_b(k) \end{bmatrix} = \begin{bmatrix} \Phi_{11} & \Phi_{1b} \\ \Phi_{b1} & \Phi_{bb} \end{bmatrix} \begin{bmatrix} x_1(k-1) \\ X_b(k-1) \end{bmatrix} + \begin{bmatrix} \Gamma_1 \\ \Gamma_b \end{bmatrix} u(k-1) \qquad (8.25)$$

$$X_b = \Phi_{bb} X_b(k-1) + \Phi_{b1} y(k-1) + \Gamma_b u(k-1) \qquad (8.26)$$

$$u^*(k-1) = \Phi_{b1} y(k-1) + \Gamma_b u(k-1), \quad \text{known at time} \, k-1$$

$$y(k) = \Phi_{11} y(k-1) + \Phi_{b1} y(k-1) + \Gamma_b u(k-1)$$

$$y(k) - \Phi_{11} y(k-1) - \Gamma_1 u(k-1) = \Phi_{1b} X_b(k-1) \qquad (8.27)$$

$$y^*(k) = y(k) - \Phi_{11} y(k-1) - \Gamma_1 u(k-1), \quad \text{known at time} \, k$$

- Build an observer for

$$X_b(k) = \Phi_{bb} X_b(k-1) + u^*(k-1)$$

$$y^*(k) = \Phi_{1b} X_b(k-1)$$

- Is this observable? This needs to be assessed as shown in the next section on reduced-order observer design.
  - If original $\{\Phi, C\}$ is observable, then $\{\Phi_{bb}, \Phi_{1b}\}$ is also observable.

8.5.1 Reduced-Order Observer Design for X_b

$$\hat{X}_b\left(k| k-1\right) = \text{estimate of } X_b \text{ based on} \left\{y(k-1),...\right\} \triangleq \hat{X}_b^-(k)$$

$$\hat{X}_b(k| k) = \text{estimate of } X_b \text{ based on} \left\{y(k), y(k-1),...\right\} \triangleq \hat{X}_b^+(k)$$

- Propagation/prediction step $k - 1 \rightarrow k$

$$\Rightarrow \hat{X}_b(k \,|\, k - 1) = \Phi_{bb}\hat{X}_b(k - 1 \,|\, k - 1) + \Phi_{b1}y(k - 1) + \Gamma_b u(k - 1) \qquad (8.28)$$

- Update step $\left.\begin{array}{r} \hat{X}_b(k \,|\, k - 1) \\ y(k) \end{array}\right\} \rightarrow \hat{X}_b(k \,|\, k)$

  - Follow the same basic approach as the full-order case:

$$\hat{X}_b(k \,|\, k) = \hat{X}_b(k \,|\, k - 1) + L_b\left[y^*(k) - \Phi_{1b}\hat{X}_b(k - 1 \,|\, k - 1) \right]$$

where the best estimate of $y^*(k) = \Phi_{1b}\hat{X}_b(k - 1 \,|\, k - 1)$ at time k, $\{y(k - 1), \ldots\}$. Then,

$$\Rightarrow \hat{X}_b(k \,|\, k) = \hat{X}_b(k \,|\, k - 1) + L_b\left[y(k) - \Phi_{11}y(k - 1) - \Gamma_1 u(k - 1) - \Phi_{1b}\hat{X}_b(k - 1 \,|\, k - 1) \right]$$

where

$$\hat{y}(k) = \Phi_{11}y(k - 1) - \Gamma_1 u(k - 1) - \Phi_{1b}\hat{X}_b(k - 1 \,|\, k - 1) \qquad (8.29)$$

- Selection of $L_b = (n - 1)$ gain vector:
  - Obtain equation for $\tilde{e}(k|k) \triangleq X_b - \hat{X}_b(k|k)$:

$$\tilde{e}(k \,|\, k) = (\Phi_{bb} - L_b\Phi_{1b})\tilde{e}(k - 1 \,|\, k - 1) \qquad (8.30)$$

- Pick L_b so that eigenvalues of $\Phi_{bb} - L_b\Phi_{1b}$ are in the unit circle:

$$p_{er}(z) = z^{n-1} + \tilde{d}_1 z^{n-2} + \cdots + \tilde{d}_{n-1}$$

- Analogous to the earlier result $\Phi \rightarrow \Phi_{bb}, C\Phi \rightarrow \Phi 1_b$

$$L_b = p_{er}(\Phi_{bb})\begin{bmatrix} - - - & \Phi_{1b} & - - - \\ - - & \Phi_{1b}\Phi_{bb} & - - \\ & \vdots & \\ - & \Phi_{1b}\Phi_{bb}{}^{n-2} & - \end{bmatrix}^{-1}\begin{bmatrix} 0 \\ 0 \\ \vdots \\ 1 \end{bmatrix} \qquad (8.31)$$

- Here, the inverse exists if $\{\Phi_{bb}, \Phi 1_b\} = $ observable, thus $H_{o,r}' = $ observability matrix R–O system.

Compute L_b by using GAINS subroutine $\Phi \rightarrow \Phi_{bb}', \Gamma \rightarrow \Phi_{1b}'$; obtain $K \rightarrow L_b'$.

8.5.2 Implementation of Reduced-Order Observer/Controller

- Note, if the original system is of standard observable form (SOF), that is,

$$
\Phi = \begin{bmatrix}
-a_1 & 1 & 0 & \cdot & 0 \\
-a_2 & 0 & 1 & \cdot & \\
\cdot & \cdot & \cdot & \cdot & \\
\cdot & \cdot & \cdot & \cdot & \\
\cdot & \cdot & \cdot & \cdot & 1 \\
-a_n & 0 & \cdot & \cdot & 0
\end{bmatrix} \Rightarrow H_{0,r}{}' = I_{(n-1)x\,(n-1)}
$$

- The algorithm at any particular k is illustrated in the flow diagram of Figure 8.14.
  - Initialization: at step $k = 0$ we do not have $y(-1)$, $u(-1)$ or

$$
\hat{X}_b(-1\,|-1) \Rightarrow \hat{y} = 0
$$

- Let $\hat{X}_b(0\,|-1) =$ best prior estimate of $\hat{X}_b(0)$, usually $= 0$

8.5.3 Loop Gain Analysis of Reduced Order (RO) Observer/Controller

- Assume:

$$
\hat{X}_b^{-}(0) = 0, r(k) = 0 \Rightarrow u(k) = -K_1 y(k) - K_b \hat{X}_b(k|k)
$$

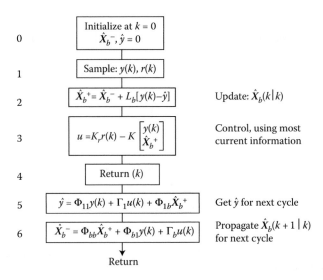

FIGURE 8.14
Algorithm flow diagram of RO observer/controller.

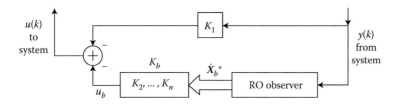

FIGURE 8.15
Block diagram of LG analysis for an RO observer/controller.

The loop-gain (LG) analysis block diagram is shown in Figure 8.15.
- Obtain the transfer function from y to $u = -H_{eq}(z)$.
  - Then $(n-1)$st-order feedback loop is given by

$$u(z) = -H_{eq}(z)y(z) = -\left[K_1 + \frac{u_b(z)}{y(z)}\right]y(z) \tag{8.32}$$

- Now obtain the dynamic equation for $\hat{X}_b(k|k)$:

$$\hat{X}_b(k|k) = \hat{X}_b(k|k-1) + L_b\left[y(k) - \Phi_{11}y(k-1) - \Gamma_1 u(k-1) - \Phi_{1b}\hat{X}_b(k-1|k-1)\right]$$

where

$$\hat{X}_b(k|k-1) = \Phi_{bb}\hat{X}_b(k-1|k-1) + \Phi_{b1}y(k-1) + \Gamma_b u(k-1)$$

$$\Rightarrow \hat{X}_b(k|k) = \left[\Phi_{bb} - L_b\Phi_{1b} - (\Gamma_b - L_bK_1)K_b\right]\hat{X}_b(k-1|k-1) + L_b y(k)$$
$$+\left[\Phi_{b1} - L_b\Phi_{11} - K_1(\Gamma_b - L_bK_1)\right]y(k-1)$$

where

$$\tilde{\Phi}_{bb} = \left[\Phi_{bb} - L_b\Phi_{1b} - (\Gamma_b - L_bK_1)K_b\right] \quad \text{and} \quad \tilde{L}_b = \left[\Phi_{b1} - L_b\Phi_{11} - K_1(\Gamma_b - L_bK_1)\right]$$
$$-u(k) = K_b\hat{X}_b(k|k) + K_1 y(k)$$

- Taking z-transforms of the above dynamic equation yields:

$$H_{eq}(z) = \left[K_b(zI - \tilde{\Phi}_{bb})^{-1}(zL_b + \tilde{L}_b) + K_1\right] = \frac{(n-1)order}{(n-1)order} \tag{8.33}$$

- Treat command inputs similarly to a full-order observer case.

The u to y in the z-domain is shown in Figure 8.16.

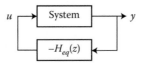

FIGURE 8.16
Block diagram of u to y in z-domain.

EXAMPLE 8.2: RADAR POSITIONING PROBLEM

$$X(k + 1) = \begin{vmatrix} 0.905 & 0 \\ 0.952 & 1 \end{vmatrix} X(k) + \begin{vmatrix} 0.095 \\ 0.048 \end{vmatrix} u(k); \quad y(k) = x_2(k) = \theta_e(k)$$

- Redefine states so that $x_1 = y(k) \Rightarrow T = \begin{bmatrix} 0 & 1 \\ 1 & 0 \end{bmatrix}; \quad T^{-1} = T.$
  Thus,

$$X(k + 1) = \begin{bmatrix} 1 & 0.952 \\ 0 & 0.905 \end{bmatrix} X(k) + \begin{bmatrix} 0.048 \\ 0.095 \end{bmatrix} u(k); \quad y(k) = x_1(k) = \theta_e(k)$$

$$K = \begin{bmatrix} 3.19 & 7.21 \end{bmatrix} \text{ from previous analysis}$$

- Select observer gain L_b ($\Phi_{bb} = 0.905$, $\Phi_{1b} = 0.952$).
  - Place the observer pole at $\tilde{\lambda}_i = 0.2$
    This yields

$$\Phi_{bb} - L_b\Phi_{1b} = 0.2 = 0.095 - L_b \, 0.952$$

$$\Rightarrow L_b = 0.740$$

- Simulation results are shown in Figure 8.17, with $x_1(0) = y(0) = -1$, $x_2(0) = 0$; $\hat{X}_b(0 \mid -1) = 0$.
  - Results are highly analogous to a full-order observer case

$$\hat{X}_b(0 \mid 0) = 0 + L_b\{y(0) - 0\} = -0.74 \Rightarrow u(0) = 0.852$$

EXAMPLE 8.3: ALTERNATE IMPLEMENTATION AND LG ANALYSIS

An alternate implementation is shown in Figure 8.18, which illustrates implementation and LG analysis.

- Feedback loop dynamics (reduced-order observer, with gain $L_b = 0.74$)
- Then the z-transform of the RO dynamic equation is given by

$$H_{eq}(z) = \left[K_b(zI - \tilde{\Phi}_{bb})^{-1}(zL_b + \tilde{L}_b) + K_1 \right] = -0.36; \quad K_1 = 3.19; \quad K_b = K_2 = 7.21$$

$$\tilde{L}_b = \left[\Phi_{b1} - L_b\Phi_{11} - K_1(\Gamma_b - L_bK_1) \right] = -0.93$$

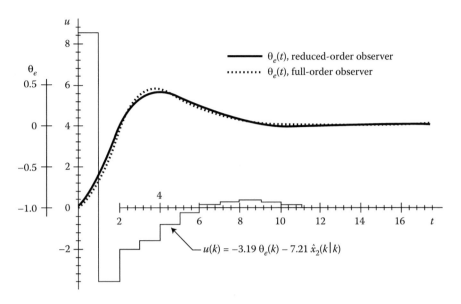

FIGURE 8.17
Simulation results of error angle with RO observer and full-order observer.

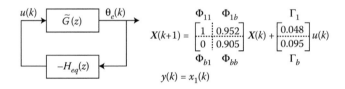

FIGURE 8.18
Alternate implementation and LG analysis.

and

$$H_{eq}(z) = \frac{8.52z - 5.55}{z + 0.362} = 8.52 \frac{z - 0.650}{z + 0.362} \tag{8.34}$$

- Examine the RL of CL system with $H_{eq} \rightarrow \beta H_{eq}$
- The RL of the CL characteristic polynomial (CP) is shown in Figure 8.19 for different values of β
  - When $\beta = 1$, CL poles are at $z_i = 0.53 \pm j\, 0.29$ and 0.2

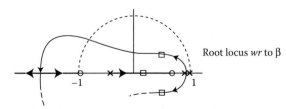

When $\beta = 1$, CL poles are at $z_i = 0.53 \pm j\, 0.29$ and 0.2

FIGURE 8.19
RL of CL CP of alternate implementation.

- In general, for an RO observer-controller
  - CL poles are observer poles plus controller poles

$$T(z) = \frac{y(z)}{r(z)} = \frac{K_r}{p_d(z)p_{er}(z)}B(z)p_{er}(z) = \frac{K_r}{p_d(z)}B(z)$$

Here, $B(z)$ = Numerator of open loop $\tilde{G}(z)$

EXAMPLE 8.4: PHASE MARGIN AND LG COMPARISONS; RADAR POSITIONING

- For SVFB, $LG(z) = K(zI - \Phi)^{-1} = \dfrac{0.84z - 0.573}{z^2 - 1.905z + 0.905}$
- For observer designs,

$$LG(z) = \tilde{G}(z)H_{eq}(z)$$

The LG Bode plot is shown in Figure 8.20 for radar positioning control systems SVFB performance.

- To compensate for the phase-shift (filtering) that they introduce, observer designs inherently reduce ω_c.

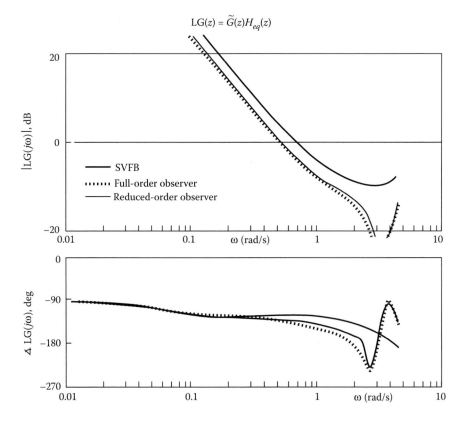

FIGURE 8.20
LG Bode plot for SVFB performance for radar positioning control system.

8.5.3.1 Continued Summary of Observer Design

- Full-order observer offers an excellent method to estimate system states from output measurement(s):
  - Can specify how fast $\hat{e}(k|k) \rightarrow 0$ *via* $\tilde{\lambda}_j$
  - Faster estimation $\Rightarrow$ higher gains L and more sensitivity to errors
  - Calculates estimator gains via use of GAINS program.
- Can obtain estimate of X between samples:

$$\hat{X}(kh + \delta) = \Phi(\delta)\hat{X}(k|k) + \Gamma(\delta)u(k) \tag{8.35}$$

$$\Phi(\delta) = e^{A\delta}; \quad \Gamma(\delta) = \int_0^\delta e^{A\delta}d\sigma B$$

- Use of $\hat{X}(k|k)$ in place of $X(k)$ in feedback:
  - Need to place observer poles closer to origin $(z = 0)$ than to primary control poles $r = p^2$ to p^3; $p = |\lambda_{\text{dom}}| = $ magnitude of dominant poles
- Implementation:
  - Requires additional computation/storage
  - Includes a "model" of the system in its structure
  - Can be implemented as an n-th order FB compensator (when $r = 0$)
- RO observer:
  - Can implement an $(n-1)$ order observer when $x_1 = y$ by setting

$$\hat{x}_1 \triangleq y$$

- Poles of CL observer/controller are set equal to

$$= \{\lambda_i, \ldots, \lambda_n, \tilde{\lambda}_1, \ldots, \tilde{\lambda}_n\}$$

- CL transfer function from r to y is the same as SVFB using actual states:
  - Observer is "transparent" in steady state
- Observer: excellent for systems that have good quality measurements, and state is subject to occasional random/deterministic changes Δx.

8.6 Modifications for Time Delay $\tau = Mh + \epsilon$

We can modify the state equation for time delay $\tau = Mh + \epsilon$; as

$$X(k + 1) = \Phi X(k) + \Gamma_1 u(k - 1 - M) + \Gamma_0 u(k - M) \tag{8.36}$$

- Control law modifications using τ-s ahead of prediction are now given by

  $u(k) = K_r r(k) - K\hat{X}(kh + \tau \mid k)$; where prediction of state at time $t + Mh + \epsilon$

  From $\hat{X}(k \mid k)$ and $u(k - 1), \ldots, u(k - M), u(k - M - 1)$

1. Obtain:

$$\hat{X}(kh + \epsilon) = e^{A\epsilon}\hat{X}(k \mid k) + \int_0^\epsilon e^{A\sigma}d\sigma Bu(k - M - 1) \tag{8.37}$$

2. M-step propagation from time $kh + \epsilon$ to $kh + \epsilon + Mh$; yields:

$$\hat{X}(kh + \tau) = \Phi^M \hat{X}(kh + \epsilon) + \sum_{i=1}^{M} \Phi^{i-1}\Gamma u(k - i) \tag{8.38}$$

- Store $\{u(k - M - 1), u(k - M), \ldots, u(k - 1)\}$ in an $M + 1$ stack
- Controller is exactly as it is for full-state FB, but with $X(k) \rightarrow \hat{X}(k \mid k)$

- Observer modifications for the "propagate" step only yield

$$\hat{X}(k + 1 \mid k) = \Phi\,\hat{X}(k \mid k) + \Gamma_1 u(k - 1 - M) + \Gamma_0 u(k - M) \tag{8.39}$$

  = prediction of state at next sample time

- Since initial estimates are incorrect, estimation error will be propagated forward in time
  - Future FB control may not be very good until $\hat{X}(\cdot) \rightarrow X(.)$
  - Response $\neq$ time shifted response with initial $X = e^{A\tau}X(0)$

EXAMPLE 8.5: SATELLITE CONTROL WITH TIME DELAY
The same problem as examined but with $M = 3$-step delay in loop ($\tau = 1.5$ s):

$$X(0) = \begin{bmatrix} -0.50 & 0.0 \end{bmatrix}', u(k) = K_r r(k) - K\hat{X}(k + M \mid k) \tag{8.40}$$

- Full-state FB, with and without delay, is shown in Figure 8.21.

8.7 Further/Advanced Topics in State Estimation

- Reformable observer pole-placement task as an optimization problem:
  - Gives more flexibility over time response $\tilde{e}(k \mid k) \rightarrow 0$

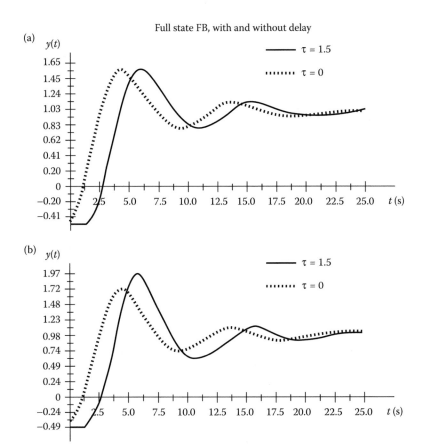

FIGURE 8.21
(a) Satellite control with and without time delay. (b) State estimation FB, with and without delay.

- Design observer for multi-output system:

$$y(k) = CX(k), \quad \text{where } C = p \times n \text{ matrix}$$

- Design reduced-order observer multi-output case:
  - (Obtain $(n - p)$th-order observer
  - Equivalent to an $(n - p)$th-order feedback compensator
- Estimation of state bias inputs:

$$X(k + 1) = \Phi X(k) + \Gamma u(k) + d \tag{8.41}$$

- d = unknown constant bias vector $(n \times 1)$
- Often $d = \Gamma_d d$; Γ_d = known n-vector, d = unknown constant
- Define $x_{n+1} = d$ with

$$x_{n+1}(k + 1) = x_{n+1}(k)$$

and estimate x_{n+1} using observer
- Yields nice formulation as an RO observer when x can be measured
- Estimation in the presence of noise:

$$X(k + 1) = \Phi X(k) + \Gamma u(k) + w(k) \tag{8.42}$$

where $w(k)$ is white noise, with covaraince W

$$y(k) = CX(k) + v(k)$$

where $v(k)$, is white noise, with covariance V
- Results in Kalman-Bucy filter for

$$\hat{X}(k \mid k), \quad \hat{X}(k \mid k - 1), \tag{8.43}$$

- Identical to observer, but with a different scheme to find gains L
- Estimation of system states and certain system parameters

8.7.1 Case Study: State Estimation in Passive Target Tracking

Multiple target tracking using multirate sampled measurements and joint probabilistic data association merged coupled filters (JPDAMCF) tracker are used; the tracking of targets is done using Probabilistic Data Association (PDA) Joint Probabilistic Data Association (JPDA) techniques.

The extensions of the JPDAMCF filters to the tracking of multiple targets using forward looking infrared (FLIR) sensors were also done by the author. These filters used the PDA technique along with a Kalman filter to track the position and velocity parameters of a target moving at a constant velocity. This procedure involves an extensive amount of computation. The multirate sampling technique suggested in this chapter is expected to alleviate these problems and release the CPU to achieve higher throughput and utilization.

In this case study, we illustrate the concept of an intelligent tracking filter called the Multirate Sampled Probabilistic Data Association (MSPDA) tracker. Here, a methodology of combining the use of the multirate sampling technique in the measurement acquisition process of a PDA filter is illustrated. It is shown that an MSPDA filter increases the computational efficiency of the tracking system.

The accuracy of the tracking algorithm is comparable with traditional PDA trackers. The computational savings is about 25% while the degradation in the accuracy is only to the extent of 5% of the mean square error (MSE) of the corresponding state vector component. Problem formulation followed by validation of the performance of the proposed is presented. Summary and conclusions follow with future extensions on the tracking of multiple targets.

The difference is to incorporate a multirate procedure that will sample velocity measurements at a different sampling rate as compared to the one where the position measurements are sampled. The process is described in Tables 8.2 and 8.3; only the first eight time intervals are included for illustration purpose.

For a single target in clutter with multirate sampled measurement model:
The state model is given as

$$X(k + 1) = FX(k) + v(k) \tag{8.44}$$

TABLE 8.2

Position and Velocity Measurements for MSPDA Filter

k	1	2	3	4	5	6	7	8
Pos	–	x	–	x	–	x	–	x
Vel	v	v	v	v	v	v	v	v

The lowercase x and v in the position and velocity rows indicate measurements taken at the corresponding time.

TABLE 8.3

Problem Formulation

k	1	2	3	4	5	6	7	8
Pos	–	x	–	x	–	x	–	x
Vel	v	v	v	v	v	v	v	v

The state vector of the target in the Cartesian coordinates is given by

$$X = \begin{bmatrix} x & \dot{x} & y & \dot{y} \end{bmatrix} \tag{8.45}$$

Measurements, which can be target originated, are as follows:

$$z(k) = HX(k) + w(k) \tag{8.46}$$

v and w are zero mean, mutually independent white-noise sequences with known covariance matrices $Q(k)$ and $R(k)$, respectively. The remaining measurements are assumed to be due to false alarms and can be modeled as independent identically distributed (iid) random variables with uniform spatial density.

T is the normal base-rate sampling interval. Choose a tracking-rate sampling interval, T_0, given by

$$T_0 = qT \tag{8.47}$$

where q is the least-common multiple of the integral multiples of the base-rate interval T associated with the tracking rate T_0 (two in this case).

The modified state and measurement models can now be written as follows:

Choose the modified state vector as

$$X_m = \begin{bmatrix} x(kT_0) \\ \dot{x}(kT_0 + T) \end{bmatrix} \tag{8.48}$$

the modified process noise vector is

$$v_m(kT_0) = \begin{bmatrix} v_k(kT_0) \\ v_{\dot{x}}(kT_0 + T) \end{bmatrix} \tag{8.49}$$

then the *modified* state model is given by

$$X_m\left[(k+1)T_0\right] = \begin{bmatrix} 1 & T \\ 0 & 1 \end{bmatrix} X_m[kT_0] + v_m(kT_0) \tag{8.50}$$

The *modified* measurement model to depict the measurement process as in Table 8.2 is given by

$$z_m(kT_0) = \begin{bmatrix} z_x(kT_0) \\ z_{\dot{x}}(kT_0 + T) \end{bmatrix} = \begin{bmatrix} \alpha & 0 \\ 0 & 1 \end{bmatrix} X_m(kT_0) + w_m(kT_0) \tag{8.51}$$

where $\alpha = 1$ for $T_0 = qT$ and zero, otherwise, $w_m(kT_0)$ is the corresponding modified measurement noise vector.

Equations 8.50 and 8.51 now represent a standard estimation problem and can be solved using the Kalman filter technique described in Section 8.7.1.

8.7.1.1 Performance Validation

The simulations can be carried out for a single target for which position, velocity, and acceleration profile (to simulate maneuver) are given as follows:

$$x(0) = 0$$

$$\dot{x}(0) = 10$$

$$\dot{x}(0) = 1 \quad \forall \quad 20 \le k \le 39$$

$$= -4 \quad \forall \quad 60 \le k \le 64$$

$$= 0, \text{otherwise}$$

The measurement noise components, as in Equation 8.51, are also Gaussian distributed with zero mean and unity variance respectively and iid. The process noise is Gaussian distributed with zero mean and variance given in Equation 8.49

$$E[v(k)v(j)] = q'\delta_{ij}; \quad q' = 0.4$$

For comparison, three filters, namely NNSF, MSPDA, and PDA, were run with the above parameters. The results of the simulations for a single run and 50 Monte Carlo runs are shown in Figures 8.22 through 8.29, respectively. The results show that the MSPDA filter does not perform as well as the PDA filter in terms of accuracy, but performs better than the NNSF filter. The advantage of the MSPDA filter exists in saving

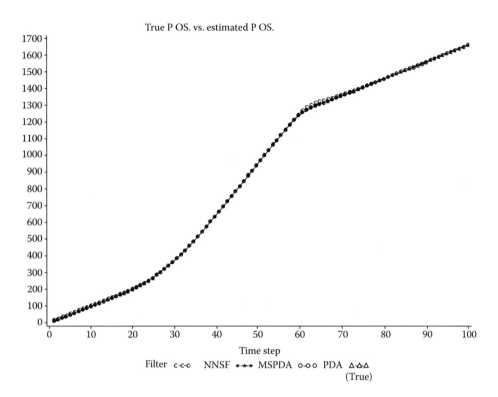

FIGURE 8.22
True position versus estimated position for NNSF, MSPDA, and PDA filters (single run), single target.

the time required to acquire measurements as compared to the PDA filter. The time savings is exactly 25%, which verifies the claims made in Section 8.7.1, Table 8.2, for the collection of measurement. The degradation of the accuracy due to MSPDA is only to the extent of 5% of the MSE.

Formulation of MS-JPDAMCF filter for two crossing targets.

Model for state vector

$$X_{mi}(kT_0) = \begin{bmatrix} X_i(kT_0) \\ \dot{X}_i(kT_0 + T) \\ X_i[k(T_0 - T)] \end{bmatrix}; \quad i = 1, 2 \tag{8.52}$$

Process noise vector

$$v_{mi}(kT_0) = \begin{bmatrix} v_i(kT_0) \\ \dot{v}_i(kT_0 + T) \\ v_i[k(T_0 - T)] \end{bmatrix}; \quad i = 1, 2 \tag{8.53}$$

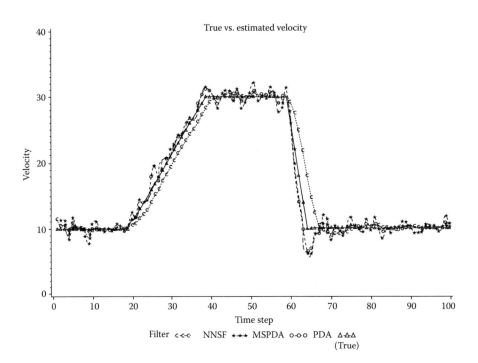

FIGURE 8.23
True velocity versus estimated velocity for NNSF, MSPDA, and PDA filters (single run), single target.

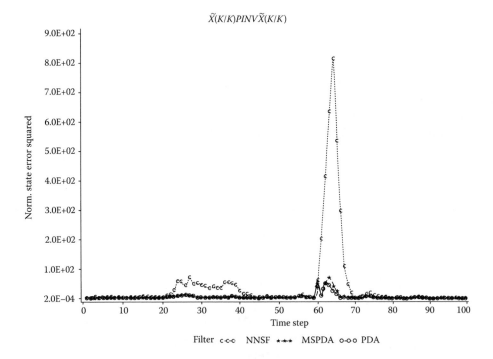

FIGURE 8.24
Normalized state error squared for NNSF, MSPDA, and PDA filters (single run), single target.

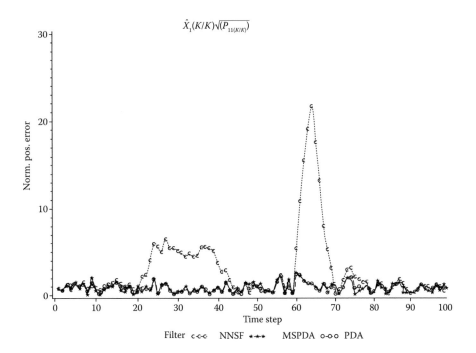

FIGURE 8.25
Normalized position error for NNSF, MSPDA, and PDA filters (single run), single target.

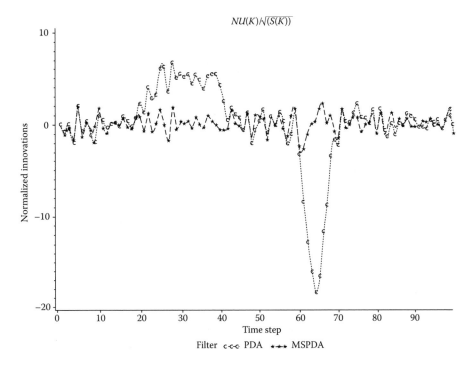

FIGURE 8.26
Normalized innovations for MSPDA and PDA filters (single run), single target.

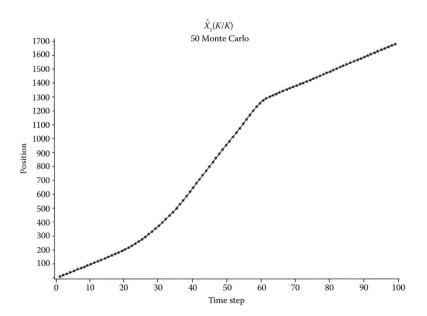

FIGURE 8.27
Estimated position evaluated by MSPDA filter.

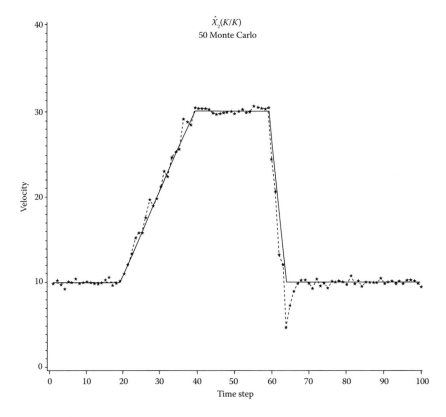

FIGURE 8.28
Estimated velocity by MSPDA filter (50 Monte Carlo runs).

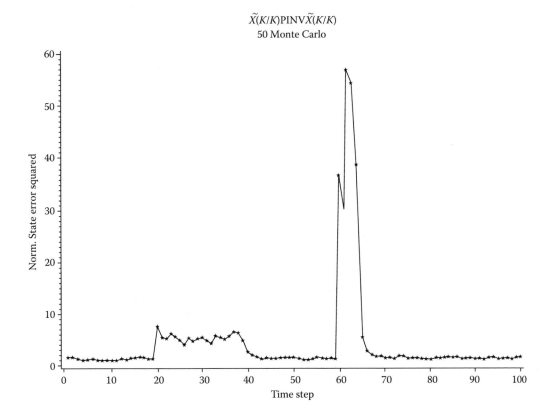

FIGURE 8.29
Normalized state error squared for MSPDA filter (50 Monte Carlo runs).

State model for each target is given by Equation 8.54 for $i = 1, 2$:

$$X_{mi}\left[(k + 1)T_0\right] = \begin{bmatrix} 1 & T & 0 \\ 0 & 1 & 0 \\ 0 & 0 & 1 \end{bmatrix} X_{mi}(kT_0) + v_{mi}(kT_0); \quad i = 1,2; \tag{8.54}$$

The measurement model to depict multirate sampling is given by

$$z_m(kT_0) = \begin{bmatrix} \beta\alpha & 0 & 0 & \beta(1 - \alpha) & 0 & 0 \\ 1 & 0 & -1 & 0 & 0 & 0 \\ 0 & 0 & 0 & 1 & 0 & -1 \end{bmatrix} X_m(kT_0) + w_m(kT_0) \tag{8.55}$$

where $\beta = 1$ for $T_0 = qT$ and zero otherwise.

$$w_m(kT_0) = \begin{bmatrix} w_c(kT_0) \\ w_{d_1}(kT_0 + T) \\ w_{d_2}[k(T_0 - T)] \end{bmatrix}; \quad \text{for two targets} \tag{8.56}$$

TABLE 8.4

Initial Conditions for Two Crossing Targets

Target 1	Target 2
$X_1(0) = 0$	$X_2(0) = 1680$
$\dot{X}_1(0) = 10$ (HEADING NORTH)	$\dot{X}_2(0) = -10$ (HEADING SOUTH)
$\ddot{X}_1(k) = 1; 20 \le k \le 39$	$\ddot{X}_2(k) = -1; 20 \le k \le 39$
$\quad = -4; 60 \le k \le 64$	$\quad = 4; 60 \le k \le 64$
$\quad = 0$; otherwise	$\quad = 0$; otherwise

TABLE 8.5

Simulation Results for Two Crossing Targets Using MS-JPDAMCF Filter ($q = 4$, and 100 Monte Carlo Runs) and the Above Initial Conditions in Table 8.4

	Target 1		Target 2	
T	Avg. Norm Pos. Error	Avg. NSES (3 States)	Avg. Norm Pos. Error	Avg. NSES (3 States)
10	−0.026	2.93	−0.033	3.32
20	0.101	2.91	−0.104	3.05
30	0.143	3.29	−0.1286	2.96
40	−0.115	3.08	0.0738	2.805
50	0.0169	2.8	−0.18	2.855
60	−0.032	2.81	0.0032	3.075
64	0.065	3.05	0.1184	3.115
80	0.241	3.36	0.0074	2.955
90	0. 032	3.15	0.0095	3.145

8.7.1.1.1 Simulation Results: MS-JPDAMCP

The formulation of MS-JPDAMCP filter is simulated using initial conditions in Table 8.4. The simulation results are illustrated for MS-JPDAMCP filter are shown in Table 8.5.

PROBLEMS

P8.1 Observability Requirement: An inverted pendulum is described in Figure P6.2. With state equations given by

$$\dot{X} = \begin{bmatrix} 0 & 1 & 0 & 0 \\ 11 & 0 & 0 & 0 \\ 0 & 0 & 0 & 1 \\ -1 & 0 & 0 & 0 \end{bmatrix} X + \begin{bmatrix} 0 \\ -1 \\ 0 \\ 1 \end{bmatrix} u$$

$$y(k) = [1\ 0\ 1\ 0]X$$

$$X(0) = \text{unknown initial condition}$$

The state vector is defined as

$X_1 = $ cart position

$X_2 = $ cart velocity

$X_3 = $ pendulum angle

$X_4 = $ pendulum angular velocity

Let an LQ controller; control signal be determined by

$$u = -(K_1 e_1 + K_2 e_2 + K_3 e_3 + K_4 e_4)$$

where

$e_1 = $ cart position error

$e_2 = $ cart velocity error

$e_3 = $ pendulum angle error

$e_4 = $ pendulum angular velocity error

For $h = 0.18$

1. Is the system observable?
2. What are the eigenvalues of Φ?
3. Find the structure of the state estimator.

P8.2 Observer Pole Placement:

1. For Problem 8.1, using Ackerman's formula select L' from $(\Phi - \Gamma K)$, and using Equations 8.9 and 8.10 then find L.

2. Use feedback control by using Equation 8.13 and write the code for the corresponding algorithm using MATLAB or suitable Simulink routines.

P8.3 Deadbeat Observer:

1. Design the deadbeat observer for $h = 0.18$ for Problem 8.2 using the above steps described therein.

P8.4 Composite CL Observer and Controller: A radar positioning system is described in Figure P8.1 for $T = 0.1$ s.

The state equations are

$$X(k + 1) = \begin{bmatrix} 1 & 0.0952 \\ 0 & 0.905 \end{bmatrix} X(k) + \begin{bmatrix} 0.00484 \\ 0.0952 \end{bmatrix} u(k)$$

$$y(k) = \begin{bmatrix} 1 & 0 \end{bmatrix} X(k)$$

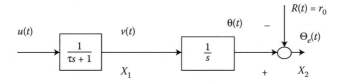

FIGURE P8.1
Composite CL observer and controller scheme.

1. Find the SVFB gains K.
2. Design the observer to estimate $X(k)$, if shaft offset $\theta_e(k) = x_2(k)$ is available for measurement.
3. Choose $C = [0\ 1]$ for the system to be observable and simulate CL response for $r = $ unit step.

P8.5 Radar Positioning: The system is shown in Figure P8.2.

The system is defined by digital control with $h = 1$:

$$X(k + 1) = \begin{bmatrix} 0.0905 & 0 \\ 0.952 & 0 \end{bmatrix} X(k) + \begin{bmatrix} 0.095 \\ 0.048 \end{bmatrix} u(k)$$

1. Find SVFB gains K.
2. Simulate the CL response for a unit step input $r = $ unit step.

P8.6 Satellite Control with Command Input:

In Problem 8.4, use

$$\dot{x}_1 = x_2;\ \dot{x}_2 = u;\quad \text{measure } y(k) = x_1(k)$$

$$h = 0.5,\quad \text{and}\quad r(k) = 1 \text{ (unit step)}$$

Show that the response of the system using the observer is identical to the response using the actual state.

P8.7 Reduced-Order Observer: In Problem 8.4 redefine states so that

$$x_1(k) = y(k) \rightarrow T = \begin{bmatrix} 0 & 1 \\ 1 & 0 \end{bmatrix};\quad T^{-1} = T$$

1. Redesign K from the previous analysis in Problem 8.4 and select the new observer gain.
2. Simulate results for $x_1(0) = y(0) = -1$; and $x_2(0) = 0$.
3. Compare results between full-order observer and reduced-order observer.

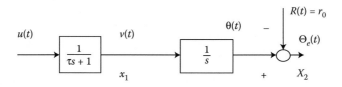

FIGURE P8.2
Satellite control with command input.

P8.8 Time Delay Modifications: In Problem 8.4, using time delay with $M = 3$ step delay in loop and $\tau = 1.5$ s,

1. Find state FB with delay
2. Find state FB without delay
3. Compare the two for $\tau = 1.5$ and $\tau = 0$

9

Implementation Issues in Digital Control

9.1 Mechanization of the Control Algorithm on Microcontrollers Motivation

- So far, the analysis and design of the digital control systems has been presented from an analytical standpoint, for example:
  - Sampling rate ← Nyquist or modified
  - Controller parameters ← performance and
  - Theoretical realizability
- However, there are physical implementations of discrete (not continuous) algorithms on real hardware.
  - A known upper limit on the sampling rate based on the hardware (microprocessor) speed limitations.
  - Finite word length characteristics of digital computers.
- Also, the control engineer may have to design the control system with available hardware.

Implementation issues are an integral part of the design process.

- Some algorithmic/implementation structures are preferable to others in minimizing the effects of error propagation throughout the system.

9.1.1 Microprocessor Implementation Structure

A microcontroller is an assembly of a clock, microprocessor, A/D and D/A converters and other large-scale integrated (LSI) chip components that are connected together by a time-sharing channel called a bus, as shown in Figure 9.1.

- Functions of a microcontroller are
  - Accept analog signals and logic interventions
  - Process information in real time
  - Deliver analog signals and logic commands
- Perform ancillary but important functions like:
  - Process other information (not in real time), but off-line, for example, periodic checking, reprogramming, software changes, etc.

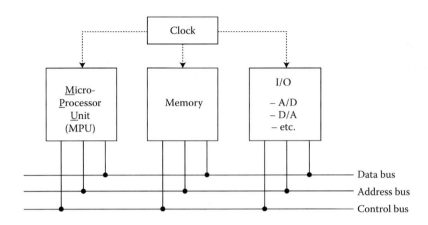

FIGURE 9.1
Microcontroller structure and its components.

- Components in a microcontroller are
  - Clock: synchronizes the whole assembly
  - MPU: (or central processing unit (CPU)) is a program-driven circuit that executes programs, arithmetic logic unit (ALU) controls the activities of the bus-organized system; and transfers data to and from registers, buffers, and so on.
  - Memory unit: stores data and programs:
    - Random access memory (RAM)
    - Read only memory (ROM)
    - Programmable ROM (PROM)
    - Erasable PROM (EPROM)
  - Input/output devices: interfaces analog to digital (A/D) and digital to analog (D/A):
    - A/D translates a single sample of an analog signal into a binary code as input to the microprocessor.
    - D/A produces an analog pulse proportional to the binary number given as its input.
  - All of the above building blocks are available as LSI chips Examples: (from 8-bit to 16-bit to 32-bit to 64-bit chips)

    Intel: 8086 → 80,286 → 80,386 → 80,486 → 860RISC → R400 by MIPS

    Motorola: 6800 → 68,000 → 68,020 → 68,030 → Power PC 74xx like PC 7448
  - Microcomputers have a finite word length that causes numerical errors that are generated and propagated through the system. Before getting to the analysis of these errors, we shall review briefly some elements of binary arithmetic with finite word length.

9.1.2 Binary Representation of Quantized Numbers

- We assume that word length of $C + 1$ bits is chosen to represent a number (C bits for numerical value or mantissa, 1 bit for sign) as shown in Table 9.1:

TABLE 9.1

$C + 1$ Bits for Numerical Value or Mantissa

s.	a_1	a_2	a_3	a_4	...	a_c
Sign bit $0 \rightarrow$ positive $1 \rightarrow$ negative	$\longleftarrow$		Value bits correspond to $\sum_{i=1}^{C} a_i 2^{-i}$			$\longrightarrow$

$$a_i; \quad i = 1, 2, \ldots, n; \quad n\text{-bits.}$$

- 2^C different nonnegative numbers may be represented with a C-bit word (and $2^C - 1$ different negative numbers).
- $2^{-C} = q$ is the least-significant bit (LSB) of the binary number (represents smallest number or resolution).
- Fixed-point representation ~ fixed scaling.

  The binary point is at a fixed position, that is, there is a fixed multiplier 2^M associated with every number $\Rightarrow$ allows one to cover a fixed range:

  $$[-(1 - 2^{-C}), 1 - 2^{-C}]*2^M$$

  (~M bit shift in the binary point)

  Example: $C = 4$, $M = 2$ (~an imaginary $C + 1 = 5$ bit μ-processor) $0.1001_b * 2^2 = 010.01b$

  $$= 1 \times 2^1 + 0 \times 2^0 + 0 \times 2^{-1} + 1 \times 2^{-2}$$
  $$= 2 + 0 + 0 + 0.25 = 2.25_d$$

- Floating-point representation:

  $N + 1$ additional bits are allocated to a signed integer exponent, M.

  Range of M is from—$(2^N - 1)$ to $(2^N - 1)$.

  (e.g., $N = 7 \Rightarrow -127 \leq M \leq 127$ and $2^M \sim 10^{-38}$ to 10^{38})

9.1.2.1 Representation of Negative Numbers

- Sign magnitude representation: the earliest form for number representation, still used some what:
  - The leading bit represents the sign
  - The value bits represent the magnitude

    Example: $0.1001_b = +2^{-1} + 2^{-4} = 0.5625_d$

    $1.1001_b = -0.5625_d$

  (*Note*: 0 has 2 representations: 0.0000 and 1.0000.)
- Two's complement representation

This is the most-used representation of binary numbers in μ-processors.

- Positive numbers are the same as above $\rightarrow$ first bit = 0.
- A number is negated by complementing bits and adding 1 to the LSB.

EXAMPLE 9.1

−[0.1001] = [1.0110] + [0.0001] = [1.0111]
And—[1.0111] = [0.1000] + [0.0001] = [0.1001]
Note: In two's complement representation

$$\text{if } x: 0. a_1 \, a_2 \, a_3 \cdots a_c \Rightarrow x = \sum_{i=1}^{C} a_i 2^{-i}$$

$$\text{Then, } -x = 1 - \sum_{i=1}^{C} a_i 2^{-i}.$$

Advantages:

1. 0 and −0 are same
2. -2^{-C} is $1.111 \cdots 1$
3. Addition carries through, to sign bit, that is,

$$1.111 \cdots 1 + 0.00 \cdots 1 = 0.00 \cdots 0$$

4. Subtraction is via the addition of the two's complement of the number being subtracted

9.1.3 Digital Quantization of a Continuous Value

- These are two basic types of quantization:
  - Truncation: values are approximated by the nearest quantization level for which the magnitude is less than the sample magnitude.
  - Roundoff: values are approximated by the nearest quantization level.
- Truncation (t):

  All bits less than the LSB are discarded. The quantization is shown in Figure 9.2.

  The truncation error is $\varepsilon_t = Q_t(x)$:

$$-2^{-C} < \epsilon_t < 0 \quad \text{(for two's complement)}$$

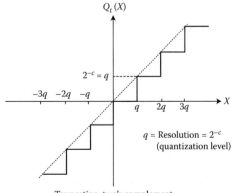

Truncation, two's complement

FIGURE 9.2
Truncation at time t.

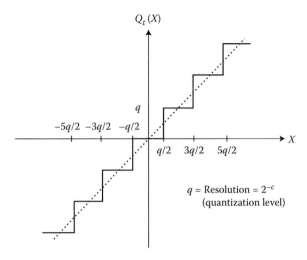

FIGURE 9.3
Round-off error in both representation methods.

- Roundoff (r)

  Rounding off a binary number to C bits is done by choosing the C-bit closest to the unrounded quantity.

  Examples: $C = 3$ $0.011|01 \rightarrow 0.011$, $0.11|10 \rightarrow 0.100$, etc.

  Round-off error is shown in Figure 9.3.

  The round-off error is $\epsilon = Q_r(x) - x$

$$\boxed{-\frac{2^{-C}}{2} \leq \epsilon_r \leq \frac{2^{-C}}{2}}$$

- Implementation of round-off in computer simulation:

  $x = $ "continuous" value; $\quad x_q = [x + \text{sign}(x)^*q/2]/q; \quad Q_r(x) = q^*\text{INT}(x_q)$

9.1.4 Sources of Numerical Errors in Digital Control

- Four primary sources of error are shown in Figure 9.4:
  - Errors in A/D converters, ϵ_{ad}: $f(t) \rightarrow \text{A/D} \rightarrow f_k = f(kh) + \epsilon_{ad}$
  - Errors in arithmetic operations, ϵ_{mi} (* and +)

EXAMPLE 9.2
Decimal arithmetic and $C + 1 = 4$ digit word length ($q = 10^{-3}$):

$$0.140 \times 0.120 = 0.0168$$

$$\text{Truncation} \rightarrow 0.016 \quad |\epsilon_t| = 0.0008$$

$$\text{Rounding} \rightarrow 0.017 \quad |\epsilon_r| = 0.0002$$

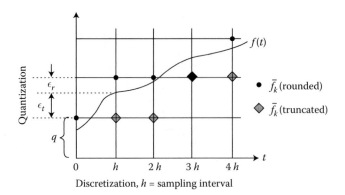

FIGURE 9.4
Sources of errors.

EXAMPLE 9.3

Binary arithmetic, $C = 4$ bits (X and Y positive), $q = 2^{-4}$,

$$\left.\begin{array}{l} x = .1110 \\ y = .1011 \end{array}\right\} \quad z = XY = .1001|1010 \quad \begin{array}{l} \nearrow Z_t = .1001 \\ \\ \searrow Z_r = .1010 \end{array}$$

- Errors in parameters and coefficient storage [e.g., α_i, β_i in $H(z)$]. The implemented $H(z)$ only approximate the $H(z)$ designed.

EXAMPLE 9.4

A coefficient in a digital compensator has been designed as $\alpha = 0.37345$; the algorithm is to be implemented on an 8-bit microcomputer ($2^{-7} = 1/128 = 0.0078125 = q$).

$$[0.0101111]_b = [0.3671875]_d$$

$$\text{true } \alpha \text{ value} = 0.37345$$

$$[0.0110000]_b = [0.375]_d$$

Truncation: $\alpha_t = 0.3671875$ with error $|\epsilon_t| = 6.2625 \times 10^{-3}$ round-off: $\alpha_r = 0.375$ with error $|\epsilon_r| = 1.55 \times 10^{-3}$.

The effect of such errors on the compensator's behavior must be taken into account in the design process.

EXAMPLE 9.5

Control algorithm given by

$$H(z) = \frac{u(z)}{e(z)} = \frac{z}{z - p} \quad (p \text{ is a quantized parameter})$$

p quantized $\rightarrow$ error $\Delta p \rightarrow$ shift in pole of $H(z)$.
Suppose $\Delta p/p = 0.1$ (large quantization).

- If $p_{\text{nominal}} = 0.5$ major effect
- If $p_{\text{nominal}} = 0.9$ severe effect!

Errors in the D/A converters.

These errors are usually small in comparison with other errors, which can be neglected easily.

9.1.4.1 Visualization of the Numerical Errors

For a first-order system with transfer function:

$$H(z) = \frac{u(z)}{e(z)} = \frac{\beta_0 + \beta_1 z^{-1}}{1 + \alpha_1 z^{-1}}$$

Error visualization is shown in Figure 9.5.
Important remarks:

- Fast varying round-off errors (ϵ_{ad}, ϵ_{mi}) cause noisy outputs but do not influence stability.
- Errors in coefficients ($\Delta\alpha_i$, $\Delta\beta_i$) influence $H(z)$ dynamics and stability.
- The effect of ϵ_{ad} on the output depends solely on the transfer function $H(z)$, and not on the algorithm structure.
- The generation of multiplication errors ϵ_{mi} and their propagation depends on the realization structure of the control algorithm.
- Next we study different realizations.

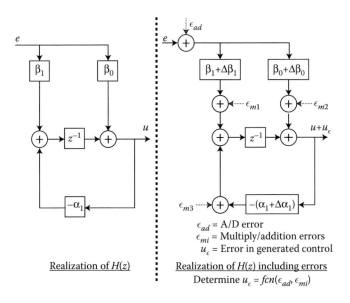

ϵ_{ad} = A/D error
ϵ_{mi} = Multiply/addition errors
u_ϵ = Error in generated control

Realization of $H(z)$

Realization of $H(z)$ including errors
Determine $u_\epsilon = fcn(\epsilon_{ad}, \epsilon_{mi})$

FIGURE 9.5
Numerical errors visualization in the realization of $H(z)$, including errors.

9.1.5 Algorithm Realization Structures

A given $H(z)$ can be realized (with hardware or software) in different ways as shown in Figure 9.5.

1. Direct form 1 (or standard observable form—SOF)
2. Direct form 2 (or standard controllable form—SCF)
3. Parallel form (or Jordan Canonical)
4. Cascade form

Different realizations should lead to different generations and propagations of numerical errors, ϵ_{mi}:

1. Direct realization 1 (SOF) is shown in Figure 9.6:

$$\frac{u(z)}{e(z)} = H(z) = \frac{\sum_{j=0}^{m} \beta_j}{1 + \sum_{j=1}^{m} \alpha_j z^{-j}} = \frac{\beta_0 + \beta_1 z^{-1} + \cdots + \beta_m z^{-m}}{1 + \alpha_1 z^{-1} + \cdots + \alpha_m z^{-m}}$$

2. Direct realization 2 (SCF):

$$\frac{u(z)}{e(z)} = H(z) = \frac{\sum_{j=0}^{m} \beta_j}{1 + \sum_{j=1}^{m} \alpha_j z^{-j}} = \frac{\beta_0 + \beta_1 z^{-1} + \cdots + \beta_m z^{-m}}{1 + \alpha_1 z^{-1} + \cdots + \alpha_m z^{-m}} \quad \text{as shown in Figure 9.7}$$

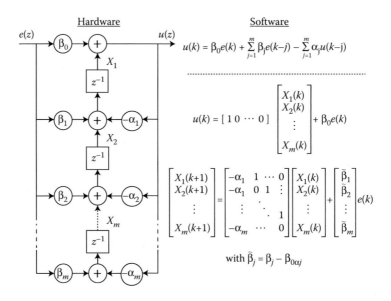

FIGURE 9.6
Hardware and software structures of SOF.

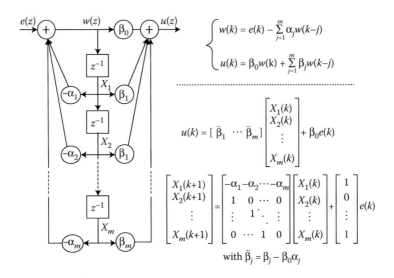

$$\begin{cases} w(k) = e(k) - \sum_{j=1}^{m} \alpha_j w(k-j) \\ u(k) = \beta_0 w(k) + \sum_{j=1}^{m} \beta_j w(k-j) \end{cases}$$

$$u(k) = [\, \bar\beta_1 \cdots \bar\beta_m \,] \begin{bmatrix} X_1(k) \\ X_2(k) \\ \vdots \\ X_m(k) \end{bmatrix} + \beta_0 e(k)$$

$$\begin{bmatrix} X_1(k+1) \\ X_2(k+1) \\ \vdots \\ X_m(k+1) \end{bmatrix} = \begin{bmatrix} -\alpha_1 & -\alpha_2 \cdots -\alpha_m \\ 1 & 0 \cdots 0 \\ \vdots & 1 \ddots \vdots \\ 0 & \cdots 1 \quad 0 \end{bmatrix} \begin{bmatrix} X_1(k) \\ X_2(k) \\ \vdots \\ X_m(k) \end{bmatrix} + \begin{bmatrix} 1 \\ 0 \\ \vdots \\ 1 \end{bmatrix} e(k)$$

with $\bar\beta_j = \beta_j - \beta_0 \alpha_j$

FIGURE 9.7
Hardware and software structure of SCF.

The hardware and software structures are shown in Figure 9.7.

3. Parallel realization (Jordan Canonical Form [JCF]):

The transfer function is decomposed using partial fraction expansion (PFE):

$$\frac{u(z)}{e(z)} = H(z) = \beta_0 + H_1(z) + \cdots + H_p(z) \quad 1 \le p \le m$$

with N_f first-order elements: $H_j(z) = \dfrac{A_j}{z + K_j} \quad j = 1,2,\ldots,N_f$

and N_s second-order elements: $H_i(z) = \dfrac{A_{i1}z + A_{i2}}{z^2 + K_{i1}z + K_{i2}} \quad i = 1,2,\ldots,N_s.$

The hardware and software structures are illustrated in Figure 9.8.

4. Cascade realization (CR):

The transfer function is expressed as a product of simple block elements:

$$\frac{u(z)}{e(z)} = H(z) = \beta_0 + H_1(z) + \cdots + H_p(z) \quad 1 \le p \le m$$

with first-order elements: $H_p(z) = \dfrac{1 + \varphi_j z^{-1}}{1 + \gamma_p z^{-1}}$

and second-order elements: $H_p(z) = \dfrac{1 + \varphi_p Z^{-1} + \varphi_{p+1} z^{-2}}{1 + \gamma_p z^{-1} + \gamma_p z^{-2}}.$

The hardware and software structures of cascade realization are illustrated in Figure 9.9.

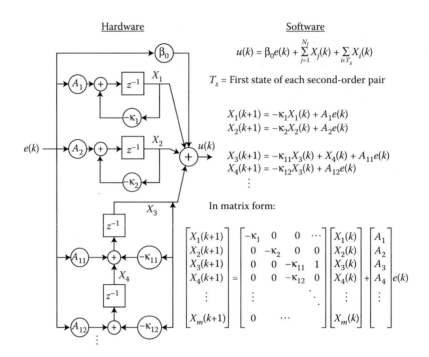

FIGURE 9.8
Hardware and software structures of JC.

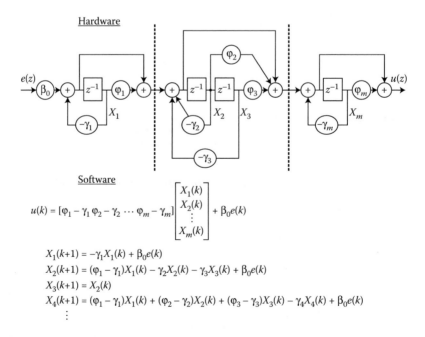

FIGURE 9.9
Hardware and Software structures of CR.

EXAMPLE 9.6: NUMERICAL EXAMPLE [10]

$$\frac{u(z)}{e(z)} = H(z) = \frac{3 + 3.6z^{-1} + 0.6z^{-2}}{1 + 0.1z^{-1} - 0.2z^{-2}}$$

1. Direct 1:

$$u(k) = 3e(k) + 3.6e(k-1) + 0.6e(k-2) - 0.1u(k-1) + 0.2u(k-2) \quad \text{or}$$

$$u(k) = \begin{bmatrix} 1 & 0 \end{bmatrix}\begin{bmatrix} x_1(k) \\ x_2(k) \end{bmatrix} + 3e(k); \quad \begin{bmatrix} x_1(k+1) \\ x_2(k+1) \end{bmatrix}$$

$$= \begin{bmatrix} -0.1 & 1 \\ 0.2 & 0 \end{bmatrix}\begin{bmatrix} x_1(k) \\ x_2(k) \end{bmatrix} + \begin{bmatrix} 3.6 - 0.3 \\ 0.6 + 0.6 \end{bmatrix}e(k)$$

2. Direct 2:

$$w'(k) = e(k) - 0.1w(k-1) + 0.2w(k-2)$$

$$u(k) = 3w(k) + 3.6w(k-1) + 0.6w(k-2) \quad \text{or}$$

$$u(k) = \begin{bmatrix} 3.6 - 0.3 & 0.6 + 0.6 \end{bmatrix}\begin{bmatrix} x_1(k) \\ x_2(k) \end{bmatrix} + 3e(k);$$

$$\begin{bmatrix} x_1(k+1) \\ x_2(k+1) \end{bmatrix} = \begin{bmatrix} -0.1 & 0.2 \\ 1 & 0 \end{bmatrix}\begin{bmatrix} x_1(k) \\ x_2(k) \end{bmatrix} + \begin{bmatrix} 1 \\ 0 \end{bmatrix}e(k)$$

3. Parallel form:

$$\frac{u(z)}{e(z)} = 3 + \frac{3.3z + 1.2}{z^2 + 0.1z - 0.2} = 3 + \frac{0.5}{z + 0.5} + \frac{2.8}{z - 0.4}$$

$$u(k) = \begin{bmatrix} 1 & 1 \end{bmatrix}\begin{bmatrix} x_1(k) \\ x_2(k) \end{bmatrix} + 3e(k);$$

$$\begin{bmatrix} x_1(k+1) \\ x_2(k+1) \end{bmatrix} = \begin{bmatrix} -0.5 & 0 \\ 0 & 0.4 \end{bmatrix}\begin{bmatrix} x_1(k) \\ x_2(k) \end{bmatrix} + \begin{bmatrix} 0.5 \\ 2.8 \end{bmatrix}e(k)$$

4. Cascade form:

$$\frac{u(z)}{e(z)} = 3\frac{(1 + 1.0z^{-1})}{(1 + 0.5z^{-1})}\frac{(1 + 02z^{-1})}{(1 - 0.4z^{-1})}$$

$$u(k) = \begin{bmatrix} 0.5 & 0.6 \end{bmatrix} \begin{bmatrix} x_1(k) \\ x_2(k) \end{bmatrix} + 3e(k);$$

$$\begin{bmatrix} x_1(k+1) \\ x_2(k+1) \end{bmatrix} = \begin{bmatrix} -0.5 & 0 \\ 0.5 & 0.4 \end{bmatrix} \begin{bmatrix} x_1(k) \\ x_2(k) \end{bmatrix} + \begin{bmatrix} 3 \\ 3 \end{bmatrix} e(k)$$

Next, we examine the effects of various realizations on the generation and propagation of numerical errors.

9.2 Analysis of Control Algorithm Implementation

A statistical model of truncation and round-off errors is given by

$$\text{error: } \epsilon = Q(x) - x; \quad Q(x) = \text{value represented in } \frac{A}{D}, CPU; \quad x = \text{exact value}$$

- If $q = 2^{-C}$ is the LSB, we model ϵ as a random variable uniformly distributed on $[-q/2 \; q/2]$ for round-off or on $[-q \; 0]$ for truncation.
- Assume no correlation between various sources of error.
- Assume no correlation between errors at different time steps.
- ϵ is modeled as a "white" noise.

A truncation PDF is shown in Figure 9.10 and round-off PDF in Figure 9.11.
Note: We will restrict our attention to the analysis of round-off errors since this error distribution has zero mean.

9.2.1 Response of Discrete Systems to White Noise

- Let $\epsilon_0 \triangleq (A/D)$ quantization noise, ϵ_{ad}

$$\epsilon_i \triangleq i\text{th} \frac{\text{multiplication}}{\text{addition}} \text{noise}, \quad \epsilon_{mi}, \quad i \geq 1$$

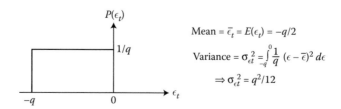

Mean $= \bar{\epsilon}_t = E(\epsilon_t) = -q/2$

Variance $= \sigma_{\epsilon t}^2 = \int_{-q}^{0} \frac{1}{q}(\epsilon - \bar{\epsilon})^2 \, d\epsilon$

$\Rightarrow \sigma_{\epsilon t}^2 = q^2/12$

FIGURE 9.10
Truncation PDF.

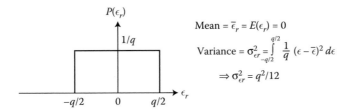

FIGURE 9.11
Round-off PDF.

$u_{\epsilon i} \triangleq$ component of u_ϵ arising from propagation of ϵ_i by superposition,

$$u_\epsilon(k) = \sum_i u_{\epsilon i}(k)$$

- The propagation of a quantization noise ϵ_i depends on the transfer function, $H_i(z)$, between the source of error and the "u" as the output:

$$\{H_0(z) = H(z)\}$$

- Given $\bar{\epsilon}_i$ and $\sigma_{\epsilon i}$, how can we find $\bar{u}_{\epsilon i}$ and $\sigma_{u_{\epsilon i}}$? $\left(\sigma_{\epsilon i} = \dfrac{q}{\sqrt{12}} \quad \text{for } i \geq 1\right)$.
- If the system is stable and linear, the response $\epsilon_i \rightarrow u$ can be described by a transfer function $H_i(z)$ or by the impulse response $h_j(k)$. Then,

$$\bar{u}_{\epsilon i} = E\{u_{\epsilon i}(k)\} = \sum_{j=0}^{\infty} h_i(j) E\{\epsilon_i(k-j)\} \tag{9.1}$$

$$\bar{u}_{\epsilon i} = \bar{\epsilon}_i \sum_{j=0}^{\infty} h_i(j) = 0 \quad \text{if rounding is used} \tag{9.2}$$

The variance of $u_{\epsilon i}$ is computed in a similar manner:

$$\sigma_{u_{\epsilon i}}^2 = \sigma_{\epsilon i}^2 \sum_{j=0}^{\infty} |h_i(j)|^2 \tag{9.3}$$

An alternate approach is via Parseval's theorem:

$$\sigma_{u_{\epsilon i}}^2 = \sigma_{\epsilon i}^2 \frac{1}{2\pi j} \oint_{\text{around}|z|=1} H_i(z) H_i(z^{-1}) z^{-1} dz \tag{9.4}$$

(shows that $H_i(z)$ and $z^{-k}H_j(z)$ will give the same results)

TABLE 9.2

h, $|\alpha|$ and Noise Amplification
Factor $1/(1 - \alpha^2)$

h	$|\alpha|$	$\dfrac{1}{1 - \alpha^2}$
0.01	0.990	50.8
0.001	0.999	500.0

EXAMPLE 9.7

$$H_1(z) = \frac{u_{\epsilon_1}(z)}{\epsilon_1(z)} = \frac{1}{1 + \alpha z^{-1}}$$

$u_{\epsilon 1}(k) = -\alpha u_{\epsilon 1}(k-1) + \epsilon_1(k), |\alpha| < 1;$ impulse response: $h_1(k) = (-\alpha)^k$

Method 1 (convolution): Use Equation 9.3 to obtain.

$$\sigma_{u_{\epsilon 1}}^2 = \sigma_{\epsilon 1}^2 \sum_{k=0}^{\infty} \alpha^{2k} = \frac{q^2}{12} \cdot \frac{1}{1 - \alpha^2}$$

Method 2 (Parseval's theorem): Use Equation 9.4:

$$\sigma_{u_{\epsilon i}}^2 = \sigma_{\epsilon i}^2 \frac{1}{2\pi j} \oint_{|z|=1} \frac{1}{1 + \alpha z^{-1}} \frac{1}{1 + \alpha z} z^{-1} dz$$

since

$$|\alpha| < 1 \Rightarrow \sigma_{u_{\epsilon i}}^2 = \frac{q^2}{12} \frac{1}{2\pi j} \oint_{|z|=1} \frac{1}{z + \alpha} \frac{1}{1 + \alpha z} dz$$

$$\frac{1}{2\pi j} \oint_{|z|=1} \frac{1}{z + \alpha} \frac{1}{1 + \alpha z} dz = \text{residue } (@z = -\alpha) = \frac{1}{1 + \alpha z} \Big|_{z=-\alpha} \Rightarrow \sigma_{u_{\epsilon i}}^2 = \frac{q^2}{12} \frac{1}{1 - \alpha^2}$$

Typically, $\alpha \sim -(1 + ah)^{-1}$ corresponding to a discretization of an $H(s)$ of the form $(s + a)^{-1}$. Assume that $a = 1$ and $h \ll 1$, as shown in Table 9.2.

Conclusion: quantization noise is amplified by fast sampling.

Due to the fast sampling rate, the difference between two successive numbers is small and only the LSBs are changing.

9.2.2 Propagation of Multiplication Errors through the Controller

There is no simple theory to help the control designer estimate the noise that will be generated by a particular realization of the control algorithm. A simple example demonstrates the phenomenon:

EXAMPLE 9.8

We assume round-off error model with $\beta_0 = 0$:

$$\frac{u(z)}{e(z)} = H(z) = \frac{A_1}{z + K_1} + \frac{A_2}{z + K_2}$$

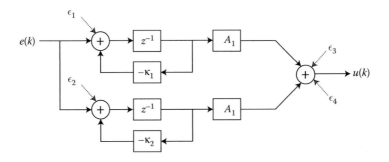

FIGURE 9.12
Parallel realization case (a).

We investigate the multiplication noise (finite word length registers).

Two cases of parallel realization will be compared first. The parallel realization as case (a) is illustrated in Figure 9.12.

- Parallel realization—case (a):
  Using Parseval's theorem and the additive property of variance (since the various errors are assumed independent), and all $\sigma_{\epsilon i}^2 = \sigma_{\epsilon}^2 = q^2/12$:

$$\sigma_{u_\epsilon}^2 = \sigma_\epsilon^2 \frac{1}{2\pi j} \sum_{i=1}^{4} \oint H_i(z)H_i(z^{-1})z^{-1}dz \tag{9.5}$$

$$\sigma_{u_\epsilon}^2 = \frac{q^2}{12}\left[2 + \frac{A_1^2}{1 - K_1^2} + \frac{A_2^2}{1 - K_2^2} \right] \tag{9.6}$$

Note: Here $H_1(z) = \dfrac{A_1 z^{-1}}{1 + K_1 z^{-1}}$; $H_2(z) = \dfrac{A_2 z^{-1}}{1 + K_2 z^{-1}}$; $H_3(z) = H_4(z) = 1$

$\left[H_i(z) = \text{transfer function from } \epsilon_i \text{ to } u \right]$
We will now change the position of the gains A_i for case (b).

- Parallel realization—case (b):
  Parallel realization case (b) is shown in Figure 9.13.

Here $H_1(z) = H_3(z) = \dfrac{z^{-1}}{1 + K_1 z^{-1}}$; $H_2(z) = H_4(z) = \dfrac{z^{-1}}{1 + K_2 z^{-1}}$

$$\sigma_{u_\epsilon}^2 = \sigma_\epsilon^2 \frac{1}{2\pi j} \sum_{i=1}^{4} 2\oint H_i(z)H_i(z^{-1})z^{-1}dz \tag{9.7}$$

$$\Rightarrow \sigma_{u_\epsilon}^2 = \frac{q^2}{12} \cdot 2\left[\frac{1}{1 - K_1^2} + \frac{1}{1 - K_2^2} \right] \tag{9.8}$$

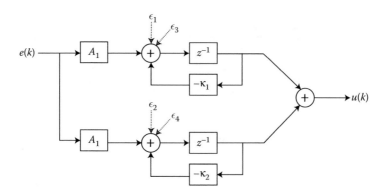

FIGURE 9.13
Structure of parallel realization case (b).

Remark

Comparing the two cases, Equations 9.6 and 9.8, we can see that for large As, case (b) generates a lower quantization noise power level.

(Case (b) has "feedback" around all of ϵ_j.)

- Direct realization:

$$H(z) = \frac{A_1 z^{-1}}{1 + K_1 z^{-1}} + \frac{A_2 z^{-1}}{1 + K_2 z^{-1}} = \frac{(A_1 + A_2)z^{-1} + (A_1 k_2 + A_2 K_1)z^{-2}}{1 + (K_1 + K_2)z^{-1} K_1 K_2 z^{-2}}$$

$$\Rightarrow H(z) = \frac{\beta_1 z^{-1} + \beta_2 z^{-2}}{1 + \alpha_1 z^{-1} + \alpha_2 z^{-2}}$$

The structure of the direct realization is shown in Figure 9.14.
The multiplication noises ε_i are generated in two different nodes:

$$H_1(z) = \frac{u(z)}{\epsilon_1(z)} = \frac{z^{-2}}{1 + \alpha_1 z^{-1} + \alpha_2 z^{-2}} = \frac{1}{(z + K_1)(z + K_2)} = \frac{1}{\Delta(z)} = H_3(z)$$

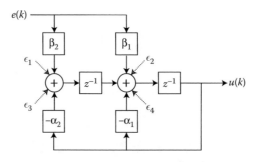

FIGURE 9.14
Structure of direct realization.

Applying Parseval's theorem and computing the variance $\sigma_{u_\epsilon}^2$:

$$\sigma_{u_\epsilon}^2 = \frac{q^2}{12} \cdot 4 \frac{1 + K_1 K_2}{(1 - K_1 K_2)(1 - K_1^2)(1 - K_2^2)} \tag{9.9}$$

Comparing this noise amplification to the one for parallel case (b):

$$r = \frac{\sigma_{u_\epsilon}^2 \text{ parallel}}{\sigma_{u_\epsilon}^2 \text{ direct}} = \frac{(2 - K_1^2 - K_2^2)(1 - K_1 K_2)}{1 + K_1 K_2}$$

At high sampling rates, $K_i \to 1$ and $r \to 0$; therefore, the parallel realization is preferable.

9.2.3 Parameter Errors and Sensitivity Analysis

- So far, we have analyzed the effects of quantization on the dynamic variables such as y (output) and u (control). However, the microcontroller must also store the equation coefficients and therefore quantize the parameter values.
- One effect of finite word length is that the controller's parameters must be chosen from a finite set of allowable values. Coefficients of arbitrary precision must be altered for implementation using digital computing devices.
- One approach is to select a filter structure that is not overly sensitive to coefficient inaccuracies. The problem is, therefore, to know how a change in parameters from α_k to $\alpha_k + \Delta\alpha_k$ produces a change in the poles of $H(z)$ from λ_j to $\lambda_j + \Delta\lambda_j$.

9.2.3.1 Solution

Define $P(z, \alpha_k)$ as the characteristic polynomial/denominator of $H(z)$, in which the values of α are the coefficients.

$$P(z, \alpha_k) = z^m + \alpha_1 z^{m-1} + \cdots + \alpha_m$$

$$P(\lambda_j, \alpha_k) = 0$$

where λ_j ($j = 1, \ldots, m$) represents the poles of $H(z)$.

A variation $\alpha_k + \Delta\alpha_k$ in parameter α_k yields a relocation of the poles of $H(z)$ from $\{\lambda_1, \lambda_2, \ldots\}$ to $\{\lambda_1 + \Delta\lambda_1, \lambda_2 + \Delta\lambda_2, \ldots\}$

$$P(\lambda j + \Delta\lambda j, \alpha k + \Delta\alpha k) = P(\lambda j, \alpha k) + \left.\frac{\partial P}{\partial z}\right|_{z=\lambda_j} \Delta\lambda_j + \left.\frac{\partial P}{\partial \alpha_k}\right|_{z=\lambda_j} \Delta\alpha_j + \cdots \tag{9.10}$$

where

$$P(\lambda j + \Delta\lambda j, \alpha k + \Delta\alpha k) = 0 \to \text{ by definition of CE;}$$

and

$$P(\lambda j, \alpha k) = 0 \to \text{ by definition of CE}$$

Therefore,

$$\Delta\lambda_j = -\frac{(\partial P / \partial \alpha_k)}{(\partial P / \partial z)}\bigg|_{z=\lambda_j} \cdot \Delta_k \tag{9.11}$$

which can be shown to be:

$$\Delta\lambda_j = \frac{\lambda_j^{m-k}}{\prod_{i \neq j}(\lambda_j - \lambda_i)} \cdot \Delta\alpha_k \tag{9.12}$$

From Equation 9.12, remarks about pole location sensitivity to parameters:

- Generally $|\lambda_j| < 1$, so the smaller the exponent $(m-k)$, the larger the variation in $\Delta\lambda_j$. The most sensitive parameter is α_m, the constant term, where $k = m$.
- However, for values of λ_j near the unit circle, the relative sensitivity decreases slowly as k gets smaller.
- From the denominator, if poles are in a cluster then the sensitivity is high; therefore, design poles should be apart in order to reduce sensitivity.

  Note: As $h \to 0$, all roots $\to 1 \Rightarrow$ high sensitivity for a small time step. Also if m increases the likelihood of two roots being close increases.

- So, keep m small and roots apart. Implement $H(z)$ as a parallel combination of first- and second-order terms.

EXAMPLE 9.9

Consider a second-order system:

$$P(z) = z^2 + \alpha_1 z + \alpha_2 = (z - re^{j\theta})(z - re^{-j\theta}) = z^2 - (2r\cos\theta)z + r^2$$

Suppose we quantize α_1 to $\bar{\alpha}_1$, and α_2 to $\bar{\alpha}_2$. Graphically, quantizing α_1 (i.e., $-2r\cos\theta$) restricts the poles to lie on a finite number of poles to of vertical lines $r\cos\bar{\alpha}_1/2$ and quantizing α_2 (i.e., r_2) further restricts the poles to lie on circles of radius $r = \sqrt{\bar{\alpha}_2}/2$.
  The possible pole locations after quantization are shown in Figure 9.15.
  To examine the phenomenon of pole migration, perform a sensitivity analysis.

$$r = \sqrt{\alpha_2}; \quad \theta = \cos^{-1}\left(-\frac{\alpha_1}{2\sqrt{\alpha_2}}\right)$$

The pole migration, due to errors $\Delta\alpha_1$ and $\Delta\alpha_2$, is given by

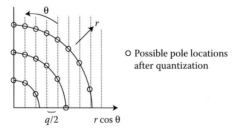

FIGURE 9.15
Possible pole locations of a second-order system after quantization.

$$\Delta r = \frac{\partial r}{\partial \alpha_1} \Delta \alpha_1 + \frac{\partial r}{\partial \alpha_2} \Delta \alpha_2; \quad \Delta \theta = \frac{\partial \theta}{\partial \alpha_1} \Delta \alpha_1 + \frac{\partial \theta}{\partial \alpha_2} \Delta \alpha_2$$

$$\frac{\partial r}{\partial \alpha_1} = 0 \quad \frac{\partial \theta}{\partial \alpha_1} = \frac{1}{2r \sin \theta}$$

$$\frac{\partial r}{\partial \alpha_2} = \frac{1}{2r} \quad \frac{\partial \theta}{\partial \alpha_2} = \frac{1}{2r^2 \tan \theta}$$

$$\Rightarrow \Delta r = \frac{\Delta \alpha_2}{2r} \quad \text{and} \quad \Delta \theta = \frac{\Delta \alpha_1}{2r \sin \theta} + \frac{\Delta \alpha_2}{2r^2 \tan \theta}$$

For a given r, as $\theta \to 0$, $\Delta \theta \to \infty$.
For a given θ as $r \to 0$, $\Delta \theta \to \infty$.
The above two variations in the parameter suggest that one should
$\to$ Expect large sensitivity in pole locations with q when θ and/or $r \sim 0$.

EXAMPLE 9.10

Second-order system with real poles λ_1 and λ_2:

a. Direct realization: $P(z) = z^2 - (\lambda_1 + \lambda_2)z + \lambda_1 \lambda_2$; here we can see that $\alpha_1 = -(\lambda_1 + \lambda_2)$ and $\alpha_2 = \lambda_1 \lambda_2$:

$$\frac{\partial P}{\partial \alpha_1} = +z \quad \frac{\partial P}{\partial z} = 2z + \alpha_1 \quad \frac{\partial P}{\partial \alpha_2} = 1$$

Using Equation 9.11:

$$\frac{\Delta \lambda_1}{\Delta \alpha_1} = \frac{z}{2z + \alpha_1}\Big|_{z=\lambda_1} = \frac{\lambda_1}{\lambda_1 - \lambda_2}$$

$$\frac{\Delta \lambda_2}{\Delta \alpha_2} = \frac{1}{2z + \alpha_1}\Big|_{z=\lambda_2} = \frac{-1}{\lambda_1 - \lambda_2}$$

$\Rightarrow$ For closely located poles $(\lambda_1 \approx \lambda_2)$, the direct realization is highly sensitive to variations in coefficients. (Do not use it!)

b. Cascade and parallel realizations: The coefficients in the algorithm are the poles themselves, $\Delta \lambda_i = \Delta \alpha_i$.

EXAMPLE 9.11

$$H(z) = \frac{u(z)}{e(z)} = \frac{z^2}{z^2 + \alpha_1 z + \alpha_2} \alpha_1 = -0.9, \quad \alpha_2 = 0.2$$

Unit pulse response with precise arithmetic and with a 3-bit quantization.

a. Precise:

$$H(z) = \frac{z^2}{z^2 - 0.9z + 0.2} = \frac{-4z}{z - 0.4} + \frac{5z}{z - 0.5}$$

$$\Rightarrow h(k) = 5(0.5)^k - 4(0.4)^k$$

$$= \{1, 0.9, 0.369, 0.2101, \ldots\}$$

b. Quantized:

$$\bar{H}(z) = \frac{z^2}{z^2 - 0.857z + 0.286}$$

$$= \frac{z^2}{(z - 0.429)^2 + 0.319^2}$$

$$\bar{h}(k) = 0.535^k \left[\cos 0.641k + 1.341 \sin 0.641k \right]$$

$$\bar{h}(k) = \{1, 0.858, 0.449, 0.140, \ldots\}$$

A Criterion for the Selection of Word-Length

- Quantization affects pole locations of $H(z)$.
  - Suppose one did an $H(s)$ design$\rightarrow \tilde{H}(z)$ via equivalents.
- $\Rightarrow \tilde{H}(z)$ implemented $\sim$ a different $H(s)$ than was designed.
  - From sensitivity analysis one learns: maximum permitted movement in pole s_1 of $H(s)$ is $s_1 \pm \Delta s$.
    Determine effective: Δs due to quantization.
- Consider a first-order $H(s)$ with pole s_1 and pole-zero mapping as shown in Figure 9.16:

$$|\Delta\lambda| = \frac{\left| e^{(s_1 + \Delta s)h} - e^{(s_1 + \Delta s)h} \right|}{2} \overset{def}{=} |s_1| \left| he^{s_1 h} \right| \left| \frac{\Delta s}{s_1} \right|$$

For a first-order $H(z)$, $|\Delta\lambda| = |\Delta\alpha_1| \leq q = 2^{-C}$ ($C + 1$ bit μ-processor, fixed-point with truncation).

Take logarithms:

$$C = -\log_2 |s_1| \left| he^{s_1 h} \right| - \log_2 \left| \frac{\Delta s}{s_1} \right| \tag{9.13}$$

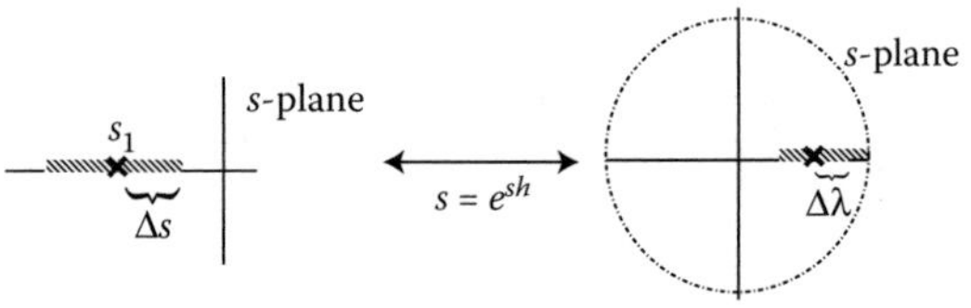

FIGURE 9.16
First order $H(s)$ with pole-zero mapping.

Determines word-length as a function of Δs.

Note: As $h \to 0$, a higher word length will be needed for the same Δs.

Example: If $H(s) = (2s + 1)/(s + 1.5)$, the system can tolerate a maximum 1% sensitivity/change in the pole $H(s)$. Find C if $h = 4$ ms.

$$C = -\log_2 1.5 \, (0.004) \, e^{-0.006} - \log_2 0.1 = 7.4 + 4.6 = 12$$

9.2.4 Nonlinear Effects

Another source of error is a slowly varying quantization error.

A recursive filter (controller) with finite word length arithmetic represents a nonlinear feedback system that generates nonlinear effects such as deadband and limit cycles. We illustrate this phenomenon by examples (Katz, 1981) [10]:

EXAMPLE 9.12

- Use decimal ($q = 1$) arithmetic with an equivalent two's complement (i.e., 8.6 truncated to the decimal point is 8 and −6.6 truncated is −9).
- $H(z) = \dfrac{u(z)}{e(z)} = \dfrac{1}{1 - 0.9z^{-1}}$; that is, $u(0) = 3$.
- Input: step function $e(z) = \dfrac{z^{-1}}{1 - z^{-1}}$; $\to$ Output: $u(k) = 0.9u(k - 1) + e(k), k > 0$.
- Compute a table of $u(k)$ using decimal arithmetic (see Table 9.3).
- For large word length, $u(\infty) = 10$.
- There is a deadband (6–15 for rounding and 0–10 for truncation) that cannot be crossed.

For a comparison of the performances, these values are plotted in Figure 9.17.

TABLE 9.3

$U(K)$ for Different k Rounded-Off and Truncated with ic $u(0) = 3$

K	e(k)	u(k)	u(k)-Rounded-Off	u(k)	u(k)-Truncated
0	0	3.0	3	3.0	⋮
1	1	3.7	4	3.7	⋮
2	1	4.6	5	3.7	⋮
3	1	5.5	⋮	3.7	3
4	1	5.5	⋮	3.7	3
5	1	5.5	6	3.7	3
6	1	5.5	6	3.7	3

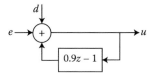

FIGURE 9.17
Nonlinear effects on output with *ic*, where $u(k) = 3$.

EXAMPLE 9.13

$$H(z) = \frac{u(z)}{e(z)} = \frac{1}{1 - 0.9z^{-1}}; \quad \text{that is } u(0) = 0.$$

- Input is a pulse of height 10 at $k = 1$, $e(z) = 10\ z^{-1}$.
- →Output: $u(k) = -0.9u(k - 1) + e(k)$, $k > 0$; steady-state value for infinite word = 0. These values are tabulated Table 9.4.
- The rounded-off value of $u(k)$ is oscillating within a limit cycle ±5.
- The truncated value of $u(k)$ converges asymptotically to the correct steady value of zero.
- Notes: Some simple numeric techniques have been developed to reduce these non-linear effects (e.g., numerical *dither*), as described in the following Example 9.14.

EXAMPLE 9.14

Use of a numerical dither (d) to avoid nonlinear effects.
We will use it here for Example 9.14: the deadband phenomenon:

$$u(z) = \left[\frac{1}{1 - 0.9z^{-1}}\right][e(z) + d]; \quad e(z) = \frac{z^{-1}}{1 - z^{-1}}; \quad u(0) = 3$$

Let $d = \pm 0.5$ be a signal included in the loop, randomly taking the values of -0.5 or 0.5 with equal probability. The nonlinear system configuration is shown in Figure 9.18.

$$u(k) = 0.9u(k - 1) + e(k) + d, \quad k > 0$$

These results are illustrated in Table 9.5 and plotted in Figure 9.19 as well.

In this case of round-off, the numerical dither helps $u(k)$ to converge to the correct steady-state value of 10. A block diagram representation of a model aircraft roll motion is shown in Figure 9.20.

TABLE 9.4

$U(K)$ for Different k Rounded-Off and Truncated with *ic* $u(0) = 0$

K	$e(k)$	$u(k)$	$u(k)$-Rounded-Off	$u(k)$	$u(k)$-Truncated
0	0	0	0	0	0
1	10	10.0	10	10.0	10
2	0	−9.0	−9	−9.0	−9
3	0	8.1	8	8.1	8
4	0	−7.2	−7	−7.2	−8
5	0	6.3	6	7.2	7
6	0	4.5	⋮	6.3	6
7	0	4.5	⋮	6.3	6
8	0	−4.5	⋮	−5.4	−6

These nonlinear effects are shown in Figure 9.18.

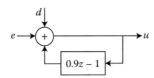

FIGURE 9.18
Nonlinear effects on output with *ic*, where $u(k) = 0$.

TABLE 9.5

Nonlinear Effects Due to Dither "*d*"

K	e(k)	d	u(k)	u(k)-Rounded-Off	u(k)	u(k)-Truncated
0	0	0	3.0	3	3.0	3
1	1	0.5	5.1	5	5.1	5
2	1	0.5	5.1	5	5.1	5
3	1	−0.5	5.0	5	5.0	5
4	1	0.5	6.0	6	6.0	6
5	1	0.5	6.9	7	6.9	6
6	1	−0.5	6.8	7	5.9	5
7	1	0.5	7.8	8	6.0	6
				.		.
				.		.
				10		?

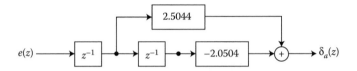

FIGURE 9.19

$H(z)$ is implemented using SCF for lateral dynamics of an aircraft (roll motion).

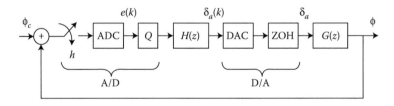

FIGURE 9.20

Model of an aircraft roll motion.

9.3 Case Study

Let us consider the simulation of a system using a relatively severe degree of quantization on the digital controller, corresponding to a short computer word length. The plant describes the simplified lateral dynamics of an aircraft (roll motion).

$$G(s) = \frac{\Phi(s)}{\delta_a(s)} = \frac{8}{s(s + 2)}$$

where Φ is the roll angle δ_a and is the aileron deflection.

Assume the model is exact and there are no errors in the model's parameters.

The design being tested is illustrated in Figure 9.21.

- The digital compensation based upon a sampling period $h = 0.1$ s is given by

$$H(z) = \frac{2.5044z - 2.0504}{z^2}$$

$H(z)$ is implemented using SCF in the following manner, as shown in Figure 9.20:

- To calculate the unit step we assume fixed-point arithmetic and a signal and parameter range [−3.75 3.75].
- Quantization: a 4-bit word length is used (1 signbit, 3 mantissa).
- Truncation with sign magnitude representation is used.

The quantization is illustrated in Figure 9.21.

$$2L = \left[2^N - 1\right]q \Rightarrow q = \frac{2L}{2^N} \begin{cases} L = 3.75 \\ N = 4 \end{cases}$$

$Qq = 0.5 = >$Quantization levels $\{q_i\} = \{0. + -0.5, - +1.0, \ldots, + -3.5\}$.
Therefore, the controller gains will be quantized as $2.5044 \rightarrow 2.5$ and $2.0504 \rightarrow 2.0$.

- We use a simulation program with simulation time step $h_1 = h/2 = 0.05$ s:

$$\dot{X}(t) = \begin{bmatrix} \dot{\Phi} \\ \ddot{\Phi} \end{bmatrix} = \begin{bmatrix} 0 & 1 \\ 0 & -2 \end{bmatrix} \begin{bmatrix} \Phi \\ \dot{\Phi} \end{bmatrix} + \begin{bmatrix} 0 \\ 8 \end{bmatrix} \delta_a(t)$$

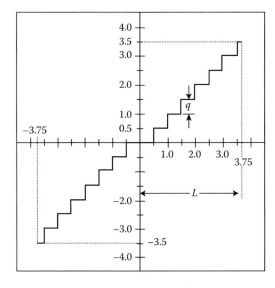

FIGURE 9.21
Quantization for fixed-point arithmetic and a signal parameter.

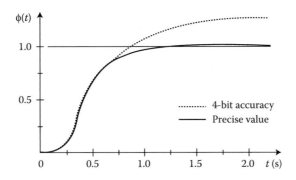

FIGURE 9.22
Step response of $\Phi(t)$ with 4-bit accuracy (dashed line) and precise value (bold).

- Discretizing $\rightarrow \Phi = e^{Ah_1}, \Gamma \ldots$
- $\rightarrow$ Control use of SCF algorithm (as in the prequel-previous page).
- Round off all control variables and parameters.
- Plot step response as shown in Figure 9.22.

Remarks:
- The limited accuracy curve is nearly the same as the ideal curve initially, except for slight variations that appear as noise.
- The steady-state value quantization is about 20% higher than the ideal curve!

9.3.1 Concluding remarks

- Amplitude quantization is a necessary evil associated with digital computers.
- With 64-bit computers becoming available, round-off is not a big issue today.
- However, great attention should be given to the issue of algorithmic realizations. Avoid direct forms that accumulate too much noise in the system and amplify numerical errors.
- At high sampling rates, poles tend to cluster around $|\lambda| = 1$. This can result in very high sensitivity to changes in controller parameters.
- Digital control implementation is yet to become an exact science at 100%; we are now up to 95% with the use of optimal control and stochastic control methodologies. Trial-and-error as well as engineering judgments are prime resources for further improvement in efficiency of digital control algorithms.

PROBLEMS

P9.1 A positive number 0.2002 needs to be converted into a digital representation with $C = 4$ bits and $M = 2$.

P9.2 A negative number 0.2002 needs to be converted into digital representation with $C = 4$ bits and $M = 2$.

P9.3 Show the 2's complement of the positive number in Problem 9.1.

P9.4   Show the 2's complement of the negative number in Problem 9.2.

P9.5   Round off the positive number in Problem 9.1 to a three bits number.

P9.6   Round off the negative number in Problem 9.2 to a three bits number.

P9.7   Perform the Direct 1 realization of the discrete function shown below:

$$\frac{u(z)}{e(z)} = H(z) = \frac{6 + 7.2z^{-1} + 1.2z^{-2}}{1 + 0.2z^{-1} - 0.4z^{-2}}$$

P9.8   Perform the Direct 2 realization of the discrete function shown below:

$$\frac{u(z)}{e(z)} = H(z) = \frac{6 + 7.2z^{-1} + 1.2z^{-2}}{1 + 0.2z^{-1} - 0.4z^{-2}}$$

P9.9   Perform the Parallel form realization of the discrete function shown below:

$$\frac{u(z)}{e(z)} = H(z) = \frac{6 + 7.2z^{-1} + 1.2z^{-2}}{1 + 0.2z^{-1} - 0.4z^{-2}}$$

P9.10 Perform the cascade form realization of the discrete function shown below:

$$\frac{u(z)}{e(z)} = H(z) = \frac{6 + 7.2z^{-1} + 1.2z^{-2}}{1 + 0.2z^{-1} - 0.4z^{-2}}$$

P9.11 For the system shown below:

$$H(z) = \frac{1}{1 + 0.9z^{-1}}$$

1. Find the response to white noise with variance $\sigma^2 = 1/12$ using the Convolution Equation 9.3.
2. Find the response to white noise with variance $\sigma^2 = 1/12$ using Parseval's Theorem in Equation 9.4.
3. Tabulate the h, $|0.9|$, and $1/(1-0.81)$ as in Table 9.2.
4. What is the effect of sampling on the results for values of $h = 0.01$ and 0.001, respectively?

P9.12 The round-off error model is given by

$$H(z) = \frac{10(z + 0.2)}{(z - 1)(z - 0.5)}$$

1. Using Parseval's theorem and the additive property of variance, investigate the propagation of the multiplicative noise through this controller as in case (a).
2. Using direct realization to investigate the propagation of the multiplicative noise through this controller.
3. Compare the two propagations for power level of quantization noise.

P9.13 For the CE given below:

$$p(z) = z^2 - 1.3z + 0.42$$

1. Find the direct realization.
2. Analyze the sensitivity of the location of poles to the variations in the coefficients.

P9.14

1. Find the unit pulse response with precise arithmetic and 3-bit quantization of the CE in Problem 9.13.
2. Use the Cascade and Parallel realizations to assess the pole location sensitivity to coefficient variation.

P9.15 $H(s)$ is given by

$$H(s) = \frac{3s + 1}{s + 2.5}$$

For a 2% sensitivity change in $H(s)$, find C if $h = 3$ ms.

P9.16 For $H(z)$ given by

$$H(z) = \frac{u(z)}{e(z)} = \frac{z^{-1}}{1 - 0.9z^{-1}}$$

With initial condition $u(0) = 0$ and input as a step function, find output $u(k)$ for $k > 0$ using decimal arithmetic.

P9.17 For $H(z)$ given by

$$H(z) = \frac{u(z)}{e(z)} = \frac{z^{-1}}{1 - 0.9z^{-1}}$$

With initial condition $u(0) = 3$ and input is a pulse function, find output $u(k)$ for $k > 0$ using k rounded off and truncated at $u(0) = 3$.

P9.18 For $d = + -0.1$ and

$$u(z) = \left[\frac{1}{1 - 0.9z^{-1}} \right] [e(z) + d]$$

$$e(z) = \frac{z^{-1}}{1 - z^{-1}}; \quad u(0) = 0$$

find the nonlinear effects.

References

1. G. F. Franklin and J. D. Powell, *Digital Control of Dynamic Systems*. Addison-Wesley Publishing Company, Reading, MA, 1980.
2. S. Almér, S. Mariéthoz, and M. Morari, Dynamic phasor model predictive control of switched mode power converters, *IEEE Transactions on Control Systems Technology*, Early Access on IEEEXplore, December, 2014.
3. T. Kailath, *Linear Systems*, Prentice Hall, Inc., New Jersey, 1980.
4. W. L. Brogan, *Modern Control Theory*, Prentice Hall, Inc., New Jersey, 1991.
5. D. L. Kleinman, On an interactive technique for Riccati equation computation, *IEEE Trans. Autom. Control*, Ac-13, 114–115, 1978.
6. S. Chapra and R. Canale, *Numerical Methods for Engineers: With Programming & Software Applications*, McGraw-Hill Companies, New York, 1998.
7. J. Ackermann, *Abtastregelung, Band I*, Springer-Verlag, Berlin, Germany, 1983.
8. J. A. Cadzow, *Discrete-Time Systems*, Prentice Hall, Inc., New Jersey, 1973.
9. K. J. Astrom and B. Wittenmark, *Computer Controlled Systems—Theory and Design*, Prentice Hall, Inc., New Jersey, 1984.
10. P. Katz, *Digital Control Using Microprocessors*, Prentice Hall, Inc., New Jersey, 1981.

The following is a collection of suggested reading material for this book on Digital Controls Applications illustrated with MATLAB®:

Further Reading

Abu-El-Haija, A. L., K. Shenoi, and A. M. Peterson, Digital filter structures having low errors and simple hardware implementation, IEEE *Trans. Circuits Syst.*, CAS-25, 593–599, August 1978.

Anderson, B. D. O. and J. B. Moore, *Linear Optimal Control*, Prentice-Hall, Englewood Cliffs, NJ, 1971.

Bar-Shalom, Y., K. C. Chang, and H. M. Shertukde, Performance evaluation of a cascaded logic for track formation in clutter, in *Proceedings 1989 IEEE SMC Conference*, Boston, MA.

Bar-Shalom, Y., K. C. Chang, and H. M. Shertukde, Performance evaluation of a cascaded logic for track formation in clutter, *IEEE Trans. Aerospace Electron. Syst.*, 25, 873–878, November 1989.

Bar-Shalom, Y., F. Palmieri, V. Kumar, and H. M. Shertukde, Analysis of wide-band cross-correlation for target detection and time delay estimation, in *Proceedings 1991 International Conference on Acoustics, Speech and Signal Processing*, Toronto, May, 1991.

Bar-Shalom, Y., F. Palmieri, A. Kumar, and H. M. Shertukde, Analysis of wide-band cross-correlation for target detection and time delay estimation, *IEEE Trans. Signal Proces.*, January 1993.

Bar-Shalom, Y., H. M. Shertukde, and K. R. Pattipati, Use of measurements from an imaging sensor for precision target tracking. *IEEE Trans. Aerospace Electron. Syst.*, 25, 863–872, November 1989.

Bar-Shalom, Y., H. M. Shertukde, and K. R. Pattipati, Precision target tracking of small extended objects, *Opt. Eng. J. (SPIE)*, 25, February 1990.

Bar-Shalom, Y., H. M. Shertukde, and K. R. Pattipati, Precision target tracking of small extended objects, in *Proceedings SPIE Conference*, Orlando, Florida, March 1989.

Bar-Shalom, Y., H. M. Shertukde, and K. R. Pattipati, Extraction of measurements from an imaging sensor for precision target tracking, in *Proceedings IEEE Conference*, ICCON, Israel, April 1989.

Benedict, T. R. and G. W. Bordner, Synthesis of an optimal set of radar track-while scan smoothing equation, *IRE Trans. Autom. Control*, 27–31, July 1962.

Boddie, J. R. et al., Digital signal processor: Architecture and performance, *Bell Syst. Technol. J.*, 60, 7, Pt. 2, 1449–1462, September 1981.

Boylestad, R. and L. Nashelsby, *Electronic Devices and Circuit Theory*, Prentice-Hall, Inc., Englewood Cliffs, NJ, 1978.

Brasch F. M. and J. B. Pearson, Pole placement using dynamic compensators, *IEEE Trans. Autom. Control*, AC-15, 34–43, February 1970.

Brockett, R. W. Poles, zeros and feedback: State space interpretation, *IEEE Trans. Autom. Control*, AC-10, 129–135, April 1965.

Brunovsky, P. A classification of linear controllable systems, *Kybernetika (Paha)*, 3, 173–187, 1970.

Bryson, A. E. and Y. C. Ho, *Applied Optimal Control*, Halsted Press, New York, 1968.

Burton, L. T. Low-sensitivity digital ladder filters, *IEEE Trans. Circuits Syst.*, CAS-22, 168–176, March 1975.

Chang, T. L. A low roundoff noise digital filter structure, *ISCAS*, 78, 1004–1008, 1978.

Chen, S. -J., S. Kulenthiran, and H. M. Shertukde, Hardware implementation of an ATM/B-ISDN switch using fuzzy-neural scheme, in *Proceedings Connecticut Symposium on Microelectronics and Optoelectronics*, Storrs, CT, March 21, 1996.

DeRusso, P. M., R. J. Roy, and C. M. Close, *State Variables for Engineers*, John Wiley and Sons, Inc., New York, 1965.

Dickinson, B. W., T. Kailath, and M. Morf, Canonical matrix fraction and state-space descriptions for deterministic and stochastic linear systems, *IEEE Trans. Autom. Control*, AC-19, 656–667, December 1974.

Feedback Instruments Limited, 5 & 6 Warren Court, Crowborough, East Sussex, TN6 2QX, UK.

Fettweis, A. Some principles of designing digital filters imitating classical filter structure, *IEEE Trans. Circuits Theory*, 314–316, March 1971.

Fortman, T. E. A matrix inversion identity, *IEEE Trans. Autom. Control*, AC-15, 599, October, 1970.

Franklin, G. F. and J. D. Powell, *Digital Control of Dynamic Systems*, Addison-Wesley Publications, New York, 1981.

Gantmacher, F. R. *Theory of Matrices*, Vols. I and II., Chelsea Publishing Company, Inc., New York, 1959.

Gardner, M. G. and J. L. Barnes, *Transients and Linear Systems*, Vol. I, John Wiley and Sons, Inc., New York, 1942.

Gilbert, E. G. Controllability and observability in multivariable control systems, *SIAM J. Control*, 1, 128–151, 1963.

Glover, K. and J. C. Willems, Parameterizations of linear dynamical systems: Canonical forms and identifiability, *IEEE Trans. Autom. Control*, AC-19, 6, 640–645, December 1974.

Godbout, L. F., H. M. Shertukde, D. Shetty, and D. Jordan, Simulation and modeling of micro-computer controlled fluid power systems, in *Proceedings ASME Winter Annual Conference*, Dallas, November 1991.

Gohberg I. C. and L. Lerer, Factorization indices and Kronecker indices of matrix polynomials, *Integ. Equations Operator Theory*, 2(2), 199–243, 1979.

Goulding, F. S. and D. A. Landis, Signal processing for semiconductor detectors, *IEEE Trans. Nucl. Sci.*, NS-29, 3, 1125–1141, June, 1982.

Gran, R. and F. Kozin, *Applied Digital Control Systems*, George Washington University Short Course Notes, Washington, DC, August 1979.

Guillermin, E. A. *The Mathematics of Circuit Analysis*, John Wiley and Sons, Inc., New York, 1949.

Ho, B. L. and R. E. Kalman, Effective construction of linear, state-variable models from input/output functions, *Regelungstechnik*, 14, 545–548, 1966.

Hoff, M. E. and M. Townsend, Single-chip n-MOS microcomputer processes signals in real time, *Electronics*, 105–110, March 1, 1979.

Houpis, C. H. and G. B. Lamont, *Digital Control Systems: Theory, Hardware, Software*, McGraw Hill, Inc., New York, 1985.

IEEE Transactions Automatic Control, Bellman Special Issue, AC-26, October 1981.

Kalman, R. E. and R. W. Koepcke, Optimal synthesis of linear sampling control systems using generalized performance indexes, *Trans. ASME*, 80, 1800–1826, 1958.

Kalman, R. E. On the general theory of control systems, in *Proceedings of the First IFAC Congress*, Vol. 1, Butterworth's, London, 1960, pp. 481–491.

Kleinman, *IEEE Transactions Automatic Control*, June 1974.

Kuo, B. C. *Digital Control Systems*, Holt, Reinhart and Winston, Inc., New York, 1980.

Lakatos, T. Adaptive digital filtering for x-ray spectroscopy, Report, Institute of Nuclear Research, Debrecen, Hungary.

Lakatos, T. Adaptive digital signal processing for X-ray spectroscopy, *Nucl. Instrum. Meth. Phys. Res.*, Section B, October 1989.

Lanczos, C. *Linear Differential Operators*, Van Norstrand Reinhold, New York, 1959.

Leondes, C. T. and L. M. Novak, Reduced-order observers for linear discrete-time systems, *IEEE Trans. Autom. Control*, AC-19, 42–46, February 1974.

Leventhal, L. A. *Introduction to Microprocessors: Software, Hardware, Programming*, Prentice-Hall, Inc., Englewood Cliffs, NJ, 1978.

Luenberger, D. G. An introduction to observers, *IEEE Trans. Autom. Control*, AC-16,596–602, December, 1971.

Martensson, K. On the matrix Riccati equation, *Inf. Sci.*, 3, 17–49, 1971.

MATLAB®, Software by MathWorks®.

Mayne, D. O. An elementary derivation of Rosenbrock's minimal realization algorithm, *IEEE Trans. Autom. Control*, AC-18(3), 306–307.

Melbert, J. G. On the filters for pulse-height and the analysis, *IEEE Trans Nucl Sci.*, NS-29(1), February, 1982.

Morf, M., G. S. Sidhu, and T. Kailath, Some new algorithms for recursive estimation in constant, linear discrete-time systems, *IEEE Trans. Autom. Control*, AC-19, 315–323, August 1974.

Moroney, P. *Issues in the Implementation of Digital Feedback Compensators*, MIT Press, Cambridge, MA, 1983.

µPD7720 *Signal Processing Interface Manual*, NEC Electronics USA, Inc.

Nagel, H. T. Jr. and V. P. Nelson, Digital filter implementation on 16-bit microcomputers, *IEEE MICRO*, 1, 1, 23–41, February 1981.

Ogata, O. *Discrete Time Control Systems*, Prentice Hall, Inc., New Jersey, 1987.

Oppenheim, A. V. and R. W. Schafer, *Digital Signal Processss*, Prentice-Hall, Inc., New Jersey, 1975.

O'Flynn, M. *Probabilities, Random Variables, and Random Processes*, Harper & Row Publishers, New York, 1982.

Paranjape, M. S. *9900 Family Systems Design*, Texas Instruments Learning Center, Dallas, Texas, 1978.

Paranjape, M. S. Microprocessor controller for a servomotor system, MS thesis, Auburn University, Auburn, AL, 1980.

Phillips, C. L. and H. Nagle Jr., *Digital Control Systems Analysis and Design*, Prentice Hall, Inc., New Jersey, 1984.

Phillips, C. L., D. L. Chenoweth, and R. K. Cavin III. z-Transform analysis of sampled-data control systems without reference to impulse functions, *IEEE Trans. Educ.*, E-11, 141–144, June 1968.

Pugh, A. C. Transmission and system zeros, *Int. J. Control*, 26, 315–324, August 1977.

Radeka, V. Signal filtering—High counting rate problems, Report—Brookhaven National Laboratory, Upton, NY.

Rosenbrock, H. H. Distinctive problems of process control, *Chem. Eng. Prog.*, 58, 43–50, September, 1962.

Schultz, D. G. and J. L. Melsa, *State Functions and Linear Control Systems*, McGraw-Hill, New York, 1967.

Shertukde, H. M., Applications of data acquisition techniques to electric machinery laboratory procedures, in *Proceedings ASEE Conference*, Toledo, Ohio, June 1992.

Shertukde, H. et al., Guide for detection of Acoustic Emissions from Partial Discharges in oil-immersed Power Transformers, IEEE guide # c.57.127.

Shertukde, H. M., L. F. Godbout, and S. C. Elmurr, Estimation of high count rate signals with a digital filter, in *Proceedings-1991 International Conference on Systems Engineering*, Dayton, Ohio, August, 1991.

Shertukde, H. M., L. F. Godbout, and S. C. Elmurr, Estimation of high count rate signals with a digital filter, in *Proceedings 1991 International Conference on Systems Engineering*, Dayton, Ohio, August 1991.

Shertukde, H. M., S. C. Elmurr, and L. F. Godbout, Spectral estimation techniques for high count rate signals in nuclear spectroscopy, in *Proceedings 1990 International Symposium on Information Theory and Its Applications (ISITA)*, Hawaii, USA, November 27–30, 1990, pp. 251–254.

Shertukde, H. M. and K. R. Pattipati, Test sequencing in hierarchical systems: Applications to electronic and electro-mechanical systems, Proceedings International Conference on CAD/CAM Robotics and Factories of the Future, New Delhi, Deccember 1989.

Shertukde, H. M. and L. F. Godbout, Nuclear pulse processors using digital filters, in *Proceedings of ICSPAT 93*, Santa Clara, CA, September 1993.

Shertukde, H. M. and S. Sambanther, Innovative use of VHDL in the VLSI implementation of an ATM/B-ISDN switch, in *Proceedings, ICSPAT97*, San Diego, CA, September 14–17, 1997.

Shertukde, H. M. and Su-Jen Chen, Hardware implementation of an ATM/B-ISDN switch, in *Proceedings Sixth International Conference on Signal Processing Applications and Technology*, Boston, October 1995.

Shertukde, H. M. and Y. Bar-Shalom, Detection and estimation for multiple targets with two omni-directional sensors in the presence of false measurements. *IEEE Trans. Acoustic Speech Signal Proces.*, 38(5), 749–763, May 1990.

Shertukde, H. M. and Y. Bar-Shalom, Target parameter estimation in the near-field with two sensors, *IEEE Trans. Acoustic Speech Signal Proces.*, August 1988.

Shertukde, H. M. and Y. Bar-Shalom, Target Parameter Estimation in the near-field with two sensors, in *Proc. ACC*, Seattle, Washington, June 1986.

Shertukde, H. M. and Y. Bar-Shalom, Target Parameter Estimation with passive sensors in the presence of false measurements, in *Proc. ACC*, Minneapolis, Minnesota, June 1987.

Shertukde, H. M. and Y. Bar-Shalom, Track detection of multiple targets with false alarms, in *Proc. International Symposium on Electronic Devices, Circuits and Systems*, Kharagpur, India, December 1987.

Shertukde, H. M. and Y. Bar-Shalom, Tracking of crossing targets with forward looking infrared sensors, *Proceedings 1989 IEEE CDC*, Tampa, Florida. (Nominated for "Best student paper" award).

Shertukde, H. M. and Y. Bar-Shalom, Tracking of crossing targets with imaging sensors, *IEEE Trans. Aerospace Electron. Syst.*, 27, 4, 582–592, July 1991.

Shertukde, H. M. and Y. Bar-Shalom, Use of track before detect approach in tracking multiple targets in low SNR environment, in *Proceedings IEEE International Conference on Systems Engineering*, Pittsburgh, August 1990.

Shertukde, H. M. Efficient call-connection model for multipoint connections in an ATM/B-ISDN network using fuzzy neural networks, *IEEE International Conference on Industrial Automation and Control; Emerging Technology Applications at Taipei*, Taiwan, May 22–27, 1995.

Shertukde, H. M. Efficient wide-band cross-correlation tracking using wavelet transforms, in *Proceedings ICSPAT '94*, Dallas, Texas, October 1994.

Shertukde, H. M. Improvements in performance of nuclear pulse processors using wavelet transforms, in *Proceedings, ICSPAT '94*, Dallas, Texas, October 1994.

Shertukde, H. M. Multiple target tracking using multirate sampled measurements and a JPDAMCF tracker, in *Proceedings Army Research Office Information Fusion Workshop*, Harper's Ferry, West Virginia, June 1–3, 1993.

Shertukde, H. M. Multirate-PDA tracker, in *Proceedings ICSE92 Conference*, Kobe, Japan, June 1992.

Shertukde, H. Tracking of crossing targets with passive sensors: Tracking of crossing targets using forward looking infrared red sensors as passive sensors and Track-Before … Detect (TBD), Methodology, VDM Mueller. ISBN # 978-3-639-29119-3, September 3, 2010.

Sokolov, R. T., J. C. Rogers, and H. M. Shertukde, Removing harmonic signal nonstationarity by dynamic resampling, in *Proceedings IEEE International Symposium on Industrial Electronics*, ISIE'95, Athens, Greece, July 10–14, 1995.

Sripad, A. B. and D. L. Snyder, Quantization errors in floating-point arithmetic, *IEEE Trans. Acoustic Speech Signal Proces.*, ASSP-26, 456–463, October 1978.

Steiglitz, K. and B. Liu, An improved algorithm for ordering poles and zeros of fixed-point recursive digital filters, *IEEE Trans. Acoustic Speech Signal Proces.*, ASSP-24, 341–343, August, 1976.

Strang, G. *Linear Algebra and Its Applications*, Academic Press, Inc., New York, 1976.

Truxal, J. G. *Control System Synthesis*, McGraw-Hill, New York, 1955.

Unpublished report from U.K. December, 1987.

VanLandingham, H. F. *Introduction of Digital Control Systems*, McMillan Publishing, New York, 1985.

Verghese, G., P. Van Dooren, and T. Kailath, Properties of the system matrix of a generalized state-space system, *Int. J. Control*, 30, 235–243, August 1979.

Wilson, R. F. An observer based aircraft automatic landing system, MS thesis, Auburn University, Auburn, AL, 1981.

Wylie, C. R. Jr., *Advanced Engineering Mathematics*, McGraw-Hill Book Company, New York, 1951.

Appendix I: MATLAB® Primer

AI.1 MATLAB for Controls: State-Space Analysis

State-space modeling of dynamic LTI systems allows the control system designer to bring the vast array of tools from linear system theory to bear on the design problem.

In addition to MATLAB's standard selection of linear systems tools, a number of specialized state-space design and analysis tools are available through the Control Systems Toolbox.

Key commands:

ss	-	conversion to state-space
ssdata	-	extraction of state-space data from a SYSTEM (SYS) model
rss	-	random stable state-space models
ss2ss	-	state transformations
canon	-	canonical state-space realizations
ctrb	-	controllability matrix
obsv	-	observability matrix

AI.1.1 Model Creation Using ss ( ), ssdata ( ), and rss ( )

To begin, consider the spring–mass–damper system as shown in Figure AI.1:

The differential equation for this simple system is

$$m\ddot{x} + c\dot{x} + kx = F \tag{AI.1}$$

An easy state-space form to convert this system is the controllability canonical form (CCF). As you learned in earlier chapters, this conversion is done using the following state definitions:

$$x_1 = x$$

$$x_2 = \frac{dx}{dt}$$

In this manner, the CCF form of the system becomes

$$\begin{bmatrix} \dot{x}_1 \\ \dot{x}_2 \end{bmatrix} = \begin{bmatrix} 0 & 1 \\ -2/3 & -8/3 \end{bmatrix} \begin{bmatrix} x_1 \\ x_2 \end{bmatrix} + \begin{bmatrix} 0 \\ 1/3 \end{bmatrix} F \tag{AI.2}$$

$$y = \begin{bmatrix} 1 & 0 \end{bmatrix} \begin{bmatrix} x_1 \\ x_2 \end{bmatrix} + [0]F \tag{AI.3}$$

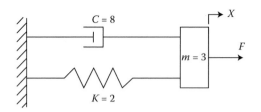

FIGURE AI.1
Spring–mass–damper system.

The output equation in this state-space model assumes the system output, which is simply the mass position, $x(t)$.

To create this state-space system within MATLAB, use the ss ( ) function, which generates a SYS object, just like the tf ( ) command for transfer function system representations:

$$A = \begin{bmatrix} 0 & 1; & -2/3 & -8/3 \end{bmatrix} \tag{AI.4}$$

$$B = \begin{bmatrix} 0; & 1/3 \end{bmatrix} \tag{AI.5}$$

$$C = \begin{bmatrix} 1 & 0 \end{bmatrix} \tag{AI.6}$$

$$D = 0 \tag{AI.7}$$

$$sys = ss\,(A,\ B,\ C,\ D) \tag{AI.8}$$

Now use

STEP(SYS)

Using this SYS object, all the MATLAB system response tools as step ( ), lsim ( ), and so on can be used on the state-space model of this system.

Given a SYS object, the component A, B, C, D matrices of the state-space model may be extracted using the ssdata ( ) command, which has the following syntax:

[A,B,C,D] = SSDATA(SYS)

retrieves the matrix data A, B, C, D respectively for the state-space model SYS.

If SYS is not a state-space model, it is first converted into the state-space representation.

Finally, the rss ( ) command may be used to generate a random state-space model. For full sysntax, enter "helprss" within MATLAB.

AI.1.2 State Transformations Using ss2ss ( ), Cannon ( )

State transformations are important for converting between various canonical state-space forms, and for reconfiguring a given state-space model into a transformed model with controllable, uncontrollable, observable, and unobservable components decoupled.

Consider a state transformation $z = Lx$, where x is the original state vector, L is a linear transformation matrix, and z is the transformed state vector. Under this transformation, the resulting system becomes

$$\dot{z} = \left[LAL^{-1} \right] z + \left[LB \right] u \tag{AI.9}$$

$$y = \left[CL^{-1} \right] z + Du \tag{AI.10}$$

The ss2ss ( ) function performs this transformation directly. For example, the system created in part A can be converted into diagonal canonical form using the following transformation discussed in the sequel:

$$x = Tz$$

$$T = \left[p_1 \quad p_2 \quad \dots \quad p_n \right], \quad \text{where } p_i = i\text{th eigen vector of } A$$

First, note that MATLAB's transformation is slightly different from the one presented earlier in Chapters 1 and 2, where the transformation is defined as $x = Tz$ (well, actually we used "X").

Thus, the MATLAB transformation matrix L is the inverse of the transform T defined above.

Next, form the transformation matrix T using the "eig ( )" function, which is defined as

$$[V,D] = \text{EIG}(X)$$

This produces a diagonal matrix D of eigenvalues and a full matrix V whose columns are the corresponding eigenvectors. This is the desired transformation matrix.

AI.2 Computing and Plotting Routines

Observe that the MATLAB steps match closely to the solution steps from the hand.

EXAMPLE AI.1

With the MATLAB code

```
% Compute average temperature and
% plot the temperature data.
%
Time = [0.0, 0.5, 1.0];
Temps = [105, 126, 119];
```

```
Average = mean(temps)
plot (time,temps),title('Temperature Measurements'),
xlabel ('Time, minutes'),
ylabel ('Temperature, degrees F'),grid
```

The words that follow % signs are comments to help us in reading MATLAB statements. If a MATLAB statement assigns or computes a value, it will also print the value on the screen if the statement does not end in a semicolon.

Thus, the values of time and temperature will not be printed, because the statements that assign them values end with semicolons.

The value of the average will be computed and printed on the screen, because the statement that computes it does not end with a semicolon. Finally, a plot of the time and temperature data will be generated.

AI.2.1 Testing

The final step in our problem-solving process is testing the solution.

We should first test the solution with the data from the hand example, because we have already computed its solution.

When the previous statements are executed, the computer displays the following output:

Average = 116.6667

A plot of the data points is also shown on the screen. Because the value of the average computed by the program matches the value from the hand example, we now replace the data from the hand example with the data from the physics experiment, yielding the following program:

```
% Compute average temperature and
% plot the temperature data.
%
time = [0.0, 0.5, 1.0, 1.5, 2.0, 2.5, 3.0, ..., 3.5, 4.0, 4.5, 5.0];
Temps = [105, 126, 119, 129, 132, 128, 131, ..., 135, 136, 132, 137];
average = mean(temps)
plot (time,temps),title('Temperature Measurements'),
xlabel ('Time, minutes')
ylabel ('Temperature, degrees F'),grid
```

When these commands are executed, the computer displays the following output:

Average = 128.1818

The plot in Figure AI.2 is also shown on the screen.

The set of steps demonstrated in this Example is used in developing the programs in the "Problem Solving Applied" sections throughout the rest of this appendix.

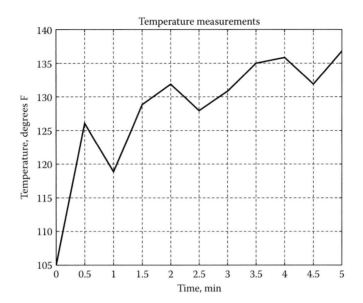

FIGURE AI.2
Temperature in °F versus time in minutes.

AI.3 Summary

Figure AI.2 shows temperatures collected in the physics experiment as a set of grand chal-lenges were presented to illustrate some of the exciting and difficult problems that cur-rently face engineers and scientists. Since the solutions to most engineering problems, including the grand challenges, involve the use of computers, we also presented a sum-mary of the components of a computer system, from computer hardware to computer soft-ware. We then introduced a five-step problem-solving methodology that we will use to develop computer solutions to problems. The five steps are as follows:

1. State the problem clearly.
2. Describe the input and output information.
3. Work the problem by hand (or with a calculator) for a simple set of data.
4. Develop a MATLAB solution.
5. Test the solution with a variety of data.

This process will be used throughout the book as we develop solutions to problems.

Key terms:
- hardware
- electronic copy
- machine language

- syntax
- compile errors
- software
- soft copy
- assembly language
- compiler
- debugging
- hard copy
- utilities
- high-level languages
- bugs
- logic errors
- algorithm

EXAMPLE AI.2

For example, to compute the square root of 9, type the following command:

sqrt (9)

The following output will be displayed:

ans = 3

AI.3.1 Some Other Commonly Used Commands

The graphics window is used to display plots and graphs. For example, to see the graphics window, type

plot([Cl,2,4,9,16],[1,2,3,4,51])

at the prompt. The graphics window will automatically appear, and a plot of the two vectors will be displayed. If you want to see some of the advanced graphics capabilities of MATLAB, as well as see a demonstration of the graphics window, enter demo at the prompt. This initiates the MATLAB Demo Window, a graphical demonstration environment that illustrates some of the different types of operations that can be performed with MATLAB. The edit window is used to create and modify M-files, which are files that contain a program or script of MATLAB commands. The edit window appears as shown in Figure AI.3.

Files are created or loaded. To view the M-File Editor/Debugger window.

type
  edit
  at the prompt.

AI.3.2 Managing the Environment

The MATLAB runtime environment consists of the variables that are placed in memory
  During a single session:
  To see the current runtime environment, type who or whos at the prompt.

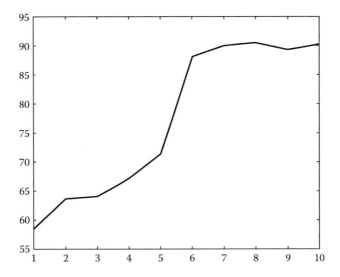

FIGURE I.3
Some plotting capabilities.

EXAMPLE AI.3

For example, enter the following commands to create and initialize the scalar X and the vector Y:

$$X = 5;$$

$$Y = [4, 5];$$

Then type *whos* to see the environment. The following should be displayed:

Name	Size	Bytes	Class
X	1×1	8	Double array
Y	1×2	16	Double array
Grand total is 3 elements using 24 bytes			

An alternative way to view the environment is to choose View and then ShowWork space from the menu bar.

The command Clear removes all variables from memory.

Type clear and then whos to see that the environment is cleared;

    Clear;
    Whos;

There are several commands for clearing windows. The clc command clears the command window, while the clf command clears the current figure and thus clears the graph window. In general, it is a good idea to start programs with clear and clf commands, to be sure that the memory has been cleared and that the graph window has been cleared and reset.

To preserve the contents of the workspace environment between sessions, you must save the environment to a file. The default format is a binary file called a MAT-file.

TABLE AI.1

MATLAB Disk Operations

MATLAB Command	Description	DOS Equivalent	UNIX Equivalent
dir	List directory	DIR	ls
delete	Delete file	DEL	rm
cd	Change directory	CD	cd
path	Show current path	PATH	printenv

To save the workspace to a file named myspace.mat, type save *myspace* at the prompt or choose File and then Save Workspace from the menu bar. To restore a workspace from a disk file, use the load command.

load myspace

The commands listed in Table AI.1 are helpful for locating and manipulating MATLAB disk files.

You can run external commands from within the MATLAB environment by the exclamation-point character (!). The particular external commands will vary depending on the operating system that you are using.

For example, in the Windows/DOS environment, the ver command shows the version of the operating system:

!ver

The results of executing the ver command on your operating system will be displayed.

For example, you might see something like:

Microsoft Windows 2000. [Version 5.00.2195] (or any latest version, for that matter.)

You can use the shell escape to execute external editors.

This capability will be useful when you are writing MATLAB programs and wish to use your favorite editor.

It is important to know how to abort a command in MATLAB. For example, there may be times when your commands cause the computer to print seemingly endless lists of numbers or when the computer goes into an infinite loop. In these cases, hold down the CTRL key and press the C key to generate a local abort within MATLAB.

AI.4 Display Format

When elements of a matrix are printed, integers are always printed as integers. Values with decimal fractions are printed using a default.format (called a short format) that shows four decimal digits. MATLAB allows you to specify other formats (see Table AI.2) that show more significant digits.

For example, to specify that we want values to be displayed in a decimal format with 14 decimal digits, one uses the command "format long."

TABLE AI.2

Numeric Display Format

MATLAB Command	Display	Example
Format short	4 decimal digit	15.2345
Format long	14 decimal digits	15.23453333333333
Format short e	4 decimal digits	1.52345e+01
Format long e	14 decimal digits	1.523453333333333e+01
Format bank	2 decimal digits	15.23
Format +	+, −, blank	+

We can return the format to a decimal format with four decimal digits by using the command "format short."

Two decimal digits are displayed when the format is specified as "format blank."

Finally, the command "format compact" suppresses many of the line feeds that appear between matrix displays and allows more lines of information to be seen together on the screen.

In our example output, we will assume that this command has been executed. The command "format loose" will return to the less compact display mode.

AI.5 Printing Text or Matrices

The disp command can be used to display text enclosed in single quotation marks. It can also be used to print the contents of a matrix without printing the matrix's name. Thus, if a scalar temp contains a temperature value in degrees Fahrenheit, we could print the value on one line and the units on the next line by using the following commands:

$$\text{disp(temp);} \quad \text{disp ('degrees F')}$$

If the value of temp is 78, then the output will be the following:

$$78$$

Degrees F

Note that the two disp commands were entered on the same line, so that they would be executed together. To display the values on the same line, you create one string from the two parts:

$$\text{disp([num2str(temp) 'degrees F'])}$$

AI.6 Formatted Output

The "fprintf" command gives you even more control over the output than you have with the "disp" command.

In addition to printing both text and matrix values, you specify the format to be used in printing the values, and you can specify when to skip to a new line.

If you are a C programmer, you will be familiar with the syntax of the command. With few exceptions, the MATLAB "fprintf" command uses the same formatting specifications as the C fprintf ( ) function. We will not cover all of the format options in this appendix.

Please refer to an ANSI C textbook for a complete description of the fprintf ( ) options. The general form of this command is

fprintf(format-string,var, …)

The format string contains format specifications to be printed. The string may contain a combination of literal text and format specifications that define how the data are to be displayed.

The remainder of the arguments is a list of names of the matrices to be printed.

The format specifiers take the following form:

%[flags] [width] [.precision] type

Each specifier begins with the % character. The flags, width, and precision format fields are optional.

Examples of type fields are

%e display in exponential notation

%f display in fixed point or decimal notation

%g display using %e or %f, depending on which is shorter

Strings are displayed literally. The % character may be displayed by using %%. The tab and newline characters are displayed by using \ t and \ n, respectively.

EXAMPLE AI.4

As an example, enter the following commands:

Temp = 98.6;

fprintf('The temperature is %f degrees F.\n', temp);

The following should be displayed:

The temperature is 98.600000 degrees F.

The width field controls the minimum number of characters to be printed. It must be a nonnegative decimal integer. The precision field is preceded by a period (.) and specifies the number of decimal places after the decimal point for exponential and fixed-point types.

Note in the preceeding example that the default precision is six digits. The precision field has a different meaning for other types.

EXAMPLE AI.5

For example, enter the following statement:

fprintf('The temperature Is %3.1f degrees.\n', temp);

The following output will be produced:

The temperature is 98.6 degrees.

The flags field is used to designate the padding characters and to format numerical signs. The zero (0) flag with a fixed-point type will cause the leading digits to be padded with zeros instead of spaces. A minus flag (–) left-justifies the converted argument in its field. A plus flag (+) always prints a sign character (+ or –).

EXAMPLE AI.6

For example, enter the following statement:

fprintf('The temperature is %08.1f degrees.\n', temp);

The following output will be produced:

The temperature is 000098.6 degrees.

The fprintf statement allows you to have a great deal of control over the form of the output.
We will use it frequently in our examples to help you become familiar with it.

EXAMPLE AI.7

Try this example to see power of the fprintf command when it is used with a matrix. Enter these two statements:

X = 0:0.2:2;

fprintf('%3.1f * 5 = %4.1f \n',X,X)

The following output will be displayed:

0.0 * 5 = 0.2

0.4 * 5 = 0.6

0.8 * 5 = 1.0

1.2 * 5 = 1.4

1.6 * 5 = 1.8

2.0 * 5 = 0.0

0.2 * 5 = 0.4

0.6 * 5 = 0.8

1.0 * 5 = 1.2

1.4 * 5 = 1.6

1.8 * 5 = 2.0

TABLE AI.3

Distance in Feet versus Trial Number

Trial	Distance (ft)
1	58.5
2	63.8
3	64.2
4	67.3
5	71.5
6	88.3
7	90.1
8	90.6
9	89.5
10	90.4

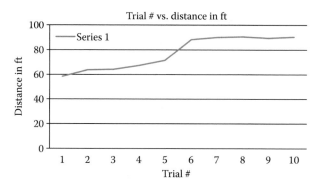

FIGURE AI.4
Simple plot of distances for 10 trials.

AI.6.1 *xy* Plots

In this section, we show you how to generate a simple *xy* plot from data stored in two vectors.

Assume that we want to plot the data shown in the accompanying table, collected from an experiment with a remote-controlled model car. The experiment is repeated 10 times, and we have measured the distance that the car travels for each trial.

Also assume that the time values are stored in a vector called *x* and that the distance values are stored in a vector called *y*, as in Table AI.3.

To plot these points, we use the plot command, with *x* and *y* as arguments: **plot(*x,y*)**

The plot in Figure AI.4 is automatically generated.

(Slight variations in the scaling of the plot may occur, depending on the computer type and the size of the graphics window.)

Appendix II: FEEDBACK≪® Guide for Applications in the Text

The following booklets published by FEEDBACK≪® have been referred to in this appendix:

1. *MATLAB® Guide for Feedback Control Instrumentation*

   FEEDBACK≪® uses the high-level technical computing language MATLAB for algorithm development, data analysis, and visualization. Together with Simulink® and an additional control toolbox, it aids in control system design and analysis that can be implemented in real time using a suitable data acquisition PCI card. A real-time model is shown in a snapshot of the manual in Figure AII.1.

   A few development steps are described in this manual, namely:

   a. Process models—first principle models designed in Simulink for the user to test. An example is shown in Figure AII.2.

   b. Dynamic analysis—model linearization is explained with Bode diagram and pole-zero plots.

   c. Discrete model identification—step-by-step model identification is described.

   d. Controller design—PID control is explained.

   e. Controller tests—RL technique helps to improve performance.

2. Coupled Tanks Installation and Commissioning—33-041-IC

   A connection diagram is shown in Figure AII.3.

   A Simulink model menu palette is shown in Figure AII.4.

   A control system is shown in Figure AII.5.

3. Digital Pendulum Installation and Commissioning—33-936IC

   This manual describes the installation and testing of the inverted pendulum experiment. Digital Pendulum Mechanical Unit is shown in Figure AII.6.

   A connection diagram of the schematic is shown in Figure AII.7.

   A control system schematic is shown in Figure AII.8.

   A PID Cart Control diagram is shown in Figure AII.9.

4. Digital Pendulum Control Experiments—33-936S

   In this manual a model of the Pendulum is introduced and the linearization is conducted. Digital algorithms are proposed and performance comparisons are assessed using test sequences. Figures AII.6 through AII.9 illustrate some aspects, with the last figure illustrating the testing process and some results shown in Figure AII.10 with PID control.

   With cart position response I, real time is shown in Figure AII.11.

5. Magnetic Levitation System—33-006-PCI

   Several chapters in this book refer to these documents to illustrate several examples and related theory.

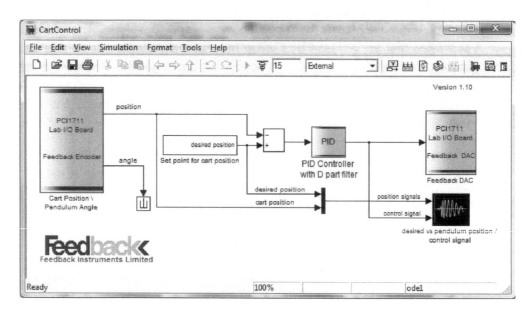

FIGURE AII.1
(See color insert.) Implementation of real-time model using FEEDBACK≪® instrumentation.

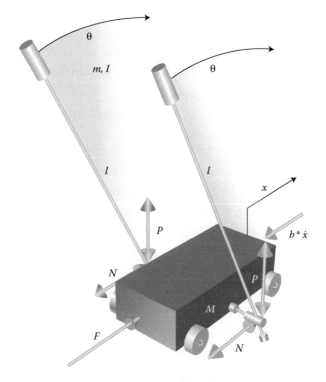

FIGURE AII.2
(See color insert.) Model description example of an inverted pendulum.

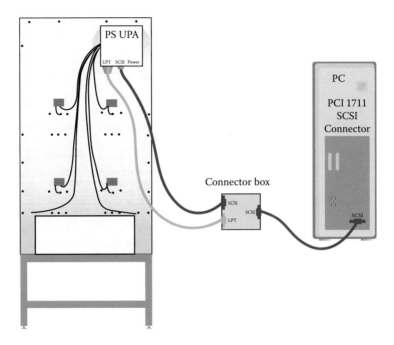

FIGURE AII.3
(See color insert.) Coupled tanks connection diagram.

FIGURE AII.4
(See color insert.) Simulink model menu for coupled tanks.

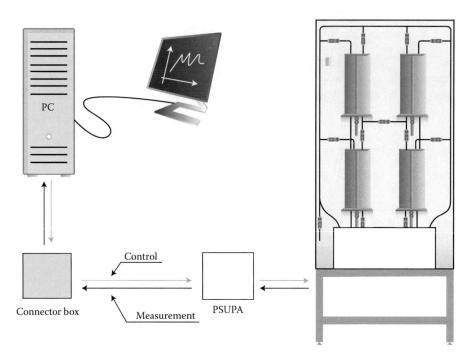

FIGURE AII.5
(See color insert.) Control system schematic for coupled tanks.

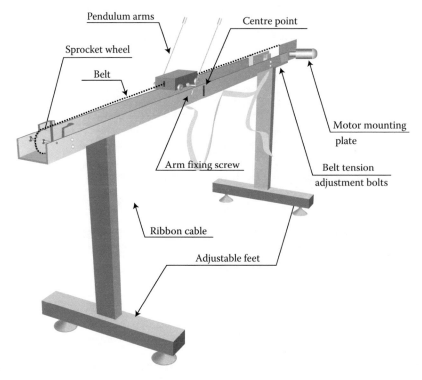

FIGURE AII.6
(See color insert.) Inverted pendulum unit.

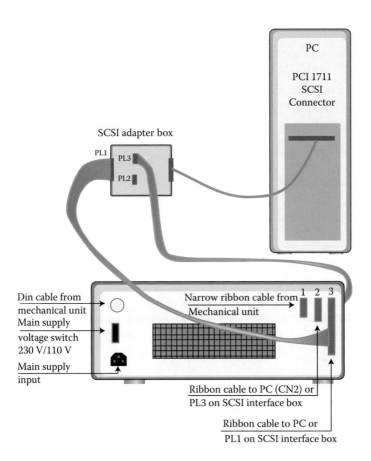

FIGURE AII.7
(See color insert.) Connection diagram for inverted pendulum.

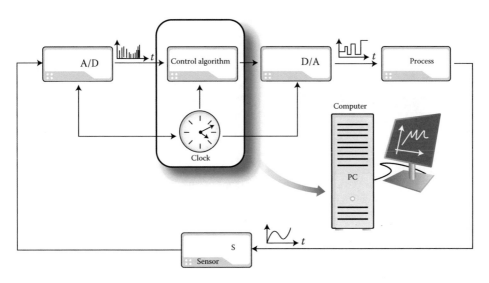

FIGURE AII.8
(See color insert.) Computer control system diagram.

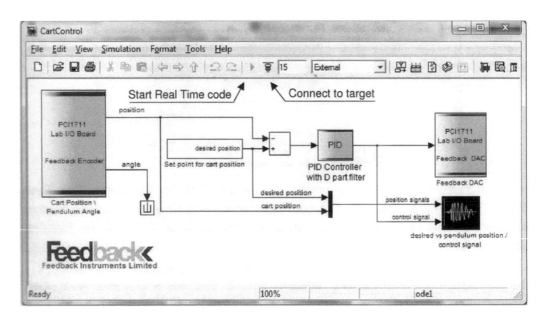

FIGURE AII.9
(See color insert.) PID cart control.

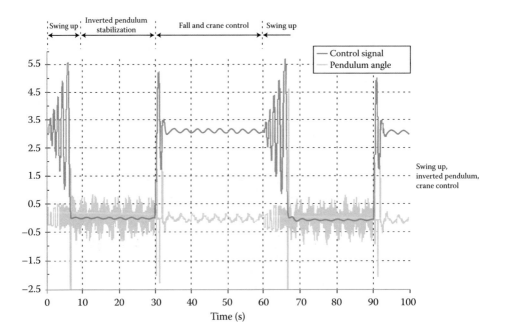

FIGURE AII.10
(See color insert.) Pendulum PID control response.

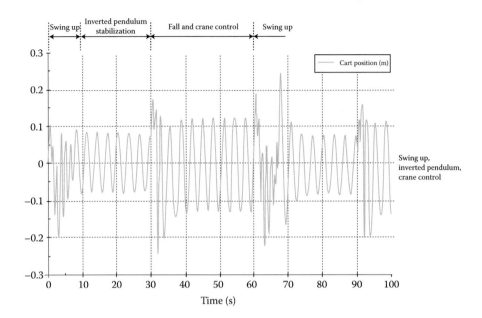

FIGURE AII.11
(See color insert.) Pendulum cart position; PID control response.

Appendix III: Suggested MATLAB® Code and Additional Examples from FEEDBACK≪®

Note: Suggested MATLAB code is provided below for several algorithms illustrated in different chapters of this book. There is no guarantee that this code will work as required by the final user depending on the version of MATLAB they use. User is advised to verify their code for suitable applications.

AIII.1 Bode Disc—Bode Plot for Different 'OPT' Values

```
% Custom Bode Plot
% BodeDisc

% [Fm,Fp,w] = BodeDisc(bi,ai,h,p,OPT);
% the values bi and ai come from the Leverier function, must run that
% before running this function
% h, p, and OPT are new values and must be decided by user

% IMPORTANT NOTE:
% although the value of h does not impact the calculations for the Bode
% plot when choosing OPT = 1 for continuous
% the value h still effects wmax, which is the maximum frequency that
  is to
% be evaluated
% wmax is calculated to by the Nyquist frequency of h, so for OPT = 1
% just choose a value h to view up to a specific frequency on the plot

% outputs are Fm, Fp, and w, are arrays
% w is an array of frequencies that are were chosen for calculation
% Fm is an array of the Magnitudes corresponding to the frequencies in w
% Fa is an array of the Phases corresponding to the frequencies in w

function [Fm, Fp, w] = BodeDisc(varargin)

bi = varargin{1}; % numerator coeffs
ai = varargin{2}; % denominator coeffs
h = varargin{3}; % sampling time
p = varargin{4}; % number of points to evaluate (chosen by user),
recommend 100 or greater
OPT = varargin{5}; % choose 1 for continuous, choose 2 (or anything else)
for discrete
wmin = 10^(-2); % a good min value for w
wmax = 2*pi*((1/h)/2); % Nyquist frequency (rad/sec)
```

```
N = wmax - wmin;
ps = N/p; % specific frequencies to be evaluated

% initialize values
w = [];
F = [];
Fm = [];
Fp = [];

% the main loop calculates the Mag and Phase at a specific frequency
% the number of frequencies that are calculated are decided by the user
% input p
% if p = 100, then the loop calculates for 100 evenly spaced values
% between the frequencies wmin and wmax
% the output array w are those frequencies

for loop = 0:p
    w(loop+1) = wmin + ps*loop;
    if OPT == 1
        X = j*w(loop+1); % for continuous transfer functions
    else
        X = cos(w(loop+1)*h) + j*sin(w(loop+1)*h); % for continuous transfer
        functions
    end
    AA = ai(1);
    BB = bi(1);

    for n = 2:length(ai)
        AA = ai(n) + X*AA;
        BB = bi(n) + X*BB;
    end

    F(loop+1) = BB/AA;
    Fm(loop+1) = 20*log10(abs(F(loop+1)));
    % Fp(loop+1) = atan(imag(F(loop+1))/real(F(loop+1))); % algorithm from
    notes, doesn't work that great visually
    Fp(loop+1) = (180/pi)*angle(F(loop+1)); % visually works better than
    above function, performs same task

    % change the Fp to see the difference
end

% plot magnitude and phase
% plots the values of Fm and Fp for the corresponding values in w

subplot(2,1,1)
semilogx(w,Fm);grid
Title('Bode Diagram')
ylabel('Magnitude (dB)')
subplot(2,1,2)
semilogx(w,Fp);grid
ylabel('Phase (deg)')
xlabel('Frequency (rad/sec)')
```

AIII.2 Dscrt Routine

```
% Dscrt

% [Phi,Gam,W] = Dscrt(n,A,B,h);
% the values A and B are matrices from the state space model
% n is the size of A (ie A is a n by n matrix), found from length(A)
% h is the sampling time

% outputs are Phi, Gam, and W
% Phi and Gam are discrete state space model matrices

function [Phi,Gam,W] = Dscrt(varargin)

n = varargin{1}; % A is a n by n matrix, thus n = length(A), be sure this
is correct
A = varargin{2}; % A matrix of state space model
B = varargin{3}; % B matrix of state space model
h = varargin{4}; % sampling time

% Select M
% First Compute ||A|| = Am
Am = 0;
for j = 1:n
    for k = 1:n
        Am = Am + A(j,k)^2;
    end
end
Am = sqrt((1/n)*Am);

% then compute how many terms in M, given h

c = Am*h/2;
for k = 2:20
    c = c*(Am*h/(k+1));
    if c < 10^(-6)
        M = k;
        break, end
end

if M < 4
    M = 4;
end

% Initialize N and Phi
N = M + 1;
Phi = A;

% Loop to make Taylor Series
for i = 1:M
    W = eye(n) + Phi*(h/N);
    N = N - 1;
    Phi = A*W;
end
```

```
% Finalize W, Phi, and Gam
W = h*W;
Phi = eye(n) + h*Phi;
Gam = W*B;
```

AIII.3 Leverier Algorithm

```
% Leverier call

% [bi,ai,Pm,Am] = Leverier(Phi,Gam,C,d);
% the values Phi and Gam come from the Dscrt function, which must be run
% before this one
% the values C and d are from the state space model

% outputs are bi, ai, Pm, and Am
% bi is an array representing the numerator coefficients of the discrete
% transfer function
% ai is an array representing the denominator coefficients of the
discrete
% transfer function
% Pm is ||P||
% Am is ||A||

function [bi, ai, Pm, Am] = Leverier(varargin)

Phi = varargin{1};
Gam = varargin{2};
C = varargin{3};
d = varargin{4};

n = length(Phi);

ai = [];
bi = [];

ai(1) = 1;
bi(1) = d;
P = eye(n);

% algorithm for ai and bi from notes
for k = 1:n
    X = Phi*P;
    ai(k+1) = (-trace(X))/k;
    bi(k+1) = C*P*Gam;
    P = X + ai(k+1)*eye(n);
end

% Compute ||P|| = Pm
Pm = 0;
for l = 1:n
```

```
    for k = 1:n
        Pm = Pm + P(1,k)^2;
    end
end
Pm = sqrt((1/n)*Pm);

% Compute ||A|| = Am
Am = 0;
for l = 1:n
    for k = 1:n
        Am = Am + Phi(l,k)^2;
    end
end
Am = sqrt((1/n)*Am);
```

AIII.4 Final Project: State Space Description of Motor-Spring— Two Rotating Masses Plant

Hints:

I	Continuous System
II	Start with the open loop system
III	Choose alpha = 2
IV	Find desired poles
V	Save them
VI	Find control gains as follows
VII	Find difference between the original open loop and the desired poles
VIII	This should give KC
IX	Then build t3, t2 and t1
X	State feedback gains are
XI	Say K2 = Tinv * KC
XII	Find LC
XIII	
XIV	Discrete System
XV	Choose appropriate h
XVI	Find phi, psi, gama using proper M from discrt routine
XVII	Use leverier to get proper transfer function coeffs
XVIII	Find HC dicrete
XIX	Then select
XX	R1 = gama
XXI	R2 = Phi * gama
XXII	R3 = phi^2*gama
XXIII	Find Hc
XXIV	Find det of HC
XXV	Find discrete gains
XXVI	Load original K and LC from above
XXVII	$K^\sim = K' + K'(A - B*K')*h/2$

XXVIII   LC⁻ = (1 − K′B*(h/2))*LC
XXIX   Use these gains for discrete design

```
%Final project for Digital Controls
 close all
format compact

%State Space description of Motor-Spring - two rotating masses Plant
%Previously imported to the workspace the Continuous Plant model for
Simulink Model - "Plant"
%unity overall gain nd alpha = 2
load plant_c2 p_c_2;
plan=p_c_2;
[A,B,C,D,TS] = ssdata(plan)
%find eigen values
[n,d]=ss2tf(A,B,C,D,1);
[Vector Value]=eig(A)
%create and save num,den of plant
[ac,bc,norm_Ac,norm_Pc,inv_zI_Ac]=leverier(A,B(:,1),C,D);
save plant_c ac bc

%Find Hc_continuous
r1=B(:,1);
r2=A*B(:,1);
r3=A^2*B(:,1);
Hc=cat(2,r1,r2,r3)
det_Hc_c=det(Hc)
cond_Hc_c=cond(Hc)
display('Large condition numbers indicate a nearly singular matrix')

%Discretize the system with input gain to make unity output
h=.15; %from calculations in report
[phi,psi,gamma,M]=Dscrt(A,B,h);
%create and save num,den of plant
[ad,bd,norm_Ad,norm_Pd,inv_zI_Ad]=leverier(phi,gamma,C,D);
save plant_d ad bd
%Find Hc_discrete
r1=gamma(:,1);
r2=phi*gamma(:,1);
r3=phi^2*gamma(:,1);
Hc=cat(2,r1,r2,r3);
det_Hc_d=det(Hc);
cond_Hc_d=cond(Hc);
display('Large condition numbers indicate a nearly singular matrix')
inv_Hc=inv(Hc);

%Describe the desired dominate pole placement for alpha = 0
Wn=.5                   %desired natural frequency
Del=.7                  %desired damping ratio
alpha=0                 %third pole
P_0=[1 2*Del*Wn+Wn*alpha Wn^2+2*Del*Wn*Wn*alpha Wn^2*Wn*alpha] %desired
dominate poles
roots([P_0])
```

```
%save desired P_2
P_0_desired=[-(2*Del*Wn+Wn*alpha) -(Wn^2+2*Del*Wn*Wn*alpha)
-(Wn^2*Wn*alpha);1 0 0 ;0 1 0]
save desired_0 P_0 P_0_desired

%Find control gains for alpha =0
%First find the difference between the plant and the desired response
%a = plant and P_0 = desired

Kc=flipud((ac(2:4)-P_0(2:4))')
%then build up the T matrix
t3=B(:,1);
t2=A*t3 + ac(2)*B(:,1);
t1=A*t2 + ac(1)*B(:,1);
T=cat(2,t1,t2,t3);
T_inv_t=inv(T)';

%Test for controllability
%build Hc matrix
Hc=cat(2,B(:,1),A*B(:,1),A^2*B(:,1))
inv(Hc);
inv(Hc');
cond_Hc_d=cond(Hc)
display('Large condition numbers indicate a nearly singular matrix')

%then, find State feedback Gains
%K_0=T_inv_t*fliplr(Kc)'
K_0=T_inv_t*Kc;
%check results of simulation with desired system specification
K=K_0;
Lc=1/.36;

%run model with continuous SVFB gains
[Ac,Bc,Cc,Dc]=linmod('p');
%create Matlab System to allow use of step function
sys=ss(Ac,Bc,Cc,Dc);
%step(sys); %removed since system with calculated K's not stable

%Test for controllability
%build Hc matrix
%Hc=cat(2,Bc(:,1),Ac*Bc(:,1),Ac^2*Bc(:,1));
%inv(Hc)
%inv(Hc');
%cond_Hc_d=cond(Hc)
%display('Large condition numbers indicate a nearly singular matrix')

%check damping and natural frequency
%from stored data to check
display('check system performance with alpha = 0 gains')
damp(sys)

%Describe the desired dominate pole placement for alpha = 2
alpha=2                                    %third pole
```

```
P_2=[1 2*Del*Wn+Wn*alpha Wn^2+2*Del*Wn*Wn*alpha Wn^2*Wn*alpha] %desired
dominate poles
roots([P_2])
%save desired P_2
P_2_desired=[-(2*Del*Wn+Wn*alpha) -(Wn^2+2*Del*Wn*Wn*alpha)
-(Wn^2*Wn*alpha);1 0 0 ;0 1 0]
save desired_2 P_2 P_2_desired

%Find control gains for alpha =2
%First find the difference between the plant and the desired response
Kc=fliplr((ac(2:4)-P_2(2:4)))'
%then build up the T matrix
t3=B(:,1)
t2=A*t3 + ac(2)*B(:,1)
t1=A*t2 + ac(1)*B(:,1)
T=cat(2,t1,t2,t3)
T_inv_t=inv(T)'
%then, find State feedback Gains
K_2=T_inv_t*Kc
%check results of simulation with desired system specification
K=K_2
Lc=1/.36

save gains_cont K Lc

%run model with continuous SVFB gains (alpha = 2)
[Ac,Bc,Cc,Dc]=linmod('p');
%create Matlab System to allow use of step function
sys=ss(Ac,Bc,Cc,Dc);
%step(sys);              Use the Simulink model to get time domain plots
save closed_loop_cont_sys sys

%Test for controllability
%build Hc matrix
%Hc=cat(2,Bc(:,1),Ac*Bc(:,1),Ac^2*Bc(:,1));
%inv(Hc)
%inv(Hc');
%cond_Hc_d=cond(Hc)
%display('Large condition numbers indicate a nearly singular matrix')

%check damping and natural frequency
%from stored data to check
%display('check system performance with alpha = 2 gains')
%damp(sys)

%Chose a sample time based on the norm of A

A
n=3;
h_norm_A=abs(sqrt((1/n)*sum(sum(A^2))))

[v d]=eig(A);
h_max_eig=max(abs(abs(d)))

norm_max=max(A);
h_max_A=1./norm_max
```

```
h_norm_max_eig=1./max(abs(eig(A)))

h_abs_A_BK=1./abs(eig(A-B*K'))

h_norm_1=1./norm(A,1)

h_norm_2=1./norm(A,1)

h_norm_inf=1./norm(A,1)

%with continuous control A
Ac
h_norm_Ac=abs(sqrt((1/n)*sum(sum(Ac^2))))

[v d]=eig(Ac);
h_max_eig=max(abs(abs(d)))

norm_maxc=max(Ac);
h_max_Ac=1./norm_max

h_norm_max_eigc=1./max(abs(eig(Ac)))

%h_abs_A_BKc=1./abs(eig(Ac-Bc*K'))

h_norm_1c=1./norm(Ac,1)

h_norm_2c=1./norm(Ac,1)

h_norm_infc=1./norm(Ac,1)

stop

%Find Gains
%K = acker(A,B,P)
%K = place(Ac,Bc,P_2(2:4))

h=.15; %from calculations in report
h=.75; %from calculations in report
[phi,psi,gamma,M]=Dscrt(Ac,Bc,h);
%Discretize the system with input gain to make unity output
h=.79; %from calculations in report
[phi,psi,gamma,M]=Dscrt(A,B,h);
load desired_2
P_2
A=P_2_desired
B=[ 1 0 0]'
[phi,psi,gamma,M]=Dscrt(A,B,h)
Hc=cat(2,gamma(:,1),phi*gamma(:,1),phi^2*gamma);
%inv(Hc)
%Hc_condition=cond(Hc)

%Evaluate X
%init
```

```
X=eye(3)
I=X
k=1
RA=real(roots([P_2(2:4)]))
RB=imag(roots([P_2(2:4)]))
n=2
if RB(k)==0
   X=X * (phi - RA(k)*I)
   k=k+1
else
   X=X * (phi^2 - 2*RA(k)*phi + (RA(k)^2 + RB(k)^2)*I)
   k=k+2
end
if k<=n
   if RB(k)==0
   X = X * (phi - RA(k)*I)
   k=k+1
else
   X = X * (phi^2 - 2*RA(k)*phi + (RA(k)^2 + RB(k)^2)*I)
   k=k+2
end
end

%calculate K_direct
Q=inv(Hc')*[0 0 1]'
K_direct = Q'*X
K_real=K_direct
K=-K_real
Lc= 8.63

save direct_gains K Lc
stop

K = place(P_2_desired,[1 0 0 ]',P_2(2:4))
%load s
load closed_loop_cont_sys
[A,B,C,D,TS] = ssdata(sys)
[phi,psi,gamma,M]=Dscrt(A,B,h)
Hc=cat(2,gamma(:,1),phi*B,phi^2*gamma)
inv(Hc)
cond(Hc)

%Evaluate X
%init
X=eye(3)
I=X
k=1
RA=real(roots([P_2(2:4)]))
RB=imag(roots([P_2(2:4)]))
n=2
if RB(k)==0
   X = X * (phi - RA(k)*I)
   k=k+1
end
```

```
X = X * (phi^2 - 2*RA(k)*phi + (RA(k)^2 + RB(k)^2)*I)
k=k+2
%calculate K_direct
Q=inv(Hc')*[0 0 1]'
K_direct = Q'*X
K_real=K_direct

stop

sysd=c2d(sys,.15);
[A,B,C,D,TS] = ssdata(sysd)
Hd=cat(2,B(:,1),A*B,A^2*B)
inv(Hd)
cond(Hd)
K = place(A,B,P_2(2:4))


[A,B,C,D,TS] = ssdata(sys)
Hc=cat(2,B(:,1),A*B,A^2*B)
[phi,psi,gamma,M]=Dscrt(A,B,h)
Hc=cat(2,gamma(:,1),phi*gamma,phi^2*B)
inv(Hc)
cond(Hc)

%load desired_2
%sys=P_2_desired
%[A,B,C,D,TS] = ssdata(sys)
%Hc=cat(2,B(:,1),A*B,A^2*B)
%[phi,psi,gamma,M]=Dscrt(A,B,h)

clc
A = [0 1 -1;3 -2 1;0 2 -1];
c = deepu(A);
h = 0.05;
c1 = (c*h)/2;
for m = 2: 20,
    c1 = c1 * (c *h)/(m+1);
    if(c1<0.000001)break;
end
end
if(m<4)
    m=4;
else
    m=m;
end
B = [1;1;0];
[phi,shi,gamma] = Dscrt(A,B,h,m);
k = [0.5 2 1];
phibar = phi - gamma*k;
C = [1 0 2];
[ai,bi] = level(phibar,gamma,C,3);
d = 0;
a0=1;b0=d;
ai1(1)=a0;bi1(1)=b0;
ai1(2:4)=ai(1:3);
```

```
bi1(2:4)=bi(1:3);
ai=ai1
bi=bi1

clc
A = [0 1 -1;3 -2 1;0 2 -1];
c = norm(A);
h = 0.05;
c1 = (c*h)/2;
for m = 2: 20,
    c1 = c1 * (c *h)/(m+1);
    if(c1<0.000001)break;
end
end
if (m<4)
    m=4;
else
    m=m;
end
B = [1;1;0];
[phi,shi,gamma] = Dscrt(A,B,h,m,3)
k = [0.5 2 1];
phibar = phi - gamma*k;
C = [1 0 2];
[ai,bi] = lev(phibar,gamma,C,3);
[ai3,bi3] = lev(phi,gamma,k,3);
d = 0;
a0=1;b0=d;
ai1(1)=a0;bi1(1)=b0;
ai2(1)=a0;bi2(1)=b0;
ai1(2:4)=ai(1:3);
bi1(2:4)=bi(1:3);
ai2(2:4)=ai3(1:3);
bi2(2:4)=bi3(1:3);
ai=ai1
bi=bi1
ai3=ai2
bi3=bi2
num1=bi
den1=ai
num2=bi3
den2=ai3
hd1=tf(num1,den1,0.05)
hd2=tf(num2,den2,0.05)
bode(num2,den2)
[Gm,Pm,wcg,Wcp]=margin(hd2)

clc
A = 0.5;
c = norm(A);
h = 0.2;
c1 = (c*h)/2;
for m = 2: 20,
    c1 = c1 * (c *h)/(m+1);
```

```
    if(c1<0.000001)break;
end
end
if(m<4)
    m=4;
else
    m=m;
end
B = 0.95;
k=1;
[phi,shi,gamma] = Dscrt(A,B,h,m,1)
[ai,bi] = lev(phi,gamma,k,1);
s=phi-(gamma*k)
d = 0;
a0=1;b0=d;
ai1(1)=a0;bi1(1)=b0;
ai1(2)=ai(1);bi1(2)=bi(1);
num=bi1
den=ai1
hd1=tf(num,den,0.2)
bode(num,den)
[Gm,Pm,wcg,Wcp]=margin(hd1);
[Gm,Pm,wcg,Wcp]


%Previously imported to the workspace the Continuous Plant model for
Simulink Model - "Plant"
%unity overall gain nd alpha = 2
load plant_c2 p_c_2;
plan=p_c_2;
[A,B,C,D,TS] = ssdata(plan)
[n,d]=ss2tf(A,B,C,D,1);
[Vector Value]=eig(A)
[ac,bc,norm_Ac,norm_Pc,inv_zI_Ac]=leverier(A,B(:,1),C,D);

%Find Hc_continuous
r1=B(:,1);
r2=A*B(:,1);
r3=A^2*B(:,1);
Hc=cat(2,r1,r2,r3)
det_Hc_c=det(Hc)
cond(Hc)

%Discretize the system with input gain to make unity output
h=.3; %from calculations in report
h=.15; %from calculations in report

[phi,psi,gamma,M]=Dscrt(A,B,h)
[ad,bd,norm_Ad,norm_Pd,inv_zI_Ad]=leverier(phi,gamma,C,D);
ad
bd
%Find Hc_discrete
r1=gamma(:,1);
r2=phi*gamma(:,1);
r3=phi^2*gamma(:,1);
```

```
Hc=cat(2,r1,r2,r3)
det_Hc_d=det(Hc)
cond(Hc)
inv_Hc=inv(Hc)

%Find discrete Gains using the equivalent method
load gains K Lc
K
Lc
K_squiggle = K' + K'*(A - B*K') *h/2
Lc_squiggle = (1 - K'*B*(h/2))*Lc
K=K_squiggle'
Lc=Lc_squiggle
save gains_equiv K Lc

stop
```

Examples provided for illustration use in the book:

```
>> num=[1,0.97];
>> num=[0.1];
>> den=[1,0.1,0];
>> sys=tf(num,den)

Transfer function:
0.1
_____ -
s^2 + 0.1 s

>> sysd=c2d(sys,1,'zoh')

Transfer function:
0.04837 z + 0.04679
_____

z^2 - 1.905 z + 0.9048

Sampling time: 1
>> sysd

Transfer function:
0.04837 z + 0.04679
_____

z^2 - 1.905 z + 0.9048

Sampling time: 1
>> num1=[1,-0.87];
>> den1=[,0.35];
>> den1=[1,0.35];
>> hd=tf(num,den,1)

Transfer function:
0.1
_____ -
z^2 + 0.1 z
```

```
Sampling time: 1
>> hd=10.5*hd

Transfer function:
1.05
--------
z^2 + 0.1 z

Sampling time: 1
>> hd=10.5*tf(num1,den,1)

Transfer function:
10.5 z - 9.135
--------
z^2 + 0.1 z

Sampling time: 1
>> hd=10.5*tf(num1,den1,1)

Transfer function:
10.5 z - 9.135
--------
z + 0.35

Sampling time: 1
>> hd

Transfer function:
10.5 z - 9.135
--------
z + 0.35

Sampling time: 1
>> sysd*hd

Transfer function:
0.5079 z^2 + 0.04938 z - 0.4274
---------------------------
z^3 - 1.555 z^2 + 0.2381 z + 0.3167

Sampling time: 1
>> s=sysd*hd

Transfer function:
0.5079 z^2 + 0.04938 z - 0.4274
---------------------------
z^3 - 1.555 z^2 + 0.2381 z + 0.3167

Sampling time: 1
>> s

Transfer function:
0.5079 z^2 + 0.04938 z - 0.4274
---------------------------
z^3 - 1.555 z^2 + 0.2381 z + 0.3167
```

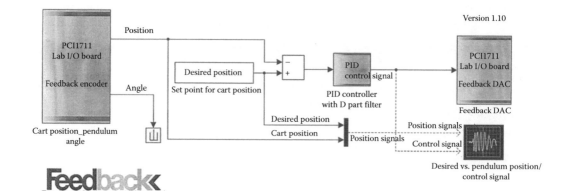

FIGURE AIII.1
PID Controller using FEEDBACK≪® PCI1171 I/O Board.

Additional Example from FEEDBACK≪® on PID Controller using PCI1711 I/O Board is given in Figure AIII.1.

Index